Transforming Urban Food Systems in Secondary Cities in Africa

Liam Riley · Jonathan Crush
Editors

Transforming Urban Food Systems in Secondary Cities in Africa

palgrave
macmillan

Editors
Liam Riley
Balsillie School of International Affairs
Waterloo, ON, Canada

Jonathan Crush
Balsillie School of International Affairs
Waterloo, ON, Canada

ISBN 978-3-030-93071-4 ISBN 978-3-030-93072-1 (eBook)
https://doi.org/10.1007/978-3-030-93072-1

This Palgrave Macmillan imprint is published by the registered company Springer Nature Switzerland AG
The registered company address is: Gewerbestrasse 11, 6330 Cham, Switzerland

Acknowledgements

This book is a product of the Food, Urbanization, Environment and Livelihoods (FUEL) Project of the African Food Security Urban Network (AFSUN) funded by an Insight Grant from the Social Sciences and Research Council of Canada (SSHRC). The publication of the book was made possible by a Connection Grant from SSHRC, which also funded a virtual workshop attended by the authors on May 13 and 14, 2021 and a series of panels at the Canadian Association of African Studies Annual Conference on June 7, 2021. Several of the contributors to this book received funding for their research from the FUEL Project and the Queen Elizabeth Advanced Scholars Program of the Hungry Cities Partnership and Universities Canada. Supplemental funding for the preparation of the volume and organizing of the workshop and conference panel was received from the Balsillie School of International Affairs and the Office of Research Services at Wilfrid Laurier University. The editors wish to acknowledge the important contributions of Maria Salamone, Nicole Langlois, Zack Ahmed, and Joan Baker in preparing the manuscript. The editors are also grateful to Joanne Weston, Mary Caesar, and Maria Salamone for their assistance in organizing the workshop.

Contents

List of Contributors

Beatrix Alsanius Department of Biosystems and Technology, Microbial Horticulture Division, Swedish University of Agricultural Sciences, Lomma, Sweden

Jordan Blekking Department of Geography, Indiana University, Bloomington, IN, USA

Andrea M. Brown Political Science, Wilfrid Laurier University, Waterloo, ON, Canada

Mary Caesar Balsillie School of International Affairs, Wilfrid Laurier University, Waterloo, ON, Canada

Gamuchirai Chakona Environmental Learning Research Centre (ELRC), Department of Education, Rhodes University, Makhanda/Grahamstown, South Africa

Emmanuel Chilanga McGill University, Montreal, QC, Canada

Jonathan Crush Balsillie School of International Affairs, Wilfrid Laurier University, Waterloo, ON, Canada

Anil Dhakal Department of Geography and Environmental Management, University of Waterloo, Waterloo, ON, Canada

Tom Evans School of Geography, Development and Environment, University of Arizona, Tucson, AZ, USA

Issahaka Fuseini Department of Adult Education and Human Resource Studies, University of Ghana, Accra, Ghana

Stacey Giroux Ostrom Workshop, Indiana University, Bloomington, IN, USA

Miriam Grant Department of Community, Culture and Global Studies, University of British Columbia Okanagan, Kelowna, BC, Canada

Gareth Haysom African Centre for Cities, University of Cape Town, Cape Town, South Africa

Idiong Christopher Idiong University of Calabar, Calabar, Nigeria

Magnus Jirström Department of Human Geography, Lund University, Lund, Sweden

Yanick Borel Kamga Department of Plant Biology, Faculty of Science, University of Dschang, Dschang, Cameroon

Lawrence Kazembe Department of Computing, Mathematical and Statistical Science, School of Science, Faculty of Agriculture, Engineering and Natural Science, University of Namibia, Windhoek, Namibia

Alexander Legwegoh London, ON, Canada

Heather Mackay Department of Geography, Umeå University, Umeå, Sweden;
Department of Human Geography, Lund University, Lund, Sweden

Ndeyapo Nickanor Faculty of Agriculture, Engineering and Natural Science, University of Namibia, Windhoek, Namibia

Michael Akindele Ojo Federal Technical University of Minna, Minna, Nigeria

Samuel Onyango Omondi Department of Agricultural Economics, University of Nairobi, Nairobi, Kenya

Inês Macamo Raimundo Faculty of Arts and Social Sciences, Department of Geography and Center for Policy Analysis, Eduardo Mondlane University, Maputo, Mozambique

Danielle Resnick Brookings Institution, Washington, DC, USA

Liam Riley Balsillie School of International Affairs, Wilfrid Laurier University, Waterloo, ON, Canada

Bhavna Sivasubramanian Department of City and Regional Planning, Cornell University, Ithaca, NY, USA

Likita Tanko Federal Technical University of Minna, Minna, Nigeria

Godfrey Tawodzera Department of Environmental Science, University of Namibia, Windhoek, Namibia

Daniel Tevera Department of Geography, Environmental Studies and Tourism, University of the Western Cape, Cape Town, South Africa

Kurt Waldman Department of Geography, Indiana University, Bloomington, IN, USA

Stephanie A. White Michigan Department of Health and Human Services, Division of Environmental Health, Lansing, MI, USA

Lovemore Itai Zuze Department of Food Security and Nutrition, University of Livingstonia (Laws Campus), Livingstonia, Malawi

List of Figures

List of Tables

CHAPTER 1

Introduction: African Secondary City Food Systems in Context

Liam Riley and Jonathan Crush

The sub-field of urban food systems in sub-Saharan Africa has expanded rapidly over the past decade with empirical research and theoretical reflection in multiple fields (Battersby & Watson, 2019; Crush et al., 2020; Frayne et al., 2018; Hovorka, 2013). This body of work dispels the myth that food systems are primarily rural development issues and that food availability in urban areas means that food security is not a problem (Crush & Frayne, 2014; Crush & Riley, 2019). The collection of research chapters in this book focuses on a specific type of urban environment

L. Riley (✉) · J. Crush
Balsillie School of International Affairs, Wilfrid Laurier University, Waterloo, ON, Canada
e-mail: lriley@balsillieschool.ca

J. Crush
University of the Western Cape, Cape Town, South Africa
e-mail: jcrush@balsillieschool.ca

L. Riley and J. Crush (eds.), *Transforming Urban Food Systems in Secondary Cities in Africa*,
https://doi.org/10.1007/978-3-030-93072-1_1

in Africa, that of secondary cities, and provides new knowledge on a wide range of topics related to their food systems. This introductory chapter draws several through lines of thematic concepts, shared experiences, and intersecting agendas. We first outline four main reasons for this book. Second, we review the definitions of "urban," "secondary cities," and "food security," with reference to the 12 countries covered in these chapters. Finally, we sketch key themes developed through the chapters. These themes illustrate the multifaceted ways in which food systems research supports the agenda for sustainable and inclusive development throughout sub-Saharan Africa.

Researching Food Systems in Secondary African Cities

The main purpose of this book is to bring together research that can draw a fuller picture of how food systems are transforming in secondary African cities and the impacts of these changes for residents. There are at least four main reasons why this is an important and timely project: first, because Africa is rapidly urbanizing; second, because food insecurity is widespread in African cities; third, because localizing global development requires knowledge about secondary city food systems; and fourth, because food systems governance is a complex challenge and information about the situation is scant and dispersed across various fields of research and policy.

Africa Is Rapidly Urbanizing

Most of the anticipated three billion increase in the human population by mid-century will occur in cities of the Global South. Cities in low- and middle-income countries will absorb 94% of urban growth in the next two decades, and, by 2050, will be home to 5.5 billion people or 83% of the world's urban population (UN-DESA, 2018). The significance of this urban transition is recognized in Sustainable Development Goal (SDG) 11: Sustainable Cities and Communities, which aims to "make cities and human settlements inclusive, safe, resilient and sustainable" (UN-DESA, 2021). James Duminy et al. (2013, p. 153) note that with rapidly expanding populations, protracted economic malaise, and governance capacity constraints, "the urban challenge in Africa is undoubtedly more serious than in any other part of the world." Sub-Saharan Africa is projected to have 1.3 billion people living in its cities by 2050, nearly trebling its urban population in 30 years.

One of the distinguishing features of rapid urbanization in sub-Saharan Africa is that it is increasingly taking place in smaller towns and cities. These urban centers currently absorb two-thirds of all urban population growth, but receive less infrastructure investment, policy focus, and academic attention than large primary cities (UN-Habitat, 2008, 2014). Cities with populations of less than 500,000 now account for a higher share of the urban population in low-income countries than in both high- and middle-income countries (Fig. 1.1). In sub-Saharan Africa, 54% of the urban population lives in secondary cities. Secondary urbanization in

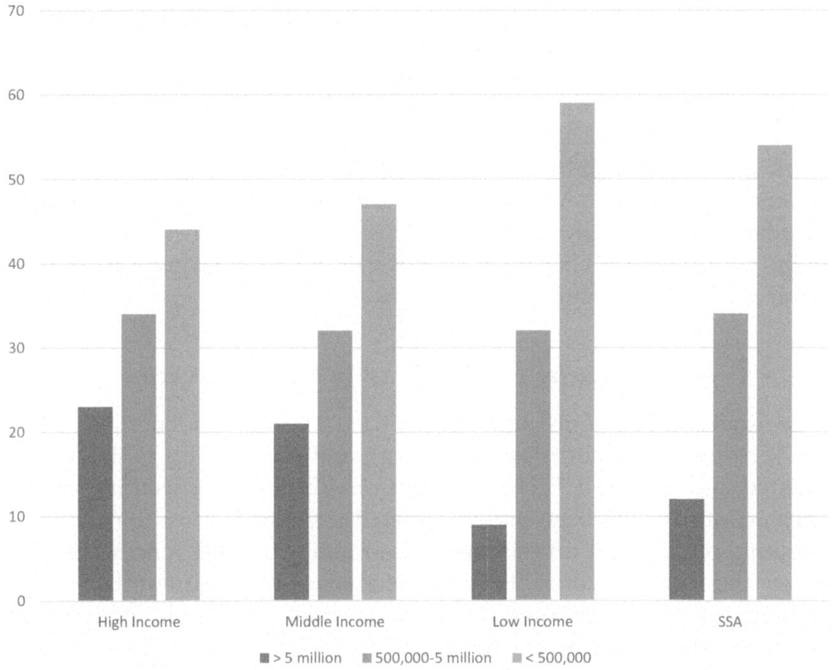

Fig. 1.1 Urban Population Distribution by City Size in High-, Middle- and Low-Income Countries Globally and in Sub-Saharan Africa Regionally (*Source* UN-DESA [2018])

Africa has many implications for development planning, which is limited by knowledge gaps both in descriptive studies of how these urban environments function and in theoretical reflection on the role of secondary urbanization in development processes. Several projects have been raising awareness of this knowledge gap in recent years (Battersby & Watson, 2019; Marais et al., 2016; Roberts, 2014; Roberts & Hohmann, 2014).

Cities Alliance (2021, p. 16) has recently noted how the impact of COVID-19 has further exposed the underlying vulnerability of African secondary cities:

> The impact of COVID-19 compounds existing crises being experienced in many secondary cities, especially in Sub-Saharan Africa. These cities face the prolonged effects of climate change, high informal sector employment and poor endowment of public goods and services. The tendency of large cities and metropolitan regions to receive a disproportionate share of public resources, investment and jobs further exacerbates the situation of inequity facing the recovery efforts and the future development of secondary cities in many countries.

A lack of resources, infrastructure, and "autonomy to pursue strategic long-term planning" (Assane, 2019) are some of the key constraints secondary cities face in addressing newer challenges such as COVID-19 and longstanding challenges of poverty, population growth, climate change, and food insecurity.

Food Insecurity Is Widespread in Secondary Cities

Marie Ruel (2020) notes that "the current state of evidence on urban food security, diets, nutrition, and health and their drivers is shockingly outdated and scattered." The problem of urban food insecurity on the African continent has been underestimated for years relative to rural food insecurity, often based on misconceptions about the differences between hunger and food insecurity. Even in the absence of consistent data disaggregated by urban and rural, there is growing case study evidence of food insecurity in urban environments across Africa. The African Food Security Urban Network (AFSUN) baseline survey found most households were severely food insecure in low-income urban areas in Southern African cities (Crush & Frayne, 2014; Frayne et al., 2018). These findings are corroborated by several empirical studies that reaffirm the widespread presence of hunger, malnutrition, and vulnerability in cities across sub-Saharan Africa (Battersby & Watson, 2019; Berlie, 2020; Chilanga et al., 2017; IPC, 2021; Van Wesenbeeck, 2018). This book provides further evidence of the high prevalence of urban food insecurity, while focusing on the secondary city context and offering specific insights into its drivers and manifestations.

The results of two interconnected research projects provide some indication that secondary cities are home to higher levels of severe household food insecurity than primary cities. The Hungry Cities Partnership conducted city-wide household surveys in three primary cities: Cape Town, South Africa (Crush et al., 2018); Maputo, Mozambique (Raimundo et al., 2018); and Nairobi, Kenya (Owuor, 2018). AFSUN's Food, Urbanization, Environment, and Livelihoods (FUEL) project conducted city-wide household surveys in Mzuzu, Malawi (Riley et al., 2018), Oshakati-Ondangwa-Ongwediva, Namibia (Nickanor et al., 2019) and Dschang, Cameroon (Legwegoh et al., 2020). Both studies used similar sampling techniques and the Household Food Insecurity Access Prevalence tool to assess household food insecurity levels. Comparing the percentage of households who were severely food insecure, the secondary cities had higher rates than the primary cities (Fig. 1.2). In Oshakati-Ondangwa-Ongwediva and Dschang, most households were severely food insecure. It is likely that other dimensions of food insecurity are also different in primary and secondary cities; for example, where the relative diversity of food retail sources creates advantages in primary cities (Mackay, 2019) and where proximity to rural areas and agricultural production creates advantages in secondary cities (Abwe & Daniel, 2021).

Localizing Global Development Requires Knowledge About Secondary City Food Systems

The large and growing population in African secondary cities provides an impetus to spur research into how to meet the SDGs in these contexts. Moreover, food systems intersect with many of the SDGs, such that efforts to understand and address food systems challenges in African secondary cities can advance progress on several SDGs for millions of people (UN-DESA, 2021). The following are six key SDGs related to the research in this book:

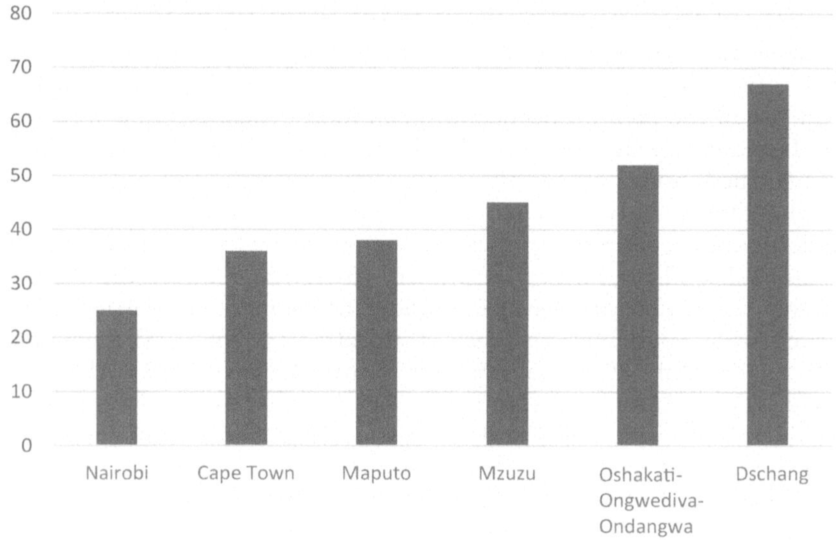

Fig. 1.2 Percentage of Severely Food Insecure Households by City (*Source* African Food Security Urban Network and Hungry Cities Partnership)

- SDG 1 (No Poverty)—Food systems are major sources of livelihoods and engines of local economic development for many secondary cities. Informal activities within urban food systems are particularly important sources of food and livelihoods for the urban poor.
- SDG 2 (Zero Hunger)—Secondary city food systems must be strengthened in order to ensure no one goes hungry. There are also relevant issues about urban food security data and subjective meanings of food insecurity in various cities.
- SDG 3 (Good Health and Well-being)—The goal of food and nutrition security underpins good health and well-being. It also relates to specific issues, such as food hygiene in shops, markets, and homes, and the cultural significance of well-being embedded in food.
- SDG 5 (Gender Equality)—The responsibility of feeding households continues to be gendered in most contexts and is often a burden for women, especially in impoverished communities. The experience of food insecurity is also generally more severe for women.

- SDG 11 (Sustainable Cities)—The goal of sustainable food systems underpins environmental, social, and economic sustainability for urban communities.
- SDG 12 (Responsible Production and Consumption)—African secondary city food systems include productive and consumptive activities that can be improved. These systems should also be studied to inform global discourse about sustainable food system models.

In the context of urban development challenges, the SDGs are usually paired with the New Urban Agenda (NUA), which sets out a vision and agenda for sustainable urbanization as part of the UN-Habitat system (Habitat III, 2017; Parnell, 2016). Jane Battersby and Vanessa Watson (2020) note that food is frequently linked in the NUA to what they call the "local trap," which over-emphasizes spatially bounded city regions and local issues of agriculture, rural–urban linkages, and small-scale farmers. While the NUA undoubtedly falls into this trap, it is plausible to think that localism may be more applicable to secondary cities than to their larger primary city counterparts, especially in agriculturally productive areas.

There is increasing recognition of the need for strong local and regional governments to influence global policy development and to implement these global visions (UCLG, 2014). The UN75 vision for post-COVID development places local and regional government in a central role for "localizing the SDGs and adapting them to the realities of the world" (UN75, 2020, p. 18). Through a "governance of proximity," local and regional governments will "bring legitimacy to the global agenda" and lead to a "renewal of the system" that has tended to be top-down. In most African contexts, and especially in African secondary cities, realizing this vision will require much deeper understanding of local governments and much more investment in their capacity to play this central role. The tendency to overestimate the capacity of local governments in Africa and to imagine their roles in unrealistic terms has been detrimental to the success of past development initiatives (Battersby & Watson, 2019). This book will aid in the effort to build knowledge about local contexts that are needed for the grand UN75 vision to be realized.

Food Systems Governance Is a Complex Challenge

The potential for local governments to make progress toward sustainable, healthy, and inclusive communities through food system improvements can only be realized through much more research into the ways various local governments function and the barriers they face. The contribution of this book is not to provide a roadmap toward sustainable urban food systems, but rather to document critically important and interrelated issues for the purposes of learning and reflection. Because of the diversity of African secondary cities and the complexity of their challenges, it is critically important to understand the context in which specific problems emerge and the vantage point from which they are analyzed and interpreted. David Satterthwaite and Cecilia Tacoli (2003, p. i) have argued that generalizations are "problematic and unhelpful in policy formulation" and that case studies can help to address the diversity of local factors, which include the very definition of "urban" areas in the local context. The multiple case studies in this book highlight the diversity of contexts and the multiple ways that food systems are being transformed through demographic growth, the actions of residents, environmental change, and economic transitions. The various methodologies, disciplinary backgrounds, and focal points of these case studies create a broadly inclusive picture of this change and the threats, opportunities, and challenges it entails. The case study approach facilitates a view of food systems in African secondary cities that situates the governance challenges within a realistic context and provides a ground-up view of the situation.

What Are Secondary Cities?

The chapters in this book discuss case studies of urban areas that are not major economic or political centers on the national scale and have fewer than 500,000 residents. Various reports and studies use different population cut-offs, sometimes with a minimum population to distinguish between secondary and large cities (Satterthwaite, 2017). In many parts of Africa, urban population statistics are highly questionable due to inadequate census records, mobile populations, informal urban development, and differing definitions of city boundaries. Coupled with rapid growth, these factors can create wide ranges in the estimated population of a particular city. Brian Roberts (2014, p. 37) provides a "reworked" definition of a secondary city, which recognizes that "population size still has a significant influence on the importance and classification of cities, but that function, specialty, logistics and trade importance and competitiveness are other determinants of a secondary city's status." He proposes a hybrid definition that integrates size, function, and role within a network of national, regional, and global systems of cities. In the first chapter of this book, Gareth Haysom expands on the relevance of Roberts' typology to research on food governance in African secondary cities.

The demographic trends in sub-Saharan Africa demonstrate the urgent need to address the food system and food security challenges that come with secondary urbanization, even as the specific definitions of secondary cities are being debated. This book is organized around case studies from 12 countries in sub-Saharan Africa, which form a unique collection of viewpoints that showcase the diversity of African urban environments. Among these 12 countries are examples of societies at varying levels of urbanization, but all are projected to be much *more* urban by 2050 (Fig. 1.3). Among the dozen countries with case study cities featured in this collection, three are currently majority urban (Cameroon, Ghana, and South Africa) and two are half urban/half rural (Namibia and Nigeria). In 2050, it is projected that eight of these countries will be majority urban.

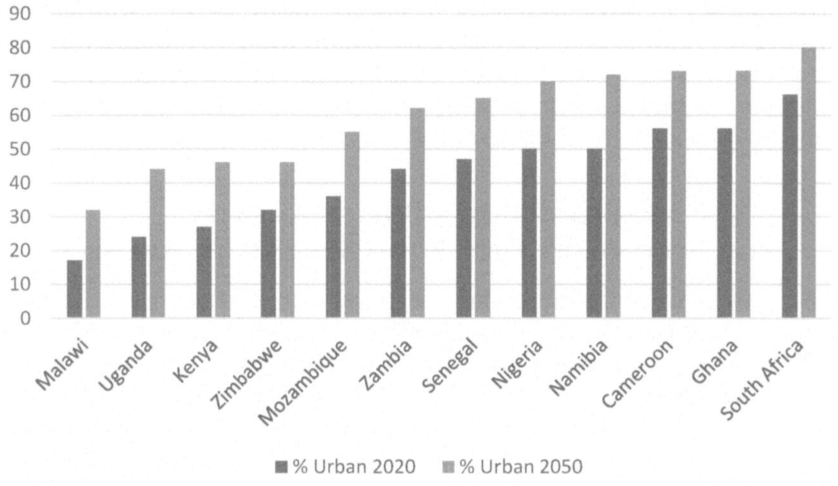

Fig. 1.3 Urban Population Projections in Selected African Countries (*Source* UN-DESA [2018])

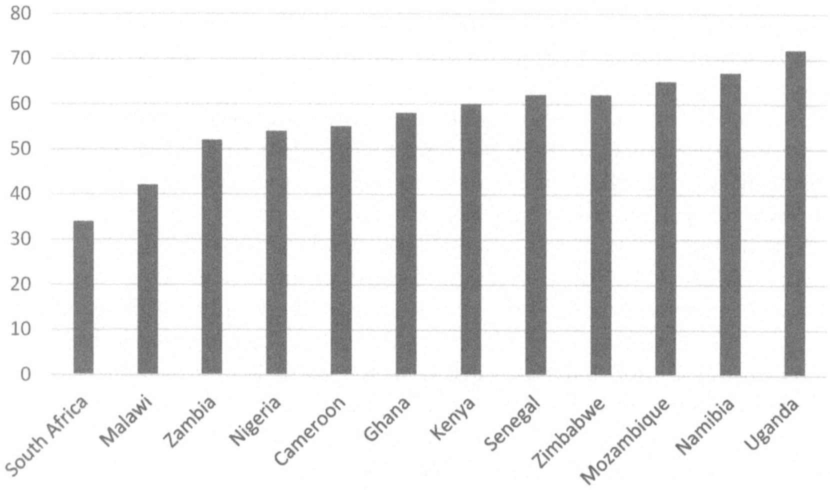

Fig. 1.4 Percentage of the Urban Population in Secondary Cities in Selected African Countries (*Source* UN-DESA [2018])

In all but two countries (Malawi and South Africa), the urban population is mostly in secondary cities (Fig. 1.4). There are different factors shaping the rates of secondary urbanization in these countries: in South Africa and Malawi, multiple major cities house most of the urban population, whereas on the other end of the spectrum, in Namibia and Uganda, only the capital city is a non-secondary city.

Elements of a Policy Agenda for Secondary City Food Systems in Africa

This book is organized into three sections. Part I, *Food System Actors, Concepts and Governance,* includes six chapters that draw attention to the theoretical and conceptual issues that underpin their case studies. Part II, *Food Security, Poverty, and Livelihoods,* includes six empirical case studies related to food and nutrition security, poverty, livelihoods, and household food strategies. Part III, *Environments, Linkages, and Mobilities,* includes six additional case studies that cover environmental and geographical factors shaping food systems and food security outcomes. These sections form the outline of a collection of chapters that converge in multiple ways and, taken together, reveal several key issues and policy dilemmas. Key issues are outlined below and orient the reader to where they are discussed in detail in the individual chapters.

Regulating Informal Food Systems

The informal food sector "plays a critical role in urban food system functioning as a whole" (Young & Crush, 2020, p. 198), and yet by its nature of operating outside of regulations it is inherently difficult to govern (Bromley, 2000; Skinner, 2008). The question of regulating informal traders raises critical concerns about the rights of the poor to participate in urban lifestyles and urban economies (Bénit-Gbaffou, 2016; Young, 2018), even as practical concerns about traffic congestion, labor standards, and food safety compel governments to attempt to control the informal food sector.

Several chapters in this volume offer empirical research that is useful for constructive policy and theoretical discourses. Godfrey Tawodzera reviews key scholarly debates about the value of informal food traders in African cities, a debate that he traces back to the 1970s with the introduction of the "informal sector" as an analytical concept for describing the unregulated areas of economic activity. In her chapter, Stephanie A. White critically examines the concept of informality and explains how its significance goes beyond the descriptive value of the degree to which activities are regulated by the state. She notes that the concept permeates the placemaking process for African urban environments, to the extent that it is a key characteristic of African urbanism.

The informal sector dominates the food system in many secondary cities, as demonstrated by Tawodzera's case study in Epworth, Zimbabwe, and other case studies in this volume, including in Namibia (Ndeyapo Nickanor et al.), Cameroon (Alexander Legwegoh and Liam Riley), Malawi (Lovemore Zuze), Zambia (Jordan Blekking et al.), and Nigeria (Danielle Resnick et al.). Jordan Blekking et al. argue that the diversity of food retailers supports wider access to food, partly by providing geographical access to low income residential neighbourhoods. As Zuze's chapter shows, the informal sector can also provide an important source of income for marginalized urban residents. Although the informal sector is vital to how these food systems currently function, as Emmanuel Chilanga and Liam Riley show in their chapter, it presents challenges such as how to ensure food safety. Resnick et al. further demonstrate the governance challenge of regulating informal traders in Minna and Calabar, Nigeria, where a lack of public concern and a lack of state resources to enforce bylaws have hindered efforts to address the problems associated with informality.

Risk and Resilience of Food Systems

The ways in which urban food environments are being built in sub-Saharan Africa are creating new forms of vulnerability to climate change, for example, through the expansion of unplanned settlements in flood-prone areas and populations that exceed the availability of water during droughts (Fraser et al., 2017; Parnell & Walawege, 2011). In secondary cities, there tends to be much less capacity for local governments to mitigate these risks through planning and infrastructure development (Satterthwaite, 2017). There is also a gap in the data required to understand how urban food systems are impacted by hazards linked to climate change, insufficient infrastructure, and endemic health hazards. The research that informs the chapters in this book was conducted before the onset of COVID-19, but it is reasonable to speculate, in line with the FAO (2020), that the pandemic has exposed the underlying vulnerabilities of the food systems described in these case studies.

Yanick Borel Kamga's chapter addresses the risk to the food system in Dschang, Cameroon, posed by the effects of climate change on Cameroon's forests. Forests provide a wealth of biodiversity and are a source of many popular and culturally significant foods in Cameroon. Climate change poses yet another threat to these biodiverse ecosystems, which have suffered from human encroachment and unsustainable exploitation over several decades. Inês Macamo Raimundo and Mary Caesar recount the impact of floods on the food system in Xai-Xai, Mozambique, a coastal city where extreme weather events have become more common due to climate change. Climate change is a present threat to these cities and any efforts to improve the sustainability of their food systems must grapple with rapid changes in the natural environment.

Other risks are documented in chapters such as Chilanga and Riley's study of hygienic facilities for informal food traders, which shows that many urban consumers are at risk from foodborne pathogens. The risk of homelessness is high in Gweru, Zimbabwe, as documented in Miriam Grant's chapter on the livelihood strategies of low-income households in the tumultuous political period of the early 1990s. Her case study also highlights the resiliency of urban residents in the face of multiple challenges. Legwegoh and Riley offer political economic explanations for the high rate of food insecurity amid the agricultural abundance of Dschang, Cameroon, where structural problems undermine the efforts of households to achieve food security. These case studies show the risks and challenges faced by households and cities, and help to address the knowledge gaps that can lead policymakers to underestimate the scale of the problems and misconstrue their nature.

Strengthening Local Policy Capacity and Community Input

In light of challenges such as regulating informal parts of the food system and mitigating risk with limited resources, the question of how to strengthen the local capacity to improve food systems is raised in several chapters. Andrea M. Brown's chapter, for example, addresses this issue directly by describing the activities of Slum/Shack Dwellers International in Mbale and Jinja, Uganda. The organization facilitates community input into the development of practical and policy solutions to their everyday problems. Resnick et al. demonstrate some of the limitations of policy-driven development in Nigerian secondary cities, where instances of progressive policies toward vendors are sometimes not implemented. In the case of the 2016 Hawkers' Rights Protection Bill in Calabar, most vendors were not aware of the policy.

The chapters offer more examples of policies that, far from strengthening local food systems, had unintended consequences. In his chapter on Tamale, Ghana, Issahaka Fuseini draws on interviews with elderly residents who recall the deleterious effects of green revolution agricultural policies on their diets and sense of food security. Research participants in Xai-Xai, Mozambique, whose views are documented in Raimundo and Caesar's chapter, criticized the framing of food security by governments and researchers. Their message was one of frustration with food security research and policy that does not give them what they need: a plate of *xima* for their daily meal. In these critical discussions of failed policy, there is an important lesson relevant to all secondary cities: that the voices of residents need to be heard and their vision for food and nutrition security should inform food system planning.

Meeting the Challenges of Global Food System Integration

There has been a sustained interest in the effects of globalization on urban food systems since at least the 1990s (Atkinson, 1995; Smith, 1998). Recent research has focused on the effects of supermarket chains on the local food systems of African cities, which offer more consistent availability of imported and packaged foods but at the cost of disrupting local food networks and informal livelihoods (das Nair, 2020). Secondary cities are generally less closely linked to these supply chains than primary cities, which tend to be the core international trading centers of their countries. The case studies in this volume reflect the diversity of stages of "supermarketization" in secondary African cities. The case studies range from Dschang, Cameroon, and Epworth, Zimbabwe, which have no supermarkets, to Oshakati-Ondangwa-Ongwediva, Namibia, where nearly all households purchase some food at a supermarket. Heather Mackay et al. found some differences in the overall rate of supermarket patronage among their six case study cities in Kenya, Uganda, and Ghana, but for all six cities higher socio-economic status was associated with higher supermarket patronage, and access to food from supermarkets was not ensured by its availability.

One of the key concerns in the literature is that the expansion of supermarkets is part of a broader nutritional transition whereby urban households consume more convenience foods, with higher sugar and fat content, and therefore experience a loss of nutrition security (Bloem & de Pee, 2017). The model suggests that this process disrupts local supply chains (which are more likely to supply freshly grown produce, locally adapted varieties of grains, and local animal products) and reduces the availability of local food. The posited effect is that the most nutritious food is too expensive for the urban poor and nutrition insecurity is a poverty-related outcome different from the supermarket-centric model observed in American "food deserts" (Crush & Battersby, 2016; Crush & Si, 2019). The process in secondary cities also appears to be more complex than the city-region food system model (CRFS) suggests (Blay-Palmer et al., 2018).

Several case studies in this book do depict secondary cities fed by the farms in their immediate vicinity, which is in line with the CRFS model and could be supported with infrastructure and agricultural investment. On the other hand, the limitations of this model are apparent in the multiple connections with places beyond the immediate hinterland

and the impact of national policies beyond the purview of municipalities. Haysom's chapter, for example, analyzes supply chains for various foods in three secondary cities and finds that supermarkets are often integrated into informal food networks. They often act as suppliers for informal traders, who frequently source food over far-reaching and international networks. From the consumption point of view, the chapters by Mackay et al. and Nickanor et al. demonstrate a strong mix of food sourcing strategies for most households, even among wealthier households who could choose to rely mainly on supermarkets. Any agenda for sustainable urban food systems needs to take these dynamic geographies, which include the food chains and global social and environmental trends described above, into account.

Ensuring Gender-Equitable Food Systems

The issue of gender equality is an important aspect of any sustainable food systems agenda, not least because gender inequality is a causal factor of poverty and gender roles play a central role in how the food system is organized, especially informal food systems (Hovorka, 2013; Riley & Dodson, 2020). In her chapter, Gamuchirai Chakona highlights the situation of female-headed households in four peri-urban centers in South Africa where women cope with limited livelihood opportunities by diversifying their food sources. In Namibia, Nickanor et al. found higher rates of food insecurity and lower dietary diversity among female-headed households compared to male-headed households. However, male- and female-headed households were equally likely to rely on intensive coping strategies. In general, factors such as the use of informal food sources and living in informal housing were much more closely associated with vulnerability than the gender of the household head. These chapters demonstrate the importance of intersecting contextual factors to understand the connections between gender and food insecurity.

Gender roles are rapidly changing in African cities with the re-allocation of women's time to productive and reproductive work (Chant & McIlwaine, 2016). Some chapters demonstrate the effect of this change on food systems. In Tamale, Ghana, Fuseini documents the intertwined processes of economic change with changing food cultures and gender roles. Drawing on interviews with older men and younger women, he presents evidence of what has been lost over time from the perspectives of the older men, such as traditional crops and recipes with a profound cultural significance. The younger women are not familiar with these foods, and they have less time to cook than women in the past because of their other pursuits. Grant highlights the situation for migrant households in Zimbabwe during a period of rapid economic decline. Housing insecurity and gender roles were key factors that interacted to shape their food security outcomes, while social networks strongly influenced the food security strategies they chose. In Zuze's chapter, the vulnerability of migrant women among food vendors in Mzuzu, Malawi, illustrates one of the ways that gender-based inequality is perpetuated in secondary cities as men and women coming to the city are presented with highly unequitable livelihood opportunities.

Including Migrants in Urban Development

The rapidly changing population dynamics in African cities present a perennial challenge for social scientists and policymakers. The dominant discourse has focused on the flow of rural migrants to large cities, with some literature showing that migrants play a key role in urban food systems while also facing many forms of vulnerability (Crush, 2013; Crush & Tawodzera, 2017). One difference observed between primary and secondary cities is the heightened importance of rural–urban linkages and circular migration in the latter, for example in urban households that operate farms in rural areas or maintain close ties with rural households who send food remittances. These dynamics are evident in several chapters in this volume.

Another type of connection is urban-urban. Legwegoh and Riley note that in the university town of Dschang, Cameroon, most of the migrant population was born in another city in Cameroon and many households receive food transfers from relatives in other cities. There are also many internally displaced people in Dschang and their presence placed a strain on local households that depressed household food security scores. Secondary cities can also be a source of migrants and act in the role usually occupied by rural communities in conventional rural–urban migration models. In his chapter, for example, Anil Dhakal documents in detail how remittances from household members working in South Africa provide support for the food security of households in Mzuzu, Malawi.

The issues of migration and complex population dynamics make it difficult to implement effective local responses to transforming urban food systems. At the household scale, migration and linkages with other places are key elements of many food security strategies.

Developing Sustainable Urban and Peri-Urban Agriculture

The final issue addressed in the chapters in this volume is urban and peri-urban agriculture (or UPA), which is particularly prominent in urban food policy discussions in Africa where it is frequently framed as a key policy tool for sustainable urban food systems. The diversity of forms of UPA food production—from industrial-scale commercial production to backyard gardens—means that the topic incorporates many normative visions of its potential. In practice, however, the extent and opportunities for engaging in UPA in primary cities are quite limited (Crush, Caesar & Haysom, 2018; Frayne et al., 2016). Rates of participation in UPA and its importance to household food security do appear to be more significant, although not barrier-free, in secondary cities than in primary cities (Davies et al., 2020). Most of the chapters in this book that discuss UPA describe situations in which households use the resources available to them to produce food for their own consumption. The high-level picture is that food production on a scale that is sufficient to ensure food security is mainly available to households with relatively greater access to resources. Daniel Tevera notes this as one of several points of political and social tension that emerge from debates about UPA as a policy tool for sustainable urbanization. While there is little evidence that UPA has been a panacea for the urban poor in Africa, it continues to hold promise for strengthening local food systems such that secondary African cities will be more resilient to shocks such as the disruptions to food networks witnessed during the COVID-19 pandemic.

References

Abwe, F., & Daniel, K. (2021). Local public markets infrastructure and urban food systems in African small cities: Insights from Arusha, Tanzania. *Cities and Health*, *1*(1), 1–4.

Assane, M. (2019). Putting Africa's secondary cities first. *Project Syndicate* (online publication, 21 May). https://www.project-syndicate.org/commentary/africa-secondary-cities-inclusive-growth-by-ibrahim-assane-mayaki-2019-05

Atkinson, S. (1995). Approaches and actors in urban food security in developing countries. *Habitat International, 19*(2), 151–163.

Battersby, J., & Watson, V. (Eds.). (2019). *Urban food systems governance and poverty in African cities*. Routledge.

Battersby, J., & Watson, V. (2020). The planned "city region" in the New Urban Agenda: An appropriate framing for urban food security? In J. Crush, B. Frayne, & G. Haysom (Eds.), *Handbook on urban food security in the global south* (pp. 341–362). Elgar.

Bénit-Gbaffou, C. (2016). Do street traders have a "right to the city"? The politics of street trader organisations in inner city Johannesburg, post-Operation Clean Sweep. *Third World Quarterly, 37*(6), 1102–1129.

Berlie, A. (2020). The invisible crisis of urban food security in Amhara Regional State, Ethiopia. *Journal of Hunger and Environmental Nutrition* (published online). 1, 20.

Blay-Palmer, A., Santini, M., Dubbeling, M., Renting, H., Taguchi, M., & Giordano, T. (2018). Validating the city region food system approach: Enacting inclusive, transformational city region food systems. *Sustainability, 10*(5), 1680.

Bloem, S., & de Pee, S. (2017). Developing approaches to achieve adequate nutrition among urban populations requires an understanding of urban development. *Global Food Security, 121*(1), 80–88.

Bromley, R. (2000). Street vending and public policy: A global review. *International Journal of Sociology and Social Policy, 20*(1/2), 1–28.

Chant, S., & McIlwaine, C. (2016). *Cities, slums and gender in the global south: Towards a feminised urban future*. Routledge.

Chilanga, E., Riley, L., Ngwira, J., Chalinda, C., & Masitala, L. (2017). *Food insecurity in informal settlements in Lilongwe, Malawi*. African Food Security Urban Network (AFSUN) (Urban Food Security Series No. 25).

Cities Alliance. (2021). *Secondary cities post-COVID 19: Achieving urban sustainable and regenerative development in emerging economies*. Cities Alliance/UNOPS.

Crush, J. (2013). Linking migration, development and urban food security. *International Migration, 51*(5), 61–75.

Crush, J., & Battersby, J. (Eds.). (2016). *Rapid urbanisation, urban food deserts and food security in Africa*. Springer.

Crush, J., & Frayne, B. (2014). Feeding African cities: The growing challenge of urban food insecurity. In S. Parnell & E. Pieterse (Eds.), *Africa's urban revolution* (pp. 110–132). Zed Books.

Crush, J., & Riley, L. (2019). Rural bias and urban food security. In J. Battersby & V. Watson (Eds.), *Urban food systems governance and poverty in African cities* (pp. 42–55). Routledge.

Crush, J., & Si, Z. (Eds.). (2019). *Urban food deserts: Perspectives from the global South*. MDPI.

Crush, J., & Tawodzera, G. (2017). South-South migration and urban food security: Zimbabwean migrants in South African cities. *International Migration, 55*(4), 88–102.

Crush, J., Caesar, M., & Haysom, G. (2018). *The state of household food security in Cape Town, South Africa* (HCP Report No. 12). Hungry Cities Partnership.

Crush, J., Frayne, B., & Haysom, G. (Eds.). (2020). *Handbook on urban food security in the global south*. Edward Elgar.

das Nair, R. (2020). The "supermarket revolution" in the south. In J. Crush, B. Frayne, & G. Haysom (Eds.), *Handbook on urban food security in the global south* (pp. 113–144). Edward Elgar.

Davies, J., Hannah, C., Guido, Z., Zimmer, A., McCann, L., Battersby, J., & Evans, T. (2020). Barriers to urban agriculture in Sub-Saharan Africa. *Food Policy,* 1(1), 101999.

Duminy, J., Watson, V., & Odendaal, N. (2013). Doing research in African cities: The case study method. In P. Kresl & J. Sobrino (Eds.), *Handbook of research methods and applications in urban economies* (pp. 153–172). Edward Elgar.

FAO. (2020). *Addressing the impacts of COVID-19 in food crises*. FAO.

Fraser, A., Leck, H., Parnell, S., Pelling, M., Brown, D., & Lwasa, S. (2017). Meeting the challenge of risk-sensitive and resilient urban development in sub-Saharan Africa: Directions for future research and practice. *International Journal of Disaster Risk Reduction, 26*(C), 106–109.

Frayne, B., McCordic, C., & Shilomboleni, H. (2016). The mythology of urban agriculture. In J. Crush & J. Battersby (Eds.), *Rapid urbanisation* (pp. 19–32). Springer.

Frayne, B., Crush, J., & McCordic, C. (Eds.). (2018). *Food and nutrition security in Southern African Cities* (1st ed.). Routledge.

Habitat III. (2017). *New urban agenda*. United Nations.

Hovorka, A. (2013). The case for a feminist foodscapes framework: Lessons from research in urban Botswana. *Development, 56*(1), 123–128.

Integrated Food Security Phase Classification (IPC). (2021). *IPC acute food insecurity analysis- Urban Kenya*. http://www.ipcinfo.org/fileadmin/user_upload/ipcinfo/docs/IPC_Kenya_Acute_Food_Insecurity_2020AugDec_Urban.pdf

Legwegoh, A., Kamga, Y., Riley, L., Martin, L., & Njukeng, P. (2020). *Food security in Africa's secondary cities: No. 3. Dschang, Cameroon*. (Urban Food Security Series No. 29). African Food Security Urban Network (AFSUN).

Mackay, H. (2019). Food sources and access strategies in Ugandan secondary cities: An intersectional analysis. *Environment and Urbanization, 31*(2), 375–396.

Marais, L., Nel, E., & Donaldson, R. (2016). *Secondary cities and development* (1st ed.). Routledge.

Nickanor, N., Kazembe, L., & Crush, J. (2019). *Food security in Africa's secondary cities: No. 2. The Oshakati-Ongwediva-Ondangwa Corridor, Namibia* (Urban Food Security Series No. 28). African Food Security Urban Network (AFSUN).

Owuor, S. (2018). *The state of household food security in Nairobi, Kenya* (HCP Report No. 11). Hungry Cities Partnership.

Parnell, S. (2016). Defining a global urban development agenda. *World Development, 78(C)*, 529–540.

Parnell, S., & Walawege, R. (2011). Sub-Saharan African urbanisation and global environmental change. *Global Environmental Change, 21(S)*, S12–S20.

Raimundo, I., McCordic, C., & Chikanda, A. (2018). *The state of household food security in Maputo, Mozambique* (HCP Report No. 10). Hungry Cities Partnership.

Riley, L., Chilanga, E., Zuze, L., & Joynt, A. (2018). *Food security in Africa's secondary cities: No 1. Mzuzu, Malawi* (Urban Food Security Series No. 27). African Food Security Urban Network (AFSUN).

Riley, L., & Dodson, B. (2020). The gender-urban-food interface in the global south. In J. Crush, B. Frayne, & G. Haysom (Eds.), *Handbook on urban food security in the global south* (pp. 218–232). Edward Elgar.

Roberts, B. (2014). *Managing systems of secondary cities: Policy responses in international development.* Cities Alliance/UNOPS.

Roberts, B., & Hohmann, R. (2014). *The systems of secondary cities: The neglected drivers of urbanising economies* (CIVIS Note Series No. 7). World Bank.

Ruel, M. (2020). Growing cities, growing food insecurity: How to protect the poor during rapid urbanization. *Center for Strategic and International Studies* (published online). https://www.csis.org/analysis/growing-cities-growing-food-insecurity-how-protect-poor-during-rapid-urbanization

Satterthwaite, D. (2017). The impact of urban development on risk in Sub-Saharan Africa's cities with a focus on small and intermediate urban centres. *International Journal of Disaster Risk Reduction, 26*, 16–23.

Satterthwaite, D., & Tacoli, C. (2003). *The urban part of rural development: The role of small and intermediate urban centres in rural and regional development and poverty reduction* (Working Paper Series on Rural-Urban Interactions and Livelihood Strategies, Working Paper No. 9). IIED.

Skinner, C. (2008). The struggle for the streets: Processes of exclusion and inclusion of street traders in Durban, South Africa. *Development Southern Africa, 25*(2), 227–242.

Smith, D. (1998). Urban food systems and the poor in developing countries. *Transactions of the Institute of British Geographers, 23*(2), 207–219.

UCLG. (2014). *Building the new urban agenda for intermediary cities.* UCLG.

UN75. (2020). *Visioning report UN75: The role of the local and regional governments in the global future* (online publication, 14 October). https://issuu.com/uclgcglu/docs/lrgs_visioning_un75_report__1_

UN-Department of Economic and Social Affairs (UN-DESA). (2018). *World urbanization prospects. The 2018 revision.* The UN-Department of Economic and Social Affairs.

UN-Department of Economic and Social Affairs (UN-DESA). (2021). *Sustainable development*. The UN-Department of Economic and Social Affairs.

UN-Habitat. (2008). *The state of African Cities 2008: A framework for addressing urban challenges in Africa.* UN-Habitat.

UN-Habitat. (2014). *The state of African Cities 2014: Reimagining sustainable urban transitions.* UN-Habitat.

Van Wesenbeeck, C. (2018). *Disentangling urban and rural food security in West Africa* (West African Papers No. 15). OECD.

Young, G. (2018). De-democratisation and the rights of street vendors in Kampala, Uganda. *International Journal of Human Rights, 22*(8), 407–423.

Young, G., & Crush, J. (2020). The urban informal food sector in the Global South. In J. Crush, B. Frayne, & G. Haysom (Eds.), *Handbook on urban food security in the global south* (pp. 198–217). Edward Elgar.

PART I

Food System Actors, Concepts and Governance

CHAPTER 2

Understanding Secondary City Typologies: A Food Governance Lens

Gareth Haysom

Introduction

Sub-Saharan Africa is urbanizing rapidly, but this urbanization process is not well understood, in intermediate or secondary cities in particular. Secondary cities will be home to the majority of future urban residents (Fox & Beall, 2012; Satterthwaite & Mitlin, 2013; UN-DESA 2018). This chapter recognizes the contestation over described trends in rates of urban growth, the associated patterns of poverty and inequality and their relationship to public policy (McGranahan et al., 2008; Potts, 2012), appreciating the challenges associated with using nationally derived demographic figures. Part of the challenge with inadequate demographic data has been a general disregard for African cities in global development and national political processes, as argued previously: "until fairly recently Africanists largely ignored, or were openly hostile to, almost all aspects of a wider urban agenda, focusing instead on issues such as the peasantry,

G. Haysom (✉)
African Centre for Cities, University of Cape Town, Cape Town, South Africa

L. Riley and J. Crush (eds.), *Transforming Urban Food Systems in Secondary Cities in Africa*,
https://doi.org/10.1007/978-3-030-93072-1_2

agriculture, natural resource use or national sovereignty. The anti-urban bias is fast receding, ushered out by evidence of the rate and scale of urbanisation" (Pieterse et al., 2018, p. 3).

The urbanization trajectory and the renewed interest in the intersections between food and urbanization have resulted in "new" concepts and approaches that seek to understand these intersections. There is emerging interest in the urban food system (and not just food insecurity or hunger) that seeks to better understand the interactions between the household and the city scale (Frayne et al., 2010). Other, more universal, food system trends, such as enquiry into the nature and form of food retail in the urban environment, both formal and informal (Reardon et al., 2003), are also evident in African secondary cities. These trends all intersect with wider governance considerations that have emerged as a result of recent global governance agreements embedded within the processes of the Sustainable Development Goals and the New Urban Agenda (Battersby, 2017).

Rapidly growing African urban areas encounter multiple and, at times, mutually reinforcing development realities related to food, including food insecurity, food systems transformation, the growth of supermarkets and the nutrition transition. When these food system-related transitions intersect with other transitions, such as increased poverty and inequality, increased climatic variability, global geopolitical shifts and global economic contraction, and watershed events such as pest infestations (often driven by climatic changes) and global health crises, urban areas face significant development and governance challenges. Urbanization across Africa is not taking place in a uniform manner and the implications of these changes, and the policy solutions, vary across contexts. This chapter starts with a brief overview of Africa's urbanization trajectory, reflecting on the different stages of urbanization and the differing trends, across regions and urban typologies. The chapter then draws on food systems studies carried out in three secondary African cities to call for far greater focus on the specificities of the different types of secondary cities. The case studies reviewed represent three different secondary city typologies: the trunk city; the satellite city; and the resource city.

The case studies are from the Consuming Urban Poverty (CUP) research project (Battersby & Watson, 2019; CUP, 2021a, 2021b; Joubert et al., 2018). CUP was a three-year research project funded by the Economic and Social Research Council and the United Kingdom's Department for International Development and carried out in collaboration among the African Centre for Cities, Copperbelt University (Zambia), the University of Zimbabwe and the Kisumu Local Interaction Platform—a research and policy knowledge hub facilitating urban research in Kisumu, Kenya. The overarching objective of the project was to understand what the urban food systems in the case study cities reveal about the dynamics of urban poverty and its governance, yielding lessons for generic poverty reduction. Conceptually the CUP project viewed food poverty as being determined by household-scale conditions—a failure of entitlements, as per Amartya Sen (1981)—and wider systemic challenges in the food system (Ericksen, 2008). The CUP project approached food poverty as a form of social exclusion determined by failures and deficiencies in food market structure and food system governance priorities, and specific priorities at the urban scale (Barling et al., 2002; Haysom & Fuseini, 2019; Smit, 2016).

A central feature of the CUP work was its alignment with the typological differentiation suggested by Lily Song (2013) and Brian Roberts (2014, p. 37), which extends the debate on secondary cities by proposing different secondary city typologies. This chapter suggests that conventional assessments of secondary cities, informed by questions of size and proximity to rural areas, are insufficient measures that fail to adequately capture the diversity of secondary cities, specifically in respect of food systems governance. Understanding the different city typologies offers new ways of describing secondary cities, less by way of location or size and more by the city's function. Three specific forms, which correspond to the three case studies in the CUP project, are evident:

- subnational urban centres of administration, manufacturing, agriculture or resource development and resource extractive areas in particular (Kitwe, Zambia);
- metropolitan clustered secondary cities, which develop on the periphery of metropolitan or urban regions and take the form of new towns, spillover growth centres and linear cities that could also include migrant and refugee cities, satellite towns or cities and often labour pool areas (Epworth, Zimbabwe); and
- corridor secondary cities, which develop as growth poles along major transportation or trade corridors, and are often sites where different modes of transport intersect (Kisumu, Kenya).

As with most typology uses, the clear typological classification may not always be possible, with some cities representing more than one such typology or new typologies that would expand this list. How cities are described and discussed at a particular time of engagement or food policy formulation, and the use of more refined typologies, can aid in the development of locally appropriate and sustainable policies. While specific typological classifications will not offer governance clarity, typological classification offers a tool to bring difference and contextual variations to the fore.

Africa's Urbanization Profile

In much of the academic and policy discourse on cities in Africa, primary cities dominate. The focus on the large metropolitan areas also dominates how urbanization in Africa is typically imagined. Mentions of the African city conjure up images of traffic chaos in Cairo, the complexity of governance in Lagos, the slums of Nairobi (with Kibera holding almost mythical status) and the crime and lawlessness of downtown Johannesburg, to mention just a few. Secondary cities have not received the same level of political, governance or popular focus. A possible reason for the focus on the primary, and mostly capital, cities is political. As Simon Bekker and Göran Therborn (2012, p. 1) have pointed out, "Capital cities have always played a central role in nation building and state building. These processes are both a symbolic movement and a quest to establish and maintain power. The nation state projects its power through the urban landscape and spatial layout of the capital city. This power is manifested in the capital's architecture, in its public monuments and the names of its streets and public spaces".

In his earlier work, David Satterthwaite dispelled "popular myths" about urbanization, the first being the supposed continued expansion of large megacities (Satterthwaite, 2007). Satterthwaite also challenged the notion of an urban bias in both academic and policy positions, suggesting that there is no evidence to support such claims (Satterthwaite, 2007). Satterthwaite suggests that in the context of urban growth, the challenges experienced are not caused by growth but by the inability of national and local institutions to adapt to the challenges presented by urban growth (Satterthwaite, 2007). These calls, made more than 10 years ago, have direct implications when engaging issues of urban food security and urban food systems governance.

African secondary cities face a number of key intersecting challenges. The first is the question of internal urban growth and increased net urban residents. This urban growth is not uniform and has clear contextual and historical trajectories. This difference is evident in the fact that settlements founded as a result of earlier resource booms often have far greater endogenous growth than those linked to trends associated with consolidation in agriculture or climate- and conflict-related migrations. The second challenge is governance and the fact that city size generally determines the skills, budgets and resourcing needed to effectively govern these secondary cities. Third, and perhaps less obvious, is that base infrastructure and existing form often dictate the nature of the expansion trajectory: the original form may have enabled some efficiency, informed by the needs of that population, but as the foundation for expansion this base is woefully inadequate. This challenge is particularly evident in a number of secondary cities in Africa that were founded during colonial periods as either resource extractive centres or regional administrative and trade centres. In these instances, the infrastructure provision was focused on a small, often white, elite, disregarding the needs of the wider population, or future needs (Duminy, 2018).

UN-DESA (2018) reports that Africa is 43% urbanized, countering the often held view that Africa is a predominantly rural continent. A further contextual consideration is the fact that, regionally, Africa is also urbanizing in different ways. Africa's urban profile is varied, specifically in terms of where the bulk of Africa's urban dwellers reside. From a governance perspective, the essential question needs to be: Where will the urban growth take place? Trends suggest that this will be in areas such as West, East and Central Africa. These are also the most populous regions, accounting for more than 320 million current urban African residents (more than 70% of all urban Africans). Combined, the scale of Africa's urban growth, the regional distribution of that growth and the typology of its growing cities, mostly secondary cities, present a significant governance challenge. This variability requires greater nuance and contextual appreciation. Current approaches and classifications of secondary cities do not provide this nuance.

Classifying Secondary Cities

The absence of nuance and contextual specificity means that governance and developmental responses based on generalizations of African urbanization miss key local trends, needs and trajectories. From a planning and development perspective, such an oversight means that development plans run the risk of effectively casting current misconceptions, and flawed policy understandings about future needs, into concrete. Developments built today will impact supply chains, infrastructure profiles and governance for the next 50 to 100 years (Pieterse et al., 2015).

Different theorists apply different terms to categorize cities. In the 1980s, a primate city was defined as "the leading city in its country or region, disproportionately larger than any others in the urban hierarchy" (Goodall, 1987). These cities were, and generally remain to this day, leading political or economic hubs in a region or country (see also Bekker & Therborn, 2012). In 1983, Dennis Rondinelli, drawing on the literature of the time, argued that secondary cities were critical for national development. Rondinelli's argument was that secondary cities were necessary to diffuse the urbanization that was mostly taking place in the primary cities at that time. Rondinelli argued that policies supporting the development of secondary cities would achieve several objectives, including: reversing polarization; alleviating primate city problems; reducing regional inequalities; stimulating rural economies; and reducing poverty in the urban sector generally (Fuchs, 1983; Rondinelli, 1983). Given the timing of Rondinelli's arguments, informed by evidence collected in the years prior, many of these primate cities had experienced the removal of colonial-imposed movement restrictions for local (essentially non-European) country residents, resulting in large numbers of internal migrants to cities, generally the primate city.

In the framing of secondary cities and the intersections between these and governance questions, the historical use of the term is also important. UN-Habitat has framed secondary cities as "an urban area generally having a population of between 100,000 and 500,000" (1996, p. 13). But as Roberts (2014) argues, the population levels informing this classification require constant revision as cities grow; based on his model in 2021, a secondary city could technically be a city with a population of several million people. Song (2013) has suggested that in Southeast Asia, secondary cities range from cities with populations of between 100,000 and three million inhabitants. There is, however, no universally agreed definition for the term "secondary city" (Roberts, 2014).

Today, secondary cities play a variety of different roles in addition to those envisioned by Rondinelli. In the race for prestige aligned to primary and capital cities, secondary cities have been forgotten, or disregarded, in terms of research and governance theorization and practice, even though these are the sites of Africa's future, as many are yet to be fully built. Secondary cities will be the engines of Africa's development trajectory. Primary and capital cities will remain important, but it will be the secondary African cities that enable, or inhibit, the attainment of Africa's potential demographic dividend (Pieterse, 2008).

This issue is increasingly important for urban food systems. In much of the food systems literature, secondary cities have been largely framed as demonstrating uniform functions and facing the same developmental challenges as larger cities. Secondary cities have also been predominantly framed as rural hubs or extensions of a rural agrarian economy. Drawing largely on demographic data, James Tefft and Marketa Jonasova (2020, p. 55) present a useful starting point in a provisional typology of cities in relation to their food systems. This differentiation addresses the need for practitioners and policymakers to tailor programme and policy recommendations in accordance with socio-economic, demographic and food system characteristics. Tefft and Jonasova (2020, p. 55) suggest three categories: agricultural towns or cities (C1) are "smaller but fast-growing populations and are in agricultural production areas with a key role in the rural economy"; medium and large secondary cities (C2) are "challenged to modernize food system architecture and strengthen food businesses to cater to the needs of diverse consumers"; and global megacities (C3) are "served by vibrant modern, traditional and informal food systems that are challenged to operate in congested environments, many in need of upgrading". Within this framework, secondary cities are key sites where new regional formations may emerge, often informed by the notion that these cities have more direct local links, to production in particular (Tefft & Jonasova, 2020). This view is inadequate in light of the CUP typologies because it perpetuates a rural framing of most secondary cities, implicitly casting them all in the role of agricultural market centres. The World Bank framing of secondary cities as "agro-cities (agricultural towns and cities with under 1 million people)" (Tefft et al., 2017, p. 84) and the Sustainable Development Solutions Network framing, detailing smaller cities, generally those of less than one million, as cities with strong agricultural links, provide further evidence of this often repeated assumption.

Roberts' (2014) framework is used in this chapter as a way of overcoming the implicit ruralization of secondary cities and assumption that secondary city economies are inherently agricultural. It is also versatile in different national contexts in that it draws focus on the different types and political, economic and even structural reasons for the development of certain types of settlements and not primarily the city's location or size. Despite the recent interest in food systems governance of secondary cities, the approaches continue to perpetuate the ruralized, "provincial" view of secondary cities, missing the depth of nuance required to see how secondary cities function, across typologies, regions and political systems. Drawing on the city's function and typology suggested by Roberts (2014), the CUP research sought to understand the food systems of different types of secondary cities.

Background to the Case Study Cities

The first of the three CUP case studies, Kisumu, Kenya, started as a regional trading centre, with its name being a Dholuo derivative meaning "a place to find food" (UN-Habitat, 2005). Kisumu is located on Lake Victoria, at the intersection of different trading routes. Historically, Kisumu was an important and strategic trading point for the Maragoli, Nandi and Luo people, long before the site was selected as a colonial commercial centre in 1898 (Nodalis Conseil, 2013; UN-Habitat, 2005). Given the strategic location of Kisumu, the British colonial government established a railroad terminus and port at the existing settlement. In 2006, Kisumu had a population of about 500,000 and an estimated growth rate of 2.8% per year (Geissler, 2013; UN-Habitat 2006). It is estimated that the daily population of the city sometimes increases to around one million as a result of trade and transit processes. Kisumu is the third largest city in Kenya, after Nairobi and Mombasa, and one of the country's fastest-growing cities (Gutberlet et al., 2017).

The second city is Kitwe, Zambia. From as early as 1924, when Northern Rhodesia was declared a British Protectorate, the colonial government began to remove the Copperbelt's established residents to native reserves to make way for the mines and farms staked out by the

British South Africa Company (Siegel, 1989). For those who did not leave the land, the imposition of punitive taxes further forced the local residents into a cash economy and necessitated their search for employment. The primary economy of Kitwe was linked to its copper mines and related mining industries, with agriculture being an important secondary economic sector. The formal colonial town was founded in the 1930s, and in 1936 a town management board was established. In 1953, Kitwe's governance structure became a municipal council and in 1967 Kitwe was officially declared a city (Hampwaye, 2008). Following trade liberalization in the early 1990s, many Kitwe residents lost employment. By the early 2000s, 45% of Kitwe's labour force was unemployed (Smart et al., 2015). Since this time, new non-mining investments have diversified the city's economy and Kitwe's population has grown, with the 2010 census recording a population of 517,543 in Kitwe District (Central Statistical Office, 2012).

The third case study city is Epworth, Zimbabwe. Located outside the capital city's municipal boundary, Epworth is a satellite town of Harare. Epworth, originally the site of two villages of subsistence farmers, was claimed by the colonial British South Africa Company. This land was donated to the Wesleyan Methodist Mission Trust in 1892 (Butcher, 1988; Rakodi, 1985). Due to the Wesleyan Methodist Mission governance of the land and unlike other colonial management of urban areas, the area had no government-imposed movement or development restrictions (Dialogue on Shelter 2009; Nyamvura & Brown, 1999). The church's opening up of the area during the peak of the Zimbabwean liberation war (between 1966 and 1979) to welcome refugees fleeing conflict in the rural areas, resulted in the area's rapid growth in the late 1970s (Rakodi, 1985). With independence in 1980, racial restrictions on movement to urban areas were lifted (Tibaijuka, 2005). Epworth, with its lack of formal development, provided an easy destination for new arrivals, further increasing the settlement's population (Butcher, 1988). This sudden growth in Epworth's population was beyond the administrative capacity of the Methodist Church, which donated part of its land to the government in 1986. The Epworth Local Board was then established as a local authority (Chirisa, 2010) and governs Epworth today. At the time of the 2012 census, Epworth had a population of 167,462 (ZIMSTAT, 2012).

Food Systems and Food Security Findings

The case study cities aligned with the various secondary city typologies. Despite imagined local benefits from rural linkages, and access to local food supplies, food insecurity levels in each case study city were high. Seventy-one percent of surveyed households in Kisumu were moderately or severely food insecure, as well as 88% of the surveyed households in Epworth and 90% in the sample in Kitwe (the Kitwe sample focused on poor neighbourhoods only) (Battersby & Watson, 2019). This supports other analyses of levels of food insecurity in African cities, including the low income areas targeted by the African Food Security Urban Network (AFSUN) surveys, which, using the same food security indicator (Household Food Insecurity Access Prevalence [HFIAP]), found high levels of food insecurity across a range of cities (Crush et al., 2012). The Household Dietary Diversity Score (HDDS) is a complementary measure, where HDDS scores of six food groups or fewer serve as a proxy indicator for under-nutrition.

In all three case study cities, the household diversity scores were recorded as being well below six, pointing to current and future developmental and health-related challenges. The HDDS and HFIAP provide some sense of urban food related challenges, but measuring multidimensional poverty offers insights into how food system challenges intersect with the wider urban system. The CUP research used the Lived Poverty Index (LPI). The LPI score ranges along a five-point scale from 0 (which can be thought of as no lived poverty) to 4 (which would reflect a constant absence of all basic necessities). Table 2.1 considers respondents who scored between 0 and 1, generally those experiencing little

Table 2.1 Comparison of FANTA and LPI scores across case study cities

	Epworth n = *483*	*Kitwe* n = *871*	*Kisumu* n = *841*
LPI (Percentage between 0–1) (%)	37	42	74
HFIAS (out of 27)	12.07	13.6	7.64
HFIAP (% food secure)	8	6	20
HDDS (possible 12)	4.12	3.25	4.05

Note While Epworth and Kisumu samples generally reflect the wider settlement trends, the Kitwe was a pro-poor survey in two poor neighbourhoods

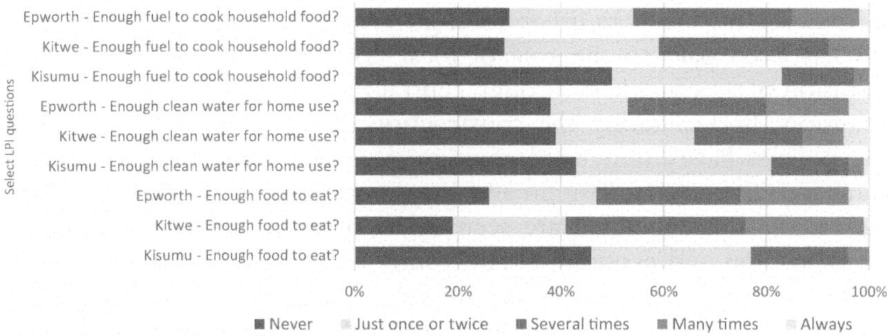

Fig. 2.1 Lived Poverty Index in case study sites

or no deprivation. By implication, this means that for Epworth, 63% experienced deprivation; for Kitwe, 58%; and for Kisumu, 26%.

Figure 2.1 provides insights into the scale and levels of lived poverty as measured through the LPI. While the absence of food and the absence of income are key drivers of lived poverty, other drivers also perpetuate lived poverty, specifically the absence of adequate energy to cook food, clean water and electricity. All impact food system outcomes. The provision of these services is predominantly the responsibility of city governments and intersects directly with the food system.

In all case study cities, poor urban residents were purchasing food from supermarkets (for further detail about Epworth, see Tawodzera in this volume). However, the main and most frequent sources of food remained markets and traders (mostly informal) within their immediate neighbourhoods. Supermarkets were far from the most important source of food. In Kisumu, more than half the consumers surveyed bought less than 25% of their food at supermarkets. This suggests that although supermarkets are an important part of the changing food retail environment, the largely informal or traditional market areas remain dominant, with frequent purchases from the local markets and less frequent large purchases from a more central location or central market.

There is evidence that supermarkets are changing the nature of the food retail sector. In the case of Kitwe, a number of smaller traders were using the supermarkets as their wholesaler. Items entering the city through the supermarket's supply chain, mostly items imported from South Africa by South African-owned supermarkets, are resold in the informal sector. From an urban food governance perspective, little attention has been paid to the role of local government in supporting the expansion of supermarkets through centralized planning that favours such expansion. By facilitating the construction of supermarkets, these planning policies have unintended consequences of altering the food supply chains into the city. In Kisumu and Kitwe, shopping malls and the supermarkets that occupy them are seen by local government as symbols of modernization and economic opportunity (Battersby & Watson, 2019). Figure 2.2, drawing on the Kisumu case specifically, demonstrates the varied food access use from the full sample of 841 households in the city. The frequency of use of local kiosks or house-shops is markedly different from the frequency of use of local markets (mostly weekly) and supermarkets. Kiosks and house-shops are used for daily purchases of key essentials, often in smaller quantities. Bulk buying of key staples (dry goods and processed foods) is often done on a monthly basis from supermarkets. On the other hand, weekly trips are made to local markets and butchers for the purchase of fresh produce.

A further contribution of the CUP project was its mapping of the sources of food coming into the city. These findings challenge some of the assumptions held within the emerging promotion of the City Region Food Systems concept led by the Food and Agriculture Organization of

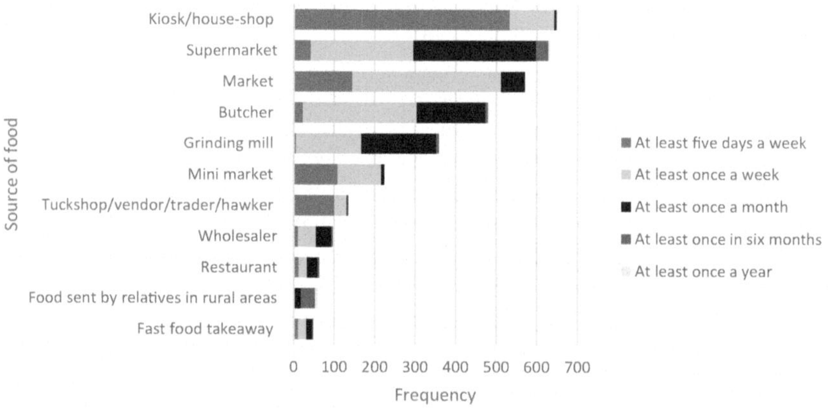

Fig. 2.2 Frequency of use by type of food retail outlet in Kisumu (*Source* Adapted from Opiyo et al. [2018, p. 26])

the United Nations (FAO), the RUAF Foundation, ICLEI and others. The food systems for the three case study secondary cities were far more globally connected than the literature suggests, with fish in Kitwe and Kisumu travelling from China, eggs at the wholesale market in Kisumu coming from Uganda, and vegetables at the Chisokone Market in Kitwe coming from South Africa. These global connections were not limited to modern, formal supply chains, but were a central part of informal sector activities. Importantly, different food types were found to have very different supply chains and import networks. Most types of food vendors were adept at negotiating these international supply chains. Importantly, during the CUP research, an FAO project sought to understand the Kitwe supply chains, asking farmers where their food was sold. The majority of this food went to Kitwe, reflecting a largely city/region food system (Mwitwa et al., 2016). The CUP research asked vendors where their foods were sourced and noted distinctly international supply chains active in the city. These different entry points to similar questions pose deeply challenging governance questions.

Food Governance Issues Arising from Typological Analysis

In the case of secondary cities, typologies provide an important link between governance structure, policy and food security outcomes. As a satellite town, Epworth was found to have very little formal food retail, with no shopping mall or other large, formal food retail outlets. However, as a dormitory type of secondary city, food purchases made by the residents of Epworth remained varied, with most formal purchases being made in the primary city of Harare and other food purchases made via a variety of informal food vending typologies (see Tawodzera, this volume). The dormitory town of Epworth required a very specific type of food system governance, one that accounted for the symbiotic relationship between Epworth and Harare, and where governance actors from both municipalities worked in conjunction with one another. Mbare Musika, the main wholesale market in downtown Harare, was a primary source of food for most informal vendors in Epworth, and thus of households from Epworth.

In Kitwe, a regional centre for commerce and services, South African supermarket chains were developing rapidly to service the city and its surrounding region. The chains were expanding their footprint in traditional business districts, but also in new shopping malls. However, most residents still frequented a wide variety of informal vendors, with the use of the traditional downtown market, Chisokone Market, still dominating purchasing patterns. Kitwe, with its extractive industry history, was highly susceptible to price and other market-related shifts in its primary resource endowment, copper. Fluctuations in the international copper price had a direct impact on the prosperity (and, as a result, food system viability) of the city. This single commodity reliance severely undermined the resilience of the city, exposing many to a highly variable economy. Planning and governing for these boom-and-bust cycles is a key food system consideration. Relying on local food systems is one thing, but when employment fails, food might be available yet not always accessible.

Kisumu, a corridor hub since its inception, was facing a similar supermarket expansion trend, with new malls being developed in a variety of locations and supermarket tenants occupying these sites as anchor tenants, despite significant challenges for the supermarket sector at the time of the research. Kisumu's role as a trunk city meant that there were direct transport routes that enabled access to produce from different regions. Eggs sold in Kisumu were found to be from Uganda, an easy trip across a nearby border, but transport was less important than the fact that poultry production was subsidized differently in Uganda than in Kenya, making eggs from Uganda more affordable. Fish caught in Lake Victoria could attract a better price in the capital, Nairobi, enabled by the trunk typology of Kisumu. However, that same trunk system ensured that fish vendors on the shores of Lake Victoria were able to import fish from China and sell it at affordable prices.

In both Kitwe and Kisumu, discussions were taking place to relocate central city markets into new "trader malls" or upgraded markets on the periphery of the city. These market upgrades were often funded by international development agencies and were seen by local officials as a vehicle to clear streets of unwanted informal traders. In the view of local officials, these traders occupy a very different position from that of shopping malls and the supermarkets that occupy them, which "hold a position in the local government as symbols of modernization and economic opportunity" (Battersby & Watson, 2019, p. 6). The role of international development agents in this process requires far greater attention and critique (Battersby & Watson, 2019).

These few examples demonstrate the central importance of urban food governance, including the importance of fully understanding the typology of the specific secondary city in applying essential urban governance approaches and determining where development attention is focused. In the CUP cities, it was found that there was generally weak governance of the urban food system, with no clear mandate for local government and limited financial and personnel capacity within local government. This meant that many governance decisions that shaped the food system, such as decisions on traders, market sites and mall development, were made outside the context of a consideration of their food system impact (Battersby & Watson, 2019). The relatively limited capacity of the state meant that key policy and planning decisions were strongly informed by national governments, large international donor agencies and private-sector actors. These governance capacity issues are particularly evident in secondary cities, urban areas with the most pressing development challenges, vulnerable economic bases and the worst access to up-to-date, accurate data on which to make governance decisions (Battersby & Watson, 2019).

Conclusion

Despite the varied food access options open to households in all three case study cities, food insecurity was still extremely high. This indicates that neither the formal supermarket system nor the informal system is enabling adequate access to food. Relying on the market systems to ensure food access is not adequate to enable health, prosperity and inclusive development in these secondary cities. The LPI results highlighted the scale and multidimensional nature of the deprivations, and the intersections between food deprivation and the wider urban system. Governing food can no longer be left to agricultural ministries and the market. City managers, politicians and officials need to play a far more active role in urban food systems governance.

The development interventions planned and decided on in the next decade will have a direct impact on the development outcomes across Africa, in both urban and rural areas, over the next 50 to 100 years. Ignoring the changes that will take place in the food systems of these urban spaces will have a profound impact on Africa's overall development outcomes. Key to ensuring that these outcomes are positive, pro-poor and health enhancing is food systems governance at the urban scale.

Africa's secondary city food system transformation is seldom considered, but requires urgent attention.

The current view of secondary cities being embedded within, and reliant on, a regional food system requires far greater interrogation and critique. Generalized approaches run the risk of missing important contextual needs and nuance. An important approach in refining urban food governance considerations in secondary African cities is to take seriously the different secondary city typologies and use them as a means to support local food systems governance planning and policy actions.

Acknowledgements The information contained in this paper was collected by the research teams from the CUP partner sites in Harare and Epworth, Kitwe and Kisumu. Thanks also to Issahaka Fuseini and Jane Battersby for their support and input into earlier drafts of this chapter.

References

Barling, D., Lang, T., & Caraher, M. (2002). Joined–up food policy? The trials of governance, public policy and the food system. *Social Policy and Administration, 36*(6), 556–574.

Battersby, J. (2017). MDGs to SDGs – new goals, same gaps: The continued absence of urban food security in the post-2015 global development agenda. *African Geographical Review, 36*(1), 115–129.

Battersby, J., & Watson, V. (Eds.). (2019). *Urban food systems governance and poverty in African Cities*. Routledge.

Bekker, S., & Therborn, G. (Eds.). (2012). *Capital cities in Africa: Power and powerlessness*. HSRC Press.

Butcher, C. (1988). *Epworth: A review of the densification policy* (Report Prepared for the Epworth Local Board).

Central Statistical Office. (2012). *Zambia 2010 census of population and Housing* (Population Summary Report). Central Statistical Office.

Chirisa, I. (2010). Prospects for the asset-based community development approach in Epworth and Ruwa, Zimbabwe: A housing and environmental perspective. *African Journal of History and Culture, 1*(2), 028–035.

Crush, J., Frayne, B., & Pendleton, W. (2012). The crisis of food insecurity in African cities. *Journal of Hunger and Environmental Nutrition, 7*(2–3), 271–292.

CUP. (2021a). https://consumingurbanpoverty.wordpress.com/?s=Urban+Poverty

CUP. (2021b). https://www.datafirst.uct.ac.za/dataportal/index.php/catalog/CUP/about

Dialogue on Shelter for the Homeless in Zimbabwe Trust (DSHZT). (2009). *Epworth profiling report*. DSHZT.

Duminy, J. (2018). Ecologizing regions; securing food: Governing scarcity, population and territory in British East and Southern Africa. *Territory, Politics, Governance, 6*(4), 429–446.

Ericksen, P. (2008). Conceptualizing food systems for global environmental change research. *Global Environmental Change, 18*(1), 234–245.

Fox, S., & Beall, J. (2012). Mitigating conflict and violence in African cities. *Environment and Planning C: Government and Policy, 30*(6), 968–981.

Frayne, B., Pendleton, W., Crush, J., Acquah, B., Battersby-Lennard, J., Bras, E., Chiweza, A., Dlamini, T., Fincham, R., Kroll, F., Leduka, C., Mosha, A., Mulenga, C., Mvula, P., Pomuti, A., Raimundo, I., Rudolph, M., Ruysenaar, S., Simelane, N., ..., Zanamwe, L. (2010). *The state of urban food insecurity in Southern Africa*. African Food Security Urban Network (AFSUN). Urban Food Security Series No. 2.

Fuchs, R. (1983). Secondary cities in developing countries: Policies for diffusing urbanization by Dennis A. Rondinelli – A review. *Economic Geography, 59*(4), 461–462.

Geissler, P. (2013). Stuck in ruins, or up and coming? The shifting geography of urban public health research in Kisumu, Kenya. *Africa, 83*(4), 539–560.

Goodall, B. (1987). *The Penguin Dictionary of Human Geography*. Puffin Books.

Gutberlet, J., Kain, J., Nyakinya, B., Oloko, M., Zapata, P., & Zapata Campos, M. (2017). Bridging weak links of solid waste management in informal settlements. *The Journal of Environment and Development, 26*(1), 106–131.

Hampwaye, G. (2008). *Decentralisation, local economic development and urban agriculture in Zambia*. PhD dissertation, Faculty of the Humanities, University of the Witwatersrand.

Haysom, G., & Fuseini, I. (2019). *Governing food systems in secondary cities in Africa* (Consuming Urban Poverty Project Working Paper No. 10). African Centre for Cities, University of Cape Town.

Joubert, L., Battersby, J., & Watson, V. (2018). *Tomatoes and taxi ranks: Running our cities to fill the food gaps*. African Centre for Cities, University of Cape Town.

McGranahan, G., Mitlin, D., & Satterthwaite, D. (2008). Land and services for the urban poor in rapidly urbanizing countries. In G. Martine, G. McGranahan, M. Montgomery, & R. Fernandez-Castilla (Eds.), *The new global frontier: Urbanization, poverty and the environment in the 21st century* (pp. 77–98). Earthscan.

Mwitwa, J., Sibajene, M., Chipoya, G., Namiluko, Y., & FAO. (2016). *City region food system situational analysis, Kitwe, Zambia* (City Region Food Systems Report). Rome, IT: Food and Agriculture Organization.

Nodalis Conseil. (2013). *Ministry of lands, housing and urban development – Kisumu County.* AFD Preparation of the integrated strategic urban development plan for Kisumu, http://www.nodalis-conseil.com/en/references.php?id_secteur=5

Nyamvura, T., & Brown, A. (1999). Epworth – An informal, unplanned and squatter settlement: Upgrading initiatives in Epworth. In A. Brown & C. Davidson (Eds.), *Responsive design and plan implementation* (pp. 107–109). Cardiff University.

Opiyo, P., Obange, N., Ogindo, H., & Wagah, G. (2018). *The characteristics, extent and drivers of urban food poverty in Kisumu, Kenya* (Consuming Urban Poverty Project Working Paper No. 4). African Centre for Cities, University of Cape Town.

Pieterse, E. (2008). *City futures: Confronting the crisis of urban development.* UCT Press.

Pieterse, E., Parnell, S., & Haysom, G. (2018). African dreams: Locating urban infrastructure in the 2030 sustainable developmental agenda. *Area Development and Policy, 3*(2), 149–169.

Pieterse, E., Parnell, S., & Haysom, G. (2015). *Towards an African urban agenda.* Nairobi: United Nations Human Settlements Programme (UN-Habitat) and United Nations Economic Commission for Africa (UNECA).

Potts, D. (2012). What do we know about urbanization in sub-Saharan Africa and does it matter? *International Development Planning Review, 34*(1), v–xxi.

Rakodi, C. (1985). Self-reliance or survival? Food Production in African Cities, with Particular Reference to Zimbabwe. *African Urban Studies, 21*, 53–63.

Reardon, T., Timmer, C., Barrett, C., & Berdegue, J. (2003). The rise of supermarkets in Africa, Asia, and Latin America. *American Journal of Agricultural Economics, 85*(5), 1140–1146.

Roberts, B. (2014). *Managing systems of secondary cities: Policy reponses in international development.* Cities Alliance.

Rondinelli, D. (1983). *Secondary cities in developing countries: Policies for diffusing urbanization.* Sage.

Satterthwaite, D. (2007). *The transition to a predominantly urban world and its underpinnings.* London: IIED. Human Settlements Discussion Paper Series.

Satterthwaite, D., & Mitlin, D. (Eds.). (2013). *Empowering squatter citizen: Local government.* Routledge.

Sen, A. (1981). *Poverty and famines: An essay on entitlement and deprivation.* Clarendon Press.

Siegel, B. (1989). The "wild" and "lazy" Lamba: Ethnic stereotypes on the Central African Copperbelt. In L. Vail (Ed.), *The creation of tribalism in Southern Africa* (pp. 350–371). James Currey and University of California Press.

Smart, J., Nel, E., & Binns, T. (2015). Economic crisis and food security in Africa: Exploring the significance of urban agriculture in Zambia's Copperbelt province. *Geoforum, 65*, 37–45.

Smit, W. (2016). Urban governance and urban food systems in Africa: Examining the linkages. *Cities, 58*, 80–86.

Song, L. (2013). *Southeast Asian secondary cities: Frontiers of opportunity and challenges*. Massachusetts Institute of Technology, Community Innovators Lab (CoLab).

Tefft, J., Jonasova, M., Adjao, R., & Morgan, A. (2017). *Food systems for an urbanizing world: Knowledge product*. FAO and The World Bank Group.

Tefft, J., & Jonasova, M. (2020). Food systems transformation in an urbanizing world. In J. Crush, B. Frayne, & G. Haysom (Eds.), *Handbook on urban food security in the global south*. Edward Elgar Publishing.

Tibaijuka, A. (2005). *Report of the fact-finding mission to Zimbabwe to assess the scope and impact of operation Murambatsvina* (UN-Habitat Report). United Nations Human Settlements Programme.

UN-Habitat. (1996). *The management of secondary cities in Southeast Asia*. United Nations Centre for Human Settlements.

UN-Habitat. (2005). *Cities development strategies for improved urban environment and poverty reduction in the Lake Victoria region: Kampala, Kisumu and Musoma* (Working Paper Series [Urban Management Programme]). United Nations Human Settlements Programme.

UN-Habitat. (2006). *Kisumu urban sector profile*. United Nations Human Settlements Programme. UN-Habitat Report prepared by R. Syrjanen & K. Sommer.

UN-Department of Economic and Social Affairs (UN-DESA). (2018). *World urbanization prospects. The 2018 revision*. UN-Department of Economic and Social Affairs.

ZIMSTAT. (2012). *Zimbabwe population census 2012* (CENSUS 2012 National Report). Population Census Office.

CHAPTER 3

Practice Theory and Informal Urban Livelihoods in M'Bour, Senegal: A Case Study of Urban Cultivation

Stephanie A. White

Introduction

Throughout sub-Saharan African, in all areas of urban life, everyday practices, such as food exchange, income-generation, transportation and infrastructure development, sustain and reproduce cities. These practices, often described as "informal", offer important entry points for research that explores how the lives of urban residents "hang together" in their efforts to produce individual and collective well-being. In turn, such research has an important contribution to make to urban planning, development, and governance, in particular as it relates to fostering just and equitable transitions towards climate-adaptive cities.

S. A. White (✉)
Michigan Department of Health and Human Services, Division of Environmental Health, Lansing, MI, USA

L. Riley and J. Crush (eds.), *Transforming Urban Food Systems in Secondary Cities in Africa*,
https://doi.org/10.1007/978-3-030-93072-1_3

The basic premise of this paper is that studies of informal livelihoods can reveal embedded social conventions, knowledge, and practices that can serve just transitions to more sustainable cities. Through a study of urban cultivation in M'Bour, Senegal, this research seeks to understand how one informal livelihood constitutes and maintains social and material worlds, and how it integrates with city processes and daily efforts to improve well-being. The conceptual frame draws on Theodore Schatzki's practice theory and is informed by scholarship in informal economies. By drawing attention to the economic, social, and spatial dimensions of urban cultivation practice in a particular place, it becomes clearer how livelihoods are social, normative, and emplaced. Exploring these livelihoods provides insight into: How the city socioecology shapes the practice of urban cultivation; the practices urban cultivators employ in response to socio-ecological conditions; and how livelihood practices shape the civic and city environment. Such a perspective on livelihood offers one way to consider the shared meanings and understandings of the urban lifeworld in relation to urban development and governance. Accordingly, this understanding of what is happening *in place* and why it is happening is useful to decision makers because, in Schatzki's words, such accounts provide direction for "shaping the organizations and trajectories of practice, creating new practices..., eliminating others" (2014, p. 22). In the language of contemporary governance, this is akin to "creating enabling environments" or shaping policy "pathways", but doing so in ways that connect to existing socio-economic systems.

The following section discusses how the prevailing framing of economic life in African cities is inherently hostile to equitable and just urban development. It then explores an alternative framing and how practice theory can more accurately situate everyday livelihood practices *in place* and in relation to the lives of urban citizens. A detailed findings section is then presented, which demonstrates the multi-dimensionality of livelihood and the ways in which people exercise agency in contingent interactions with the urban landscape and other urban residents. The paper concludes with a short discussion of the theoretical and practical implications of the research.

Conceptual Framework and Understanding Informal Urban Livelihoods

Urbanization in Africa is commonly regarded as a haphazard, unthinking, spontaneous, and problematic phenomenon, and the relative size of the informal economy as an indicator of "backwardness" (Potts, 2008). Held up against "the fantasy of (Western) urban modernity" (Robinson, 2006, p. 13), African cities and the efforts of citizens who make them are not typically examined for insights about how to do urban development in ways that improve well-being.

Prevailing perceptions of cities and the lives of urban residents influence the production of knowledge and power relationships and, likewise, urban development policy, planning, and practice. If the urban economic activity of poor people is primarily understood to be chaotic, desperate, and spontaneous, there is little inclination on the part of decision makers and researchers to refer to these processes and practices to inform urban development policies. In fact, these practices are often perceived by city authorities to be "out-of-place" in an urban environment and to adversely affect the image of the city that town leaders would like to project, i.e. planned, clean, tidy, and modern (Potts, 2008). Such perspectives underlie and enable the authoritarian actions of city governments in both the Global North and South, in which mostly poor people are persecuted in their daily efforts to make a life. Examples include the harassment of Sidi Bouzid by Tunisian authorities, who subsequently set himself alight in front of the governor's office, which provided the catalyst for the Arab Spring; "Operation Murambatsvina" in Zimbabwe; and the violent eviction of residents by the Nigerian Navy in Lagos.

Prevailing Conceptualizations of Informality and the Politics of Planning

Urban life in Africa is commonly portrayed as a "catalogue of the utterly devastating conditions that characterize the daily lives" of people living in cities (Pieterse, 2013, p. 2). Within this discourse, which portrays African cities as "all that can go wrong with urbanism", the organizing conceptual dichotomy of formality/informality is used to make sense of urban economic practice (Myers, 2011, p. 4). Although the most basic definitions of formality and informality refer simply to whether an economic activity is regulated or unregulated, the concept has come to serve as

Table 3.1 Characteristics associated with the formal/informal dichotomy

Formal Economy	*Informal Economy*
Modern	Traditional
Legitimate, legal	Illegitimate, unlawful
Growth-oriented, dynamic, innovative, technological	Small-scale, family-run, low levels of technology
Progressive	Static
Enables accumulation of capital	Pre-capitalist, survivalist, and subsistence

shorthand for a number of other characteristics that have implications for how particular economic activities are perceived by authorities. As William Maloney and Jamie Saavedra-Chanduvi (2007) note, "The term 'informality' means different things to different people but almost always bad things: unprotected workers, excessive regulation, low productivity, unfair competition, evasion of the rule of law, underpayment or nonpayment of taxes, and work 'underground' or in the shadows" (p. 21). Table 3.1 provides a summary of qualities commonly associated with each side of the dichotomy.

Several authors have discussed the dichotomy's analytical shortcomings and discredited the notion of a dual economy (see, e.g., Owusu, 2007; Potts, 2008; Roy, 2005), and those arguments are not extensively repeated here. However, there are two main problems that are important to highlight because they provide the impetus for the proposed alternative conceptual approach that results in better characterizations of urban economic life. First, although informality is overwhelmingly associated with economic activities of poor people, research shows that informal livelihoods are practised by people at all income levels who may simultaneously engage in both regulated and unregulated work, depending on available opportunities. Francis Owusu (2007) explains that working in multiple livelihoods has just become the way of doing things in African cities, while Ananya Roy (2009) points out "informality is not synonymous with poverty" (p. 82).

The second issue is more fundamental and raises serious questions about a conceptual approach that organizes economic life into two oversimplified categories that confer legitimacy on some livelihoods, but not on others. The implications for equality, social justice, and the ability to pursue livelihood become evident when one acknowledges how the dichotomy works as part of a "regime of representation" within a particular discursive context, which, in turn, affects how urban space is managed (Escobar, 2011, p. 19). When Keith Hart (1973) initially made the distinction between formality and informality, characterizing it as the difference between wage-earning and self-employment, his intent was to caution against the "unthinking transfer of western categories to the economic and social structures of African cities" and to demonstrate how people adapt their livelihood strategies when the state's macroeconomic policies fail them (p. 61). The significance of the distinction shifted as the concepts became colonized by the development apparatus and were folded into the prevailing "developmentalist" macroeconomic policy framework, thus enabling informality to emerge as an "object of state regulation" (Roy, 2005, p. 149). To his obvious chagrin, Hart (1994) subsequently noted that the "informal economy became a way of turning what is defiantly external to bureaucracy into something internal to it, incorporating the autonomous life of the people into the abstracted universe of their rulers".

A morphological reading of "informality" indicates economic activity "without form" and suggests that informal livelihoods can be distinguished from formal livelihoods according to appearance or type. Within a policy and planning context that regards parochial Western notions of modernity and capitalist economies as "formal", and thus, developed, what seems to concern planners and policymakers is not so much whether an economic activity is regulated, but whether it conforms to narrow and limited notions of (modern) form. For example, Roy (2009) demonstrates how urban governments and businesses engage in informal practices, but because outcomes from these practices accord with a "modern" sensibility and are glossed with a veneer of formality (Table 3.1), they are perceived as legitimate. In contrast, small-scale, "improvisational" livelihoods that comprise the street economy, for example, appear chaotic and are subject to authoritarian crackdowns by urban authorities.

The analytical frames used to understand economic processes must enable consideration of complexity and encourage analyses that discern how these processes differ from place to place and how they differentially affect people in those places. Roy (2005), for example, has proposed understanding informality as a *mode* of development, which, she asserts, provides a common logic that can be applied to all economic activities. Owusu (2007) proposes that a Multiple Modes of Livelihood (MML) approach can help to "indigenize planning" by recognizing that the conditions in African cities require individuals to diversify economic activities.

The next section offers a complementary approach, but argues for an additional emphasis on "place" as an important theoretical touchstone. A place-based approach entails understanding places as "constructed historical processes...in which the extralocal is as constitutive as the local" (Biersack, 2006, p. 16) and thus provides the impetus to focus on the ways in which factors at multiple scales combine to produce distinct and specific places that require different ways of pursuing livelihoods (Casey, 1996; Escobar, 2001; Massey, 2013). Such an approach compels an analysis that valorizes local level complexity, perspectives, and experience, and thus serves an agenda to decolonize city development processes and narratives.

Re-thinking Livelihood with Schatzki's Practice Ontology

Urban cultivation is commonly viewed by urban authorities to be a temporary, opportunistic, backwards use of city space. In reality, however, urban cultivation is widespread and persists even in densely populated cities. As a use of urban space that is increasingly appearing on developmentalist agendas and which is attracting more interest from African urban planners, it is important to understand how citizens use agriculture to improve urban well-being and negotiate the urban environment. In contrast to much development research, which seeks to produce universal accounts and generalizable policy recommendations, an analysis that uses practice theory can reveal how differences in different places might inform decision-making processes that are *locally* relevant and responsive. Accordingly, this research demonstrates how *one* city plays a critical and active social role in the ways people make their lives, as well as how, in making their lives, people make this city (Biersack, 2006; Escobar, 2001; Feld & Basso, 1996). The idea that the socio-material environments play an active part in the constitution of livelihood is not without precedent, although such research often concerns rural places and livelihoods (Batterbury, 2001; Bebbington & Batterbury, 2001; King, 2011).

In contrast to rational choice or behaviour change theory, which explains human activity as the pursuit of individual interests, practice theory views human activity as a means of reproducing wider social systems (Giddens, 1984). Schatzki et al. (2001) explain that, "practice accounts are joined in the belief that such phenomena as knowledge, meaning, human activity, science, power, language, social institutions and historical transformation occur within and are aspects or components of the field of practices" (p. 11).

The practice of urban livelihoods, then, is much more than simply the means for economically sustaining a household. Rather, the practice of urban livelihoods reproduces social networks that bring into being and sustain cultural, socio-economic, and physical environments (Bourdieu, 1990; Gibson-Graham, 2008; Illich, 1980). That is, urban livelihoods are fundamental in a broad and multidimensional sense to how people "make a life". Urban livelihoods, in addition to providing for material necessities, are constitutive of what Jürgen Habermas calls the lifeworld, or the "informal and unmarketized domain of social life", that serves as a repository for shared meanings and understandings and is a critical locus for social action and change (Finlayson, 2005, p. 51).

Drawing from Schatzki's social ontology (2010), this analysis examines how socio-material worlds are reproduced through urban cultivation and vice versa. Schatzki's approach roots practices of sociality firmly in place by connecting practice to the material dimensions of place. According to Schatzki (2010), social life "transpires as part of nexuses of practices and material arrangements" (p. 129). Practices are construed as organized "manifolds of human activity", shaped by the particularities of locations and relationships. In Schatzki's ontology, the physical aspects of environments are considered fundamental to how lives transpire; human social worlds are made in and from material worlds. Such a perspective compels close attention to city form and material configurations and the ways in which those configurations are implicated in the social practice of livelihood.

Schatzki (2010) proposes three ways to consider how materiality is implicated in the "hanging together of human lives": physical entities in combination with practices compose social sites; the physical–chemical composition of materials affects practice-material arrangements and affects social life; and "biological and physical flows pass through practice arrangement nexuses" (p. 137). Additionally, he proposes four types of relationships among practices-material arrangements that shape social life in various ways: causality, prefiguration, constitution, and intelligibility. Causality refers to how materiality leads to certain practices; prefiguration refers to how material environments set the parameters for possible "paths of action"; constitution refers to the idea that certain material arrangements are necessary, or strongly linked, to certain practices; and intelligibility refers to the idea that the arrangements make sense to/are given sense by the people who carry out the associated practices. Following a brief section about the research site of M'Bour, Senegal, and a description of methods, findings are presented using Schatzki's social ontology.

Site Background and Methods

M'Bour, Senegal, is a rapidly growing, ethnically diverse, coastal secondary city with a population of around 233,000 (most recent data from 2013 census, ANSD, 2014), located about 80 kilometres south of the capital city, Dakar. M'Bour draws new residents from all over Senegal and beyond, as well as expatriates and retirees from Europe. Although small business, fishing, and tourism are generally regarded by town officials as the most economically important livelihoods, many of M'Bour's citizens make their lives, either partially or wholly, with unregulated economic activities.

On the west side, M'Bour's growth is limited by the Atlantic Ocean. On the north, south, and east sides, it is limited by the territorial boundary it shares with the *communauté rurale* of Malicounda. In Senegal, each *communauté rurale* is governed by an elected set of representatives that maintains sovereignty over that territory. M'Bour has subdivided all the land within its territory and any additional land must be acquired through negotiations with the government of Malicounda, although, at the time of research in 2010, the territorial boundary was not clearly defined and disputes were common. As a result, land is in increasingly short supply and is appreciating rapidly.

Enough and diverse kinds of food are easily accessed if one has the means to purchase them. Residents commonly remark upon cost-of-living increases and how expensive food and other material necessities have become in the last decade. Daily power and water outages, petrol shortages, severely depleted fish stocks, and increases in the price of flour and cooking gas are common occurrences, with causative factors at multiple scales. Despite these difficulties, many consider M'Bour to be a city of opportunity and abundance.

The following empirical findings are drawn from interviews, participant observation, photographs of research sites, and document analysis.

Reviewed documents included "Historical Overview and Purpose of Master Plan" (no date), "Town of M'Bour" (2008), "Extract of a Planning Report from the Director of Sanitation" (2008), "Extract of an impact study on the environment for the town of M'Bour" (2008) and "A table synopsis of the needs expressed by the neighbourhood councils of the town of M'Bour" (no date). Most data were collected while studying multiple forms of small-scale dry season urban cultivation from August 2010 to May 2011. The forms of cultivation include micro-gardening, ornamental plant production, fruit tree production, and vegetable production. At the time, micro-gardening was a national government programme that was initially started as a collaborative development project with the Food and Agriculture Organization of the United Nations (FAO) in 1999. The programme operated through local agriculture offices and can be described as simplified hydroponics: plants are grown in lightweight substrates on tables that can be set up on rooftops, terraces, or within walled compounds, which, in densely built environments with limited "soil space", represents a viable option for growing herbs and vegetables close to home.

Following Seidman's guidance on interview sequencing, two semi-structured interviews with each cultivator, recorded several months apart, focused on livelihood strategies and practices, outputs of gardens, individual life histories, economic challenges, urban governance, and hopes for the future. The purpose of interview sequencing is to enable interviewees to "reconstruct and reflect upon their experience within the context of their lives" (Seidman, 2006, p. 21). It also helps to ensure that interviewers formulate questions based on realities as they are relevant to interviewees, rather than solely in terms of an external research agenda. Lastly, interview sequencing has the ancillary effect of building trust and familiarity between interviewee and interviewer, thereby better ensuring the trustworthiness of data.

Interviews were conducted with eight men and 10 women at 14 different cultivation sites, and eight officials (seven men, one woman) representing five government bureaus.[1] Interview protocols were developed following multiple visits to, and participant observation at, research participants' homes (Creswell & Clark, 2017; Lincoln & Guba, 1985).

[1] Mayor's Office, Prefect's Office, Urban Planning Office, Office of Decentralization and Local Development, and Rural Development Office, which managed the micro-gardening program.

Site visits typically lasted one to two hours. Each site was visited, on average, five times over the nine-month study period. Ongoing memo writing and informal conversations with a core group of neighbourhood friends helped develop research themes and insights, and provided a sort of locally grounded peer-review process (Glesne, 2006; Lincoln & Guba, 1985).

Systematic member-checks with interviewees were used to ensure reliability of interpretations, which was an especially important analytical aspect of the research in light of the deep cultural dimensions of the data (Glesne, 2006). Interviews were conducted and recorded in Wolof and French, transcribed by the author upon arrival back in the United States, and coded and organized using NVivo software.

Findings: Material-Practice Assemblages of Urban Cultivation

"Suñu barkaay, mingi ci suuf" ("Our fortune is in the soil").

The above quote, expressed by one of the participant farmers, demonstrates a perception that farming is the basis of well-being. This is a common sentiment in M'Bour and suggests that even among city dwellers, people remain attuned to the value of agriculture in their lives. It is thus not surprising that urban cultivation plays a significant role in shaping urban environments and civic relationships. The research findings unpack this significance by addressing the following concerns: how social and environmental conditions shape the practice of urban cultivation; the practices urban cultivators employ in responding to those conditions; and how livelihood practices shape the civic and city environment. Although the focus is on the cultivation of plants, small livestock figure into the analysis because they are a significant part of the urban cultivation practice-arrangement nexus. Findings are organized according to Schatzki's conceptual framework: physical entities and practices compose social sites; physical–chemical composition of materials affects social life; and biological and physical flows pass through assemblages.

Physical Entities and Practices Compose Social Sites

This section demonstrates how urban cultivation practice-material assemblages have formed, and how such arrangements sustain and reproduce social relationships.

Cultivation Assemblages: Land Tenure and the Construction of Wells and Walls

The basic physical requirements for a dry season garden in the city are that it has a water source and physical protection to help keep roaming livestock and would-be thieves away. Walled vacant lots with wells are common material configurations in M'Bour. That so many of these vacant lots exist, which create the conditions that prefigure urban cultivation, is the direct result of cities' formal land tenure management practices in combination with a common informal practice to safeguard one's claim to urban space. Vacant lots with wells and walls might not prefigure urban cultivation in all places, but in M'Bour, which is populated by people who regard primary production as a useful and productive livelihood, open space is commonly regarded for its agricultural potential.

Land is generally considered to be the property of the state, and urban land tenure is managed through a method called *mise en valeur des zones du terroir* (the enhancement or development of local areas). The state allocates land based on a contingency agreement in which the recipient swears to make productive use of the land within two years. Once someone holds a lease, and invests in the land, they can sell it to someone else.

Land represents a considerable investment, even for relatively wealthy Senegalese, and many people are unable to come up with the capital to both buy the land and build a house. To protect their investment and to comply with the productive use stipulation, many people build a well and wall. Their intent is to return later, when they have saved enough capital to build a house. Partially built houses are a common sight throughout Africa. Instead of "eating their money", or having it in a form that opens them up to requests for (permanent) loans, which they are obliged to respond to, people instead put their money into a physical and durable object. Thus, partially built houses represent a rational economic practice and physically symbolize how one balances social obligations and social/personal goals.

People in M'Bour commonly recount tales of land disputes and trickery. Stories of land "theft" abound, in which someone who is not the leaseholder builds a structure on a piece of land that has not been productively used. In at least some of those cases, the "squatter" apparently emerged victorious when challenged by the leaseholder in front of authorities, due to the productive use stipulation. Similarly, there are many stories of two or three different people holding papers to the same piece of land. For example, one urban cultivator told a story in which two people were sold the same piece of land. Each came to know of the other, and because a claim to land is legitimized by demonstrating productive use, each immediately hired a crew to build bricks. The two crews worked side by side until someone from the town hall arrived on the scene and ordered them to stop. The situation remained unresolved during the research period.

Protecting One's Investment Through Practices of Informality

Many people who hold a lease to a parcel of land live and work elsewhere and are unaware of what happens on their land from day to day. Although claims to land should, in theory, be secured by the lease and by the existence of the well and wall, there remains a common fear among leaseholders that, in their absence, someone will "sell" the land to an unwitting third party or claim the land and build a house. Rather than rely on the formal lease to safeguard their investment, leaseholders create informal relationships with neighbours who keep them apprised of any suspicious activity. In some of these arrangements, the neighbour will ask permission to cultivate the vacant lot. In other cases, urban cultivators seek out leaseholders to ask permission to cultivate the parcel of land. In either case, what emerges is a reciprocal relationship in which each party benefits. In these circumstances, in which the formal institution has proven insufficient to safeguard an individual's claim, people pursue an informal strategy that is rational in the local context.

The walls and wells that compose the social site of urban cultivation might be understood as a material form of resistance to the imposed productive use stipulation. Implicit in that stipulation is a particular view of how land should be managed, and an assertion by the state that it reserves the right to take back land and redistribute it. Building a well and wall, in combination with stationing a local guardian at the site, seems to be an effective and strategic form of resistance even when high-value land is in question. For example, the largest site in the study was composed of nine subdivided parcels in a prime area of M'Bour, just south of a busy intersection leading to many luxury beach hotels. One woman, an absentee leaseholder who lives in Dakar, holds the leases to all nine parcels. She obtained the leases when land was inexpensive, and since then, has successfully resisted several attempts by both the state and private individuals to claim (or in the case of the state, reclaim) the land. She initially hired a guardian to live on the premises in 2003, at which time he started cultivating the site. When the guardian observed that a "for sale" sign had been erected in front of the site, he called the owner in Dakar. She travelled to M'Bour to confront the planning department, which had erected the sign, and was told she was not allowed to hold leases to so much land. In response, she raised the wall, and built more bricks, which are stored on the land, and which, incidentally, are used to protect growing fruit trees planted by the guardian/urban cultivator. At the time of the research period, the guardian was actively cultivating the site and had plans to invest his own money to bring in electricity to run a water pump and thus expand his production capability.

Physical–chemical Composition of Materials

In many African cities, daily reminders of uncertainty provide the impetus to create urban environments that enhance the ability to navigate periodic or chronic shortages. Daily power and water outages, propane and petrol shortages, price increases and depleted fish stocks are realities that people must contend with, but which they cannot control. In that context of uncertainty, the city evolves as a series of material-practice assemblages that mitigate such insecurity.

Throughout M'Bour, there are material objects embedded with meanings that have to do with navigating uncertainty. Many people raise small livestock, just as many have papaya or mango trees growing in their compounds. Public wells exist throughout the city and are used frequently. In the materiality of a well, a chicken or a mango is basic sustenance, which provides the basis for social life.

Such practices might seem more appropriate in rural environments and, indeed, are remarked upon by both Senegalese people and visitors as being "out of place" in cities. But maintaining such practice-material assemblages reflects a rational and vital response to the inability of city governments to provide reliable infrastructure and an ongoing perception that self-provisioning and primary production are important to food and livelihood security. Although citizens in M'Bour might observe that such practices are more rural than urban, their relative acceptance reflects what Maller and Strengers (2014) conceptualize as "practice memory", the active maintenance of practices that "are able to be reinvigorated under the right conditions" (p. 148).

In addition to concerns about basic sustenance, the social dimensions of food and agriculture are well discussed and recognized, and there are infinite ways in which the physical or chemical composition of plants and animals could be demonstrated in relation to social life: taste, scent, medicinal value, religious symbolism, etc. In this section, three are discussed. First, the chemical composition of soils affects practice-material arrangements and causes cultivators to require soil amendments that they acquire by forming social relationships. Second, the chemical composition of specific plants carries a specifically urban social significance and conveys social meaning. Third, the physical composition of gardens as it occurs in an urban environment reflects a particular norm, which, in turn, has instrumental value for urban cultivators.

Composition of Soils

Soils in M'Bour are mainly composed of sand and thus have little nutrient or water-holding capacity. To make them viable as a growing medium, cultivators use considerable amounts of organic material, which serves as a nutrient source and helps to mitigate water loss. Warm ambient temperatures and daily watering provide optimal conditions for microbial activity. Thus, degradation of organic material is rapid and cultivators must amend soil regularly.

The effect that soils have on social life through the practice of urban cultivation is causative. The composition of soils, and the causal relationships between soils and, for example, heat, microbial action, and humidity, lead cultivators to perform the practice of urban cultivation in some similar ways. For example, every urban cultivator found sources of manure to amend soils on a regular basis. Cultivators choose manures according to availability, stage of plant growth/type of plant, and personal preference. To access manures, cultivators may form relationships with people who raise livestock. In some cases, they pay for manure; in others, they do not. For example, several cultivators had relationships with friends who raise chickens and acquired that manure, which is often mixed with wood shavings (used as a bedding material), for free. That manure is stockpiled and exchange occurs because the practice is intelligible to both parties. In other words, manure has a mutually recognized value to those who produce livestock and those who cultivate gardens, and this mutual recognition leads to a practice of manure exchange, which is an inherent part of the material-practice arrangement of urban cultivation in M'Bour.

Schatzki (1996) notes, "Practice theorists champion practices as the central constitutive phenomenon in social life because they view them as the site where understanding is ordered and intelligibility articulated" (p. 110). In other words, it is through practice that places are created or constituted. Intelligibility, as a socially contingent phenomenon, has spatial, environmental, and cultural dimensions; what makes places place-specific are the ways in which intelligibility evolves and articulates via materiality. Farmers everywhere, as people who engage in the practice of reorganizing and channelling forms of energy into primary production, generally understand that nutrient management practices are important to plant growth. But the relationships, significance and practice-arrangement nexuses associated with implementing these practices vary over time and space, and with place-specificity. The same is true of any practice and its associated material arrangement. It is intelligibility of the practice, or the ways in which meaning is conferred upon these practice-material arrangements, that "organize entities into the integrated nexuses that are what reality is and can be for us" (Schatzki, 1996, p. 115). Although raising animals in the city or exchanging manure might seem out-of-place or "backward" to those who attach specific normative meaning to "urban", or who attach specific spatial parameters to raising animals, it is commonsensical to the people who engage in these practices and who carry them out in ways specific to the particularities of M'Bour's practice-material arrangements.

Composition of Plants

Likewise, food, its meaning, and the ways in which it figures into the practice-material arrangements of a place, is environmentally, spatially, and culturally contingent. Specific dishes are commonly associated with specific celebrations, while certain foods are embedded with social significance. In M'Bour, lettuce is identified as a "city food", and is disparaged by people in rural areas as "food for sheep". Lettuce is a common feature of urban gardens in M'Bour; some cultivators grow only lettuce, while other cultivators grow it in association with other plants. Lettuce, of which there are several varieties, is easier to grow in the cool, dry season. It is often served in combination with fried or boiled potatoes, fresh tomatoes, onions, fried or grilled fish or grilled meat, a mustard-vinaigrette dressing, and French bread. Lettuce is in higher demand when there is plenty of fish in the markets. If fish are scarce, as happens often, people do not buy as much lettuce. As part of a dish that is unlikely to be served in rural areas, lettuce helps to constitute meals that have a specific cultural significance, or intelligibility, in the city. For example, it is in high demand around several holidays, including Christmas Eve and New Year's Eve, because it helps to signify the special occasion. Farmers plan their planting schedule to respond to the increased demand and are able to charge a higher price than usual.

Mint is another plant that is commonly grown by urban cultivators. Several varieties are grown, and cultivators tend to remark on and favour those varieties that are especially fragrant. Mint is almost requisite for the daily (or twice daily) and highly valued social ritual of brewing tea, in which friends and neighbours gather in public or private spaces for several hours to discuss politics, sports, and any number of other subjects. The material arrangement of the tea ritual (i.e. tea, mint, brewing pot, charcoal, stove, sugar and teacups) and the practice of the tea ritual (i.e. brewing, pouring, and drinking) are co-constitutive in that material entities help constitute the practice "by being pervasively involved…at particular times and places" (Schatzki, 2010, p. 140). Likewise, the ways in which the tea ritual is carried out is in a co-constitutive relationship with a particular way of gathering socially. The arrangement of the tea brewing and drinking materials serves as a focal point, and thus "holds" a group of people together for a time.

Without the tea ritual, would such daily gathering occur? Indeed, the tea culture is so pervasive in daily life and among Senegalese of all ages and ethnicities, it might be reasonable to suggest that this practice alone has enormous value in thwarting the social isolation and anonymity that are sometimes assumed to be a natural outcome of urbanization. Although tea drinking is not specifically an urban activity, it takes on an urban significance when practised among people who may be new to each other. Furthermore, cities, more so than rural environments, are sites of innovation and novelty. A novel mint variety, or the mint produced with the innovative hydroponic micro-gardening method, was regarded with enthusiasm and added to the social cachet of the mint.

The role of urban cultivators in the tea ritual is to provide a constant source of mint, and as often as they sell it around town, they also give it away to friends and neighbours, which has the effect of increasing and maintaining community cohesion. Giving away a portion of one's production, incidentally, was requisite and performed by every cultivator in the study. Many cultivators, because they bring in little income, used the gardens to satisfy cultural responsibilities. All Senegalese are expected to give away a portion of their income, a practice performed by both Christians and Muslims. In the absence of disposable income, gardens served an important social function, a reminder of Karl Polanyi's insight that economic activity safeguards social standing, social claims, and social assets (1944).

Composition of Urban Gardens

The particular physical–chemical composition of cultivated spaces produces a number of social interactions that are particular to urban cultivated spaces. First, a healthy garden visible from the road often leads to conversations between the cultivator and passersby. One woman, a recent arrival from the Casamance in southern Senegal, who knew relatively few people in her neighbourhood, cultivated a large garden of eggplant, peppers, onions, tomatoes, bananas, and okra. In the space of 90 min working in her garden, she was greeted by many passersby and engaged in two extended conversations with people who initially stopped to remark on the garden. Furthermore, this woman became familiar with people in at least two other households in the neighbourhood based on their gardening activities. In one instance, she bought a fruit tree, and in the other instance, she asked for pest management advice.

Secondly, a common practice for some entrepreneurial market women is to seek out and buy directly from urban cultivators. These women go from cultivated space to cultivated space, carrying an empty washtub that they use to carry away produce, and negotiate a wholesale price with cultivators. Though cultivators could earn more by selling their own produce directly to consumers, they often prefer to sell it to these *bana-bana*, who are more familiar with the going market prices, as well as market practices and processes.

Thus, it is the specific qualities of gardens, inherent to gardens, that provoke interaction. This is an important point of Schatzki's ontology that distinguishes it from other ontologies that perceive "the social" to be fundamentally separated from "the material". People's lives hang together in particular ways because of the particular social qualities of the garden.

Lastly, cultivators often sold produce to neighbours, who were grateful for the convenience and quality of neighbourhood gardens. A common theme to emerge was that many women disliked the expense and time required to travel to the single market in the centre of town. Furthermore, many women, both cultivators and consumers, noted that they did not trust what they bought in the market, or thought it was of lesser quality than what they could buy from neighbourhood cultivators or grow themselves. Schatzki (2010) notes that "the bearing of materiality on human activity and social life lies not just in the constitutive and causal relations that hold between individual actors and particular objects, but also in how material entities are connected with temporally and spatially extended manifolds of organized human actions" (p. 135). In M'Bour, the existence of a single market is a burden for many women, which turns out to be advantageous for urban cultivators.

Biological and Physical Flows

This section discusses the ways in which matter-energy and various organisms flow through the material-practice arrangements of urban cultivation and affect social life. Pests such as termites, ants, birds and whitefly "pass through" the material-practice arrangements and cause urban cultivators to institute certain practices or to suffer various losses. Very often, these flows are mediated by practice, while at other times, they pass through and affect the material-practice arrangement but are not managed.

The decisions about which flows to manage and which to ignore are affected by several variables, including knowledge of biological processes and interactions and prioritization of tasks given time constraints and other responsibilities. Ensuring the flow of water, for example, is the single most critical task, and also the most arduous and time-consuming. Most cultivators pulled water from wells, poured it into a basin, and then hand-watered with watering cans. Because of the poor water-holding capacity of soils, cultivators watered at least once a day, and often twice a day. Some cultivators referred to the practice of urban cultivation as *rosaay*, which is the Wolof adaptation of the French word *arroser* ("to water"). That the whole practice of urban cultivation is referred to as "watering" suggests that most of a cultivator's time and energy is spent watering.

Because a critical dimension of the material-practice arrangement of urban cultivation in M'Bour is constituted by a practice that requires a considerable amount of strength, urban cultivation is primarily identified as men's work. There was only one woman in the study who identified as the primary manager of her garden plot, and she had access to a pump and was able to water quickly. Another woman who had cultivated a plot of land in the past and in association with other women was reluctant to cultivate that same plot by herself because of the watering requirements. Micro-gardening, on the other hand, was performed mostly by women.

The task of managing water flow sometimes affected interpersonal relationships. Several of the research participants were married to each other. In all but one of these cases, men were identified as the primary manager of the garden and performed most of the watering tasks. However, women helped to water from time to time. In some cases, men noted their appreciation when their wives voluntarily take this task on. One man said that his wife could not usually help with the work of watering, but that she sometimes does it when he is out working on his occasional day-job of building bricks. He noted that she has asthma and that she should not be doing the work because it is difficult for her, but that when she does it, he regards it as an act of compassion towards him. Another man said that the work of caring for the garden brings him and his wife together, and that when she cares for it in his absence, it makes him happy. Most research participants noted the cultural imperative of husbands and wives working together in harmony and working out disagreements through discussion to set an example for the children. The garden represented a physical space where "working together" took place daily.

An observation of how practice mediates biological and physical flows, in conjunction with an understanding of the underlying ideas behind certain practices, points towards potential areas for agricultural extension education. Urban cultivators had little to no access to extension services and most information and knowledge exchange was limited to a small geographic area, among a small group of farmers. There were a few instances in which a better understanding of how to manage flows would have resulted in increased yield. For example, one cultivator had to uproot and destroy a whole plot of *bissap* (*Hibiscus sabdariffa*) because it was afflicted with a variety of fungal leaf rust, which often occurs because of over-wet conditions. By simply avoiding wetting the leaves when watering, he might have prevented such a loss and the extra labour it entailed. Another cultivator noticed how the addition of compost reduces the incidence of termite damage, but had no outlet to pass this information along to other farmers.

Conclusions

> [U]nless the complex, dynamic, highly improvising and generative actions of the urban poor are acknowledged and explored, it is foolish to come to conclusions about what is going on in a city, or what may or may not work, either from an insurrectionary perspective or from a "policy-fix" approach. (Pieterse, 2013, p. 3)

Jennifer Robinson (2006) observes, "Western modernity…is only one moment in the astonishingly diverse circulations and productions of new things and new ways of being that are assembled in distinctive ways to produce different kinds of places and ways of understanding them" (p. 20). In African cities, economic activities are improvisational, decentralized, flexible, and responsive to the highly dynamic environment of the city. This chapter has sought to demonstrate the ways in which seemingly "backwards" livelihoods are composed from fields of practice in combination with the material dimensions of the urban environment. They are enacted in relation to social and spatial position, dynamic urban processes, and normative and material concerns. The widespread practice of urban cultivation serves as an entry point into the "vernacular economy" to discover how people, through their interactions with each other and the city, are working to sustain themselves and create meaningful lives. Through livelihood, people are negotiating city and civic environments to deliberately make lives that include pleasure, innovation, and good relationships and social standing. Such an understanding of city life implicitly argues for an approach to urban planning and development that does not dismiss livelihoods because they are not indicative of modern form. As Robinson (2006) notes, "For urban studies to contribute to development strategies for ordinary cities, it will need to offer analyses that have a purchase on the diversity of economic activities, political interests and the range of needs of citizens" (p. 116).

Recent efforts to retool planning education and practice in Africa note the importance of engaging with informality differently (Odendaal, 2012). Future research and policy engagements must address the power relationships and biases embedded in informality discourse and better acknowledge how that discourse might serve to legitimate the unjust exercises of control over urban space. Furthermore, the underlying implications of informality, i.e. that these are economic activities without form, must be vigorously contested. The ontological claim to the city that is enabled by this discourse rests, in part, on the assumption that some economic activities threaten civic order because of their chaotic nature. This assumption must be challenged through research that seeks to find the organizing logic behind seemingly chaotic and spontaneous economic practices.

Practice theory offers conceptual tools that ground economic analysis in place and thus provides an opportunity to move away from conventional economic thinking that underlies the persecution of people in their efforts to make a life in the city. In coming years, it is likely that people living in growing secondary cities will need to continue to leverage all kinds of social relationships and practices to navigate uncertainty and sustain their households, and not just those that fall into narrow, externally constructed economic categories. Schatzki (2010) tells us, "Explanations of social phenomena should be sought in the specifics of pertinent practice-arrangement nexuses and the events that happen to them" (p. 146). Urban cultivation exists as a constitutive part of the city because it makes sense to the people who practise and benefit from it. Such an understanding of why people do the things they do, and a deliberate valuation of those things, offers alternative direction for urban research, planning, and development based on local imaginaries of well-being and meaning. The approach suggested in this chapter deliberately grounds livelihood in place by focusing on how the city's materiality is assembled through practice (and vice versa) and privileges the daily experiences and perspectives of urban citizens. However, for such an approach to gain legitimacy, urban planners and decision makers must reject parochial Western notions of modernity and urbanity that have obscured alternative urban imaginaries. Without valorizing the notion that development proceeds differently in different places, the prevailing approach to management of urban space will remain one that alternates between neglect and oppression, and that facilitates growing inequality and injustice.

Acknowledgements The data collection for this research took place over a period of 10 months between 2010 and 2011, and was funded by a Fulbright IIE Fellowship. It is based on one of three dissertation studies that used different place-oriented frames to understand urban cultivation in M'Bour, Senegal. The other two studies were published, one in 2015 in the journal *Gender, Place, and Culture*; the other in 2017 as a chapter in *Global Urban Agriculture: Convergence of Theory and Practice between North and South* (edited by Antoinette Winkler-Prins). The project was determined exempt by the Institutional Review Board, IRB# x09-602, Application #I033440.

References

ANSD. (2014). *Recensement Général de la Population et de l'Habitat, de l'Agriculture et de l'Elevage*. République du Sénégal, Ministère de l'Economie, des Finances et du Plan. Rapport Provisoire, mars 2014.

Batterbury, S. (2001). Landscapes of diversity: A local political ecology of livelihood diversification in south-western Niger. *Ecumene, 8*(4), 437–464.

Bebbington, A., & Batterbury, S. (2001). Transnational livelihoods and landscapes: Political ecologies of globalization. *Ecumene, 8*(4), 369–380.

Biersack, A. (2006). Reimagining political ecology: Culture/power/history/nature. In A. Biersack & J. Greenberg (Eds.), *Reimagining political ecology* (pp. 3–40). Duke University Press.

Bourdieu, P. (1990). Structures, habitus, practices. In P. Bourdieu (Ed.), *The logic of practice* (pp. 52–65). Polity.

Casey, E. (1996). How to get from space to place in a fairly short stretch of time: Phenomenological prolegomena. In S. Feld & K. Basso (Eds.), *Senses of Place* (pp. 14–51). School of American Research Press.

Creswell, J., & Clark, V. (2017). *Designing and conducting mixed methods research*. Sage.

Escobar, A. (2001). Culture sits in places: Reflections on globalism and subaltern strategies of localization. *Political Geography, 20*(2), 139–174.

Escobar, A. (2011). *Encountering development: The making and unmaking of the third world* (Vol. 1). Princeton University Press.

Feld, S., & Basso, K. (1996). *Senses of place*. School of American Research Press.

Finlayson, J. (2005). *Habermas: A very short introduction* (Vol. 125). Oxford University Press.

Gibson-Graham, J. (2008). Diverse economies: Performative practices for other worlds. *Progress in Human Geography, 32*(5), 613–632.

Giddens, A. (1984). *The constitution of society: Outline of the theory of structuration*. University of California Press.

Glesne, C. (2006). *Becoming qualitative researchers: Meeting qualitative inquiry (chapter 1)*. Pearson and Allyn and Bacon.

Hart, K. (1973). Informal income opportunities and urban employment in Ghana. *Journal of Modern African Studies, 11*(1), 61–89.

Hart, K. (1994). *African enterprise and the informal economy: An autobiographical note*. https://thememorybank.co.uk/papers/african-enterprise-and-the-informal-economy/

Illich, I. (1980). *Vernacular values*. http://www.davidtinapple.com/illich/1980_vernacular_values.html

King, B. (2011). Spatialising livelihoods: Resource access and livelihood spaces in South Africa. *Transactions of the Institute of British Geographers, 36*(2), 297–313.

Lincoln, Y., & Guba, E. (1985). *Naturalistic inquiry*. Sage.

Maller, C., & Strengers, Y. (2014). Resurrecting sustainable practices. In Y. Strengers & C. Maller (Eds.), *Social practices, intervention and sustainability: Beyond behaviour change* (pp. 147–162). Routledge.

Maloney, W., & Saavedra-Chanduvi, J. (2007). The informal sector: What is it, why do we care, and how do we measure it. In G. Perry, O. Arias, P. Fajnzylber, W. Maloney, A. Mason, & J. Saavedra-Chanduvi (Eds.), *Informality: Exit and exclusion* (pp. 21–42). World Bank.

Massey, D. (2013). *Space*. Wiley.

Myers, G. (2011). *African cities: Alternative visions of urban theory and practice*. Bloomsbury Academic.

Odendaal, N. (2012). Reality check: Planning education in the African urban century. *Cities, 29*(3), 174–182.

Owusu, F. (2007). Conceptualizing livelihood strategies in African cities: Planning and development implications of multiple livelihood strategies. *Journal of Planning Education and Research, 26*(4), 450–465.

Pieterse, E. (2013). *City futures: Confronting the crisis of urban development*. Zed Books Ltd.

Polanyi, K. (1944/2001). *The great transformation: The political and economic origins of our time*. Beacon Press.

Potts, D. (2008). The urban informal sector in sub-Saharan Africa: From bad to good (and back again?). *Development Southern Africa, 25*(2), 151–167.

Robinson, J. (2006). *Ordinary cities: Between modernity and development*. Routledge.

Roy, A. (2005). Urban informality: Toward an epistemology of planning. *Journal of the American Planning Association, 71*(2), 147–158.

Roy, A. (2009). Why India cannot plan its cities: Informality, insurgence and the idiom of urbanization. *Planning Theory, 8*(1), 76–87.

Schatzki, T. (1996). *Social practices: A Wittgensteinian approach to human activity and the social*. Cambridge University Press.

Schatzki, T. (2010). Materiality and social life. *Nature and Culture, 5*(2), 123–149.

Schatzki, T. (2014). Practices, governance and sustainability. In Y. Strengers & C. Maller (Eds.), *Social practices, intervention and sustainability: beyond behaviour change* (pp. 29–44). Routledge.

Schatzki, T., Knorr-Cetina, K., & Von Savigny, E. (Eds.). (2001). *The practice turn in contemporary theory* (Vol. 44). Routledge.

Seidman, I. (2006). *Interviewing as qualitative research: A guide for researchers in education and the social sciences*. Teachers College Press.

CHAPTER 4

Co-productive Urban Planning: Protecting and Expanding Food Security in Uganda's Secondary Cities

Andrea M. Brown

Introduction

Co-production is a strategy increasingly adopted in urban planning which, due to its mandate to empower primary stakeholders to address their own priorities, has potential for protecting and expanding urban food security. Africa's urban poor consistently rank food security high on their list of immediate priorities. Co-production goes beyond participation, with more substantive collaboration in policy design, implementation and monitoring, thus shifting some of the power associated with planning decisions and actions to primary stakeholders. Co-production is

A. M. Brown (✉)
Political Science, Wilfrid Laurier University, Waterloo, ON, Canada
e-mail: abrown@wlu.ca

L. Riley and J. Crush (eds.), *Transforming Urban Food Systems in Secondary Cities in Africa*,
https://doi.org/10.1007/978-3-030-93072-1_4

desirable for empowerment outcomes, but also on grounds of greater efficiency, cost savings, government accountability and more locally informed planning.

Present in 13 African nations, Slum/Shack Dwellers International (SDI) is a lead actor in co-production. SDI facilitates co-productive planning relationships among African states, federations of slum dwellers, community-based organizations and other partners, for example, faculty and students from local universities. They facilitate slum upgrading via targeted infrastructural improvements in informal settlements. Drawing on the case of Uganda, this research explores how co-production is engaging slum dwellers and governance actors in the secondary cities of Jinja and Mbale, through Transforming Settlements for the Urban Poor in Uganda (TSUPU), a programme supported by donors (the World Bank and the Gates Foundation) and government. This chapter looks to assess how urban food security is supported by current SDI co-productive programming, given the limitations posed by Uganda's political environment. Empirical evidence to support this research is drawn from interviews with urban planning stakeholders in Kampala, Jinja and Mbale.

Co-production

Co-production is "the process through which inputs used to produce a good or service are contributed by individuals who are not 'in' the same organization", from sectors not traditionally considered part of government, for example, citizens, the private sector and non-governmental organizations (NGOs) (Ostrom, 1996, p. 1073). The defining shift from conventional public service delivery is the role citizens play: they have active input shaping, or co-producing, services (Brandsen & Honigh, 2016, p. 428).

Co-production is valued for efficiency reasons, with arguments that cooperation with directly impacted stakeholders provides more accurate knowledge about priorities, which more removed planners do not have access to, as well as potential cost savings, depending on how non-traditional groups provide inputs into the governance process. For example, in addition to targeting priority areas for investment, primary stakeholders may contribute voluntary labour for community infrastructure, ideally fostering community ownership and incentives for upkeep as well as cost savings for governments. Co-production also has the potential for expanding accountability, as government actors are less able to operate in a closed forum and resource transfers are mandated. Further, community empowerment is a goal, as co-production offers opportunities for marginalized groups to shape development processes and voice their priorities (Mitlin, 2008).

Efficiency, accountability and empowerment may not all be fully realized, or even equally valued as goals, in every co-productive practice. Incentives to adopt co-production vary among participants, in ways that politically impact potential for successfully achieving desired outcomes. Thus, agreeing to and formally institutionalizing co-productive urban planning is likely to encounter a variety of political and bureaucratic bottlenecks and barriers.

The potential of co-production to achieve empowerment outcomes for primary stakeholders is debated. Concerns overlap with those directed at other programming that seeks to combine the interests and mindsets of international financial institutions with those of marginalized populations. For example, Poverty Reduction Strategy Papers (PRSPs) replaced Structural Adjustment Programs (SAPs) with goals to insert a pro-poor agenda in parallel with the austerity and liberalization conditions that had worsened poverty in the 1980s. The added provisions of popular participation, country ownership and pro-poverty ringfencing in budgets, represent movement from the rigid ideological top-down conditionality of SAPs. However, given the continued priority-setting dominance and values of the World Bank and the International Monetary Fund, they do not result in meaningful change to the neoliberal planning agenda (Cammack, 2004; Fraser, 2005; Soederberg, 2005). Arne Ruckert (2010) suggests, "The main goal of civil society participation might not lie in the incorporation of alternative ideas into the PRSP, but rather in the co-optation of NGO [non-governmental organization] voices, with the ultimate goal of disarming criticism and creating a stronger consensus around the content of the PRSP" (p. 119).

Similarly, co-production has been criticized as "coopting popular action" (Havassy & Yanay, 1990, p. 215). Jakub Galuszka (2019) notes that the term is frequently blurred with mainstream participatory approaches, and thus open to the same criticisms of co-optation, but when practices are deliberately structured to engage grassroots and the government on more equal footing, urban solutions that incorporate possibilities for meaningful power sharing are possible. For citizens to genuinely share in the process of urban governance—and this is more often the exception than the rule—specific conditions must be in place throughout the entire process of co-productive urban planning and implementation. These include willingness and capacity to collaborate from both government and community participants, and an institutionalized transfer of resources, including decision-making power, to non-traditional governance actors. Further, the intended outputs of co-production also matter: in the case of TSUPU's agenda, do infrastructural improvements that result from slum upgrading facilitate empowerment, or is this dependent on the rights-based community association aspects that emerge in tandem? Can a fairly limited agenda of slum improvement and upgrading impact urban hunger, a cross-cutting priority that is not the central target of urban planning for the government or NGO actors, despite its high priority for the primary stakeholders?

Urbanization, Planning and Food Security in Uganda's Secondary Cities

In 20 years' time, Africa will be predominantly urban, resulting in enormous pressures on urban infrastructure, but also marked by limited employment opportunities and rising levels of urban poverty and hunger. In Uganda, this transformation lags behind most of the continent: the nation is only 25.6% urbanized today. However, it is changing rapidly, increasing each year by more than five percent, making its urbanization rate among the highest in the world (CIA, 2021). Rising proportions of urban dwellers are living in crowded informal settlements.

Sixty-nine percent of Uganda's urban population currently live in secondary cities, with populations of 500,000 or less (World Bank, 2015). Uganda has responded to rising urbanization proactively in select secondary cities with the TSUPU and seeks to develop mechanisms and institutions to facilitate greater participation and agenda-setting by poor urban residents in cooperation with municipal governments in five urbanizing towns (Jinja and Mbale are two of these), ultimately to be scaled up nation-wide. SDI facilitates this process, by way of a co-productive urban planning strategy.

Jinja and Mbale have strong potential for planning success, given they are Uganda's only two secondary cities that benefited from an early colonial physical development strategy (Kiggundu, 2014). Although growing rapidly, informal settlements are not as established as in Kampala, so there is also scope for meaningful upgrading. Smaller populations further allow for easier facilitation of community cooperation. Limitations are posed by weak capacity in local councils, and more transient migratory patterns than in Kampala, with less established community organization and leadership.

Jinja and Mbale have similar population sizes, at 89,700 and 96,189, respectively. They are both growing steadily, fuelled by migration from rural areas and high birth rates. Rural poverty in Uganda is high, and increased land allocations for large-scale production is constraining space available for subsistence crops and sustainable land cultivation for small producers.

The land around Jinja is dominated by sugar cane production, both on large plantations and with locally owned land rented out to large sugar producers. The sugar cane industry is controlled by a handful of land owners. Most remaining locally owned land has been converted to growing sugar cane, with negative food security impacts. Eighty-seven percent of households in sugar cane-growing areas of Uganda report inadequate food to meet family needs and 21 in 25 households report sugar cane growing as the main source of food insecurity in the area (Anguyo, 2013). Jinja, in Eastern Uganda, banks on Lake Victoria, and was initially a fishing village, but today the city is more a market for fish than a centre of the fishing industry. Relocating to Jinja town from neighbouring rural areas is in part motivated by availability of services and employment, but there is poor access to both. Rural migrants settle in informal settlements, which face deficits in access to sanitation, clean water, education and employment opportunities, and safe, secure and affordable housing. Informal fishing remains central to the livelihoods of many in Jinja, but government pressures to formalize fishing, as well as market retail, make this increasingly difficult (Lince, 2011).

Around Mbale, 120 kms northeast of Jinja, production is much more diversified, but with significant concentrated land holdings devoted to rice production and coffee. Wetland areas near the city were used until recently for urban agriculture by the urban poor, growing rice, but this practice has now been prohibited, although it is unclear how this land will be used in the future. In surrounding rural and peri-urban areas of Mbale, subsistence agriculture still dominates, but population pressures and negative effects of climate change have reduced plot sizes and output, contributing to rural poverty, food insecurity and steady migration to Mbale town. Mbale is more ethnically diverse than Jinja and has long served as a transit point for migrants from Kenya as well as from more remote areas of Uganda en route to Kampala. Mbale also has a higher crime rate than other Ugandan towns and has a reputation for criminal gang activity and as a destination for those escaping the law.

Roughly 40% of the populations in Jinja and Mbale live in informal settlements. These settlements are not new, but these areas in both towns have doubled in size over the past 15 years, with economic inequality in both towns rising sharply. Slum residents rely heavily on the informal food sector to purchase and sell prepared food and foodstuffs for cooking at home. Every informal settlement has several informal business areas where food is sold, some specializing in items that attract consumers from across the city. One settlement in Mbale, for example, sells pork products unavailable elsewhere in town, and is a destination for locals of all economic status from across the city. Slum residents also sell and purchase items in the informal sector in town where informal markets—both stationary and mobile—exist near formal retail centres. Informal retail is the main source of income for poor households in Uganda, women in particular, and in turn the source for provisioning food and household goods. Locations for this retail have emerged in response to where people live and work and their daily travel routes.

Food security in urban Uganda is poorly recognized and addressed. Existing governance attention to the issue remains firmly grounded in narrow assumptions related to safeguarding and increasing food production, without attention to access or nutrition concerns of the urban poor. For example, only the capital, Kampala, has a government-supported policy on urban agriculture; both the community-driven and government-supported food security focus remains on increasing agricultural output in rural areas of the nation. Longstanding practices of urban agriculture in Mbale, such as fishing in Jinja, have been restricted to facilitate urban development in line with the municipal government's vision of modern town life. The myriad ways food security is attained in urban settings, related to stable access to nutritional food in a cash economy, are not considered. Smaller populations and less advocacy from local and international NGOs leave secondary cities further behind in addressing urban hunger than even the limited progress being made in centres such as Kampala.

Urban hunger is poorly addressed in part because the necessary responses are tied to a range of urban deficits limiting access to nutritious, culturally appropriate food, including income-generation opportunities, adequate access to water and sanitation, environmental hazards, transportation, stigmatization and marginalization of populations, and the protection of and access to informal food markets, all of which may be governed under different authorities who do not hold food security as their central priority. In contrast, food is the central priority of the urban poor, accounting for the largest portion of their spending. "For the urban poor in low and middle-income countries, food affordability and utilisation are shaped by the income and non-income dimensions of poverty that include the urban space" (Tacoli, 2017, p. 1554).

Municipal councils are the main intermediaries and most important actors in secondary city urban development planning in Uganda. It is widely recognized that municipal government capacity is low and that corruption is an immense problem. Only 60% of town positions are currently filled and funding from the national government is based on decades-old census data (Kiggundu, 2014). Also important is that town councils see their role as limited to enforcing and regulating urban development in accordance with planning processes and institutional rules that reflect prevailing visions of urban ideals. The dominant culture in local councils and among planners in Uganda is a central constraint on innovative urban development. While there is considerable attention to the need for local entrepreneurs to innovate, there is weak attention to how municipal functions could extend beyond infrastructural planning and enforcing regulations. When expanded understandings have been put forward, for example UN-Habitat's Right to the City initiatives arguing for public use of public space, these have received only short-term support from municipal governments and national ministries not interested in changes that undermine their authority. Planning culture in local government remains rigid and technocratic, and embracing the collaboration necessary for co-production will take time and a dramatic change in assumptions for actors who see one another as opponents more often than sharing mutually held goals.

Local governments in Uganda have long faced problems with planning, related to limited financial capacity, low education levels among councillors and corruption. Insufficient financial resources are available for local councils to carry out improved service delivery. Although there is authority for some tax collection, with financial transfers from the central government via a range of national ministries, this is inadequate and a considerable portion of centrally transferred resources come with detailed spending conditions that effectively leave authority centralized (Muriisa, 2008, p. 89). Local tax collection was undermined in 2005 when the central government abolished graduated personal tax, making councils even more dependent on transfers from the centre. National ministries have been unwilling to relinquish resources or decision-making to lower levels of government, in particular when a case can be made that national strategies with priority overlap with areas of local policy implementation. This became especially relevant when PRSPs were introduced.

Local governments also struggle with capacity, in the form of educated and experienced local officials. Inadequate skills in accounting, administration, planning and engineering, combined with a shortage of qualified personnel to support better health and education services, have stymied effective municipal planning. Elected local councillors have lower levels of education than central government civil services, contributing to the reluctance for ministries to truly devolve authority (Steiner, 2006, p. 14). This is an important point, highlighting that a perceived or real deficit in skills at lower levels can result in the failure of administrators with experience and formal education ceding control over planning and service delivery, and associated resources.

More serious structural problems also limit the success of local planning. Over the 30-plus years of President Yoweri Museveni's rule as leader of the National Resistance Movement, there has been a steady increase in concentrated power and authoritarianism. In practice, decentralization was utilized as a means for central authority to secure loyalty throughout the nation, by co-opting members of local government. Rather than a stakeholder model of decentralized public administration "grounded in democratic participation, accountability and empowerment" (Wenene et al., 2016, p. 170), competition among ministries for scarce state resources, alongside practices of rent-seeking, draws local councils into patron-client networks, where their limited authority and resource command are compromised by the need to be "loyal" to nation-state patrons. More devolves to district-level appointees rather than elected local councils (Green, 2015, p. 496). Corruption at local as well as national levels of government is endemic, even as enough professionalism and capacity, combined with ongoing international support, has facilitated overall economic growth. Corruption is at highest levels in local councils (Deininger & Mpuga, 2005, p. 178; Green, 2015, p. 501; Lambright, 2011, p. 1).

Co-production in Jinja and Mbale

In Uganda's co-productive urban planning, community participation is channelled through SDI, the local NGO AcTogether (which was established by SDI) and Federation of the Urban Poor (also initiated by SDI). SDI has replicated this institutional model from other Global South settings where it has been working since 1996; it is currently operating in 33 nations. In Uganda, TUSPSU launched in 2010, with detailed enumerations of slums in the initial five towns. Residents, members of the federation, provided the volunteer labour for these enumerations. SDI organized communities to identify priorities around potential upgrading of infrastructure, such as toilets, showers, water, better roads and drainage. Further, SDI and AcTogether support savings and micro-credit groups for women, which are central in community building in the federations and ensure women have key roles. Urban forums were organized, members signed up and meetings include municipal leaders and representatives from slums. Frequent networking opportunities exist to allow AcTogether leadership to share insights. Through these strategies, SDI strives to empower settlement residents so they can advocate for their rights, in particular those related to housing tenure and security. Students from Makerere University have also been brought into the process at different times—for example, to help design toilet blocks—and faculty from Makerere University participate in some planning meetings, although not in Jinja or Mbale, adding additional voices to both counter and inform municipal planning thinking and decision-making.

In Jinja and Mbale, immediate impacts of recent investments in urban sectors are evident: there is heavy construction, primarily roadwork, underway, with almost all roads in the town centres dug up, and hundreds of workers involved. In addition, both towns have had central open-air markets recently replaced by large new multi-billion dollar buildings, with new rental requirements and fees for vendors. Also visible are new toilet blocks and water pipes in slum communities, a direct result of priorities identified during the TSUPU slum upgrading co-productive process.

Cooperative planning between informal settlements and municipal authorities is challenging. Contestations between informal settlement residents and municipal authorities are frequent. Municipal police continue to remove informal vendors from the town centres and even from some of the informal settlements. Funding for new urban planning initiatives is partly supported by transfers from the national government, but also by increasing tax revenues from licensed shops and the new markets, adding pressure to the existing bias of formal over informal retail activities and spaces. Having clean, modern cities, with tidy, open sidewalks and no garbage areis the goal of urban development as expressed by municipal leaders. This narrative is particularly strong in Mbale, where the town long held a reputation of being clean and orderly, before urban growth and urban poverty rose, bringing with it the disorder and garbage of unregulated trading areas. TSUPU's slum upgrading initiatives operate alongside a larger urban planning agenda, with overlapping resources released by government authorities. Co-productive planning should, in theory, allow for more power and control of planning resources to be in the hands of community groups; given limited resources and competing understandings of upgrading and other urban development needs, however, this is not a straightforward process.

Evaluating Impacts of TSUPU

Food security is explicitly recognized by both SDI and residents of informal settlements as a central concern, but is not a direct part of the agenda of slum upgrading underway. From the SDI perspective, food security is a consequence of poverty and precarious income, itself compounded by the deficits of slum existence. The central undertakings of TSUPU involve infrastructural improvements to slums, through a process that allows targeted and cost-effective improvements in line with the sectors that residents see as most urgent. Enumerations further provide for accurate planning data on population density and size, and access and demand for housing, roads, drainage, water, toilets and retail areas.

There are several ways that food security is being addressed through TSUPU's programming, but these are indirect, and thus their impact is difficult to assess. Based on communities' own prioritizations, improvements have been made in informal settlements in Jinja and Mbale to water access and sanitation, which have both demonstrated links to food safety and nutrition. One settlement in Mbale successfully lobbied for a new school to be built in their area, since greater levels of formal education correlate with improved food security, due to lunch programmes, as well as better employment opportunities in the future. Drainage ditches have been dug to protect against flooding and the health risks of standing water. As with improvements to sanitation, better environmental conditions can have indirect benefits for nutritional security, for children in particular. These are all examples of how upgrading infrastructure has been planned and put in place with community participation and some community labour, achieving the co-productive goals of efficient and cost-saving urban planning. Early problems were identified, with some new infrastructure experiencing breakdowns and the fees for toilet access resulting in limited use, but overall AcTogether reports satisfaction with the pace and outcomes of upgrading.

While urban food security relies on safe water, healthy environments and access to services such as education, by far the most immediately impactful upgrades for food security are those connected with supporting spatial and income accessibility to food. In terms of increased income generation, support for women's savings groups and access to micro-credit are foundational to SDI's community organization, to allow for women's participation in co-production and to improve their financial autonomy and income-generating opportunities. This is one of the areas where SDI credits co-production with empowerment outcomes. Informal markets are critical to food access in urban settings. In this area, TSUPU has had mixed success with protecting and improving venues for informal trade and food access. AcTogether has helped communities advocate for water and sanitation access near informal markets, and women have been granted micro-loans in order to start or expand informal businesses. However, evidence that micro-credit is not an effective tool for improving incomes or promoting women's empowerment is mounting (Banerjee et al., 2015; Mahmud, 2003; Rankin, 2010). Some studies indicate that only a narrow band of women with established businesses benefit from micro-credit, while most women encounter unsustainable debt burdens.

Micro-credit in TSUPU's programming has also highlighted the disconnect between municipal and NGO goals. Micro-credit has been provided to women's groups in order to improve their income-generating activities, most of which involve informal sector businesses. Women are disproportionately represented in lower income informal activities, the informal food retail sector in particular. Improving women's business opportunities and income generation directly boosts food access for women and their families. However, municipal police in both Jinja and Mbale continue to harass street vendors, fining them for operating without a licence and confiscating their wares. When credit recipients in both cities reported back to AcTogether regarding these losses, they were advised to operate their businesses in the evenings, when they would be less likely to be targeted by authorities. Views on the necessity and legality of unlicensed vendors remain at odds with the larger urban planning agenda of city planners, despite co-productive arrangements to support women's informal business enterprises. A parallel example occurred in Mbale, where a toilet block constructed by TSUPU near an informal market area was destroyed by municipal police as part of a crackdown on illegal vendors in the central business area of town.

Uganda's National Urban Plan seeks to formalize food retail by developing new marketplaces in cities across the nation. New market areas have been constructed in both Jinja and Mbale. Improvements to existing central marketplaces are needed, to allow for more space, protection from the elements and safer access to clean water for vendors, but having the means to purchase licences to operate in formal markets remains out of reach for many of the urban poor, a reality largely disregarded by planning authorities.

Another mechanism SDI uses to increase community power is organizing communities more broadly into federations of the urban poor to enable them to assert their rights, in particular rights related to housing tenure. In India, SDI programming has been quite effective in this regard and the organization remains outspoken in its defence of housing rights for slum dwellers across the Global South. The goal is to have urban federations become part of a global movement, with federations able to draw support and learn from one another in their defence of rights. This has been less successful in Africa, perhaps because populations there have less established urban mobilization, but over time, as SDI expands its presence and affiliate federations continue to network, opportunities may grow. The primary obstacle to empowerment is that many African states, including Uganda's, are not on board with protecting rights to shelter or informal markets.

While urban food security is not identified as a direct planning priority by urban planners or integrated into TSUPU's approach, from a food security lens there are encouraging signs that this co-productive planning approach will respond to some urban food security needs. Improvements in communities' infrastructure, including access to clean water, sanitation, electricity and transportation, and developing safer, cleaner, more secure environments, all contribute to the broader requirements for supporting nutritional security. That communities are able to identify which upgrading needs are most pressing allows this to happen in a more efficient and accountable manner. SDI's enumerations in Asia have begun to incorporate a food security lens, in response to consistent community prioritization of hunger (Boonyabancha et al., 2019); given the extensive affiliate networking, this may inform future co-productive planning in Uganda's secondary cities.

SDI's explicit agenda to empower communities is also important. Centring women at the forefront of community organizing and leadership, by mobilizing them into savings groups, ensures those with primary community care have a strong voice in setting priorities for upgrading. Women are central to protecting food access in their homes and communities, with primary roles in food retail, preparation and responsibility. Facilitating regional networking with SDI affiliates has potential to add scope to TSUPU's agenda, as well as adapt tactics in response to a wider pool of experience and local needs. The knowledge sharing from these networks is geared towards developing a rights-based global movement, and SDI does not shy away from speaking out against governments it works with, in particular around defending tenancy rights, but more generally as well. This will take time, and at present only a small number of urban residents in Jinja and Mbale are actively involved in the forum's activities, or even aware of TSUPU or the source of new upgrades in their communities. Further, when AcTogether was formed, existing community-based organizations were not involved in the process and there was a missed opportunity to collaborate with existing community groups with overlapping concerns. These groups have longstanding relationships with communities, and their work with important marginalized populations (children, women, those with HIV/AIDS) could contribute to informing SDI's empowerment agenda if connections are fostered in the future.

Conclusions

Attention on urban food security remains weak in Uganda, with a continued bias from government actors to see food security through a productivist rural lens. Further, all levels of government continue to see long-term formalization of informal housing and employment sectors as urban planning goals. However, willingness to upgrade informal settlements indicates that governments, national and municipal, have shifted from seeing informal settlements as only urban blights to a greater recognition of the pressing needs of rising numbers of urban poor residents.

Infrastructural improvements only impact food security tangentially at present, but opening space for urban poor voices through co-production may permit the central importance of urban hunger to be heard in future. TSUPU's upgrading in Jinja and Mbale has been successful in achieving some efficiency and cost-saving goals, institutionalizing a transfer of resources and limited decision-making power to local communities. The main barriers to transferring more power to communities stem from limited resources, weak municipal governance capacity, competing and uncoordinated planning priorities and a continued planning culture that values formal urban planning over the needs of rising populations of urban poor.

References

Anguyo, I. (2013). *67% of Ugandans vulnerable to poverty*. www.newvision.co.ug/news/640813-67-of-ugandans-vulnerable-to-poverty.html

Banerjee, A., Duflo, E., Glennerster, R., & Kinnan, C. (2015). The miracle of microfinance? Evidence from a randomized evaluation. *American Economic Journal: Applied Economics, 7*(1), 22–53.

Boonyabancha, S., Kerr, T., Lumanti, J., & Tacoli, C. (2019). How the urban poor define and measure food security in Cambodia and Nepal. *Environment and Urbanization, 31*(2), 517–532.

Brandsen, T., & Honigh, M. (2016). Distinguishing different types of coproduction: A conceptual analysis based on the classical definitions. *Public Administration Review, 76*(3), 414–426.

Cammack, P. (2004). What the World Bank means by poverty reduction, and why it matters. *New Political Economy, 9*(2), 189–211.

CIA. (2021). *The world Factbook: Uganda, "Urbanization"*. www.cia.gov/the-world-factbook/countries/uganda/#people-and-society

Deininger, K., & Mpuga, P. (2005). Does greater accountability improve the quality of public service delivery? Evidence from Uganda. *World Development, 33*(1), 171–191.

Fraser, A. (2005). Poverty reduction strategy papers: Now who calls the shots? *Review of African Political Economy, 105*(5), 317–340.

Galuszka, J. (2019). What makes urban governance co-productive? Contradictions in the current debate on co-production. *Planning Theory, 18*(1), 143–160.

Green, E. (2015). Decentralization and development in contemporary Uganda. *Regional and Federal Studies, 25*(5), 491–508.

Havassy, H., & Yanay, U. (1990). Bridging between local needs and centralised services: Co-optation or co-production? *Community Development Journal, 25*(3), 215–224.

Kiggundu, A. (2014). Constraints to urban planning and management of secondary towns in Uganda. *Indonesian Journal of Geography, 46*(1), 12–21.

Lambright, G. (2011). *Decentralization in Uganda: Explaining successes and failures in local governance*. Lynne Rienner Publishers.

Lince, S. (2011). The informal sector in Jinja, Uganda: Implications of formalization and regulation. *African Studies Review, 54*(2), 73–93.

Mahmud, S. (2003). Actually how empowering is microcredit? *Development and Change, 34*(4), 577–605.

Mitlin, D. (2008). With and beyond the state—Co-production as a route to political influence, power and transformation for grassroots organizations. *Environment and Urbanization, 20*(2), 339–360.

Muriisa, R. K. (2008). Decentralisation in Uganda: Prospects for improved service delivery. *Africa Development, 33*(4), 83–95.

Ostrom, E. (1996). Crossing the great divide: Coproduction, synergy, and development. *World Development, 24*(6), 1073–1087.

Rankin, K. (2010). Governing development: Neoliberalism, microcredit, and rational economic woman. *Economy and Society, 30*(1), 18–37.

Ruckert, A. (2010). The poverty reduction strategy paper of Honduras and the transformations of neoliberalism. *Canadian Journal of Latin American and Caribbean Studies, 35*(70), 113–139.

Soederberg, S. (2005). Recasting neoliberal dominance in the global south? A critique of the monterrey consensus. *Alternatives, 30*, 325–364.

Steiner, S. (2006). *Decentralisation in Uganda: Exploring the constraints for poverty reduction*. German Institute of Global and Area Studies (GIGA Working Paper No. 31).

Tacoli, C. (2017). Food (in)security in rapidly urbanizing, low-income contexts. *International Journal of Environmental Research and Public Health, 14*(12), 1554.

Wenene, M., Steen, T., & Rutgers, M. (2016). 'Civil servants' perspectives on the role of citizens in public service delivery in Uganda. *International Review of Administrative Sciences, 82*(1), 169–189.

World Bank. (2015, March 3). *Managing rapid urbanization can help Uganda achieve sustainable and inclusive growth*. World Bank press release. http://www.worldbank.org/en/news/press-release/2015/03/03/managing-rapid-urbanization-can-help-uganda-achieve-sustainable-and-inclusive-growth

CHAPTER 5

The Role of the Informal Sector in Epworth's Food System, Zimbabwe

Godfrey Tawodzera

Introduction

The informal sector refers to economic activities that take place in unincorporated, small, or unregistered enterprises that are not assessed for taxation by a central government (ICLS, 2013; Paradza, 2009). The economic activities undertaken in this sector generally occur outside the rigid confinement of formal sector regulations and guidelines (Njaya, 2015). The greater proportion of the employment in the informal sector is without social protection. The informal sector is an important component of the global economy, as it provides employment and income to millions of people and also contributes to economic growth. The International Labor Organization (ILO, 2018) estimates that more than two billion people work in the informal sector worldwide, while Johannes

G. Tawodzera (✉)
Department of Environmental Science, University of Namibia, Windhoek, Namibia

L. Riley and J. Crush (eds.), *Transforming Urban Food Systems in Secondary Cities in Africa*,
https://doi.org/10.1007/978-3-030-93072-1_5

Jutting and Juan de Laiglesia (2009) put the proportion of the workforce engaged in informal employment at 60% globally. This proportion, however, differs between regions: 69% in South Asia, 57% in East and Southeast Asia (excluding China), 53% in sub-Saharan Africa, and 34% in Latin America (Vanek et al., 2014). The contribution of the informal sector to the economy varies globally, but Mohammed Yelwa and A. J. Adam (2017) estimate that it accounts for 40% of the gross national product of most low-income countries. The informal economy accounts for approximately 50–80% of the gross domestic product (GDP) in Africa (Steel & Snodgrass, 2008).

In Zimbabwe, the informal sector is important, especially in the absence of any noticeable growth in the formal sector in the past two decades. Leandro Medina and Friedrich Schneider (2018) posit that Zimbabwe's informal sector is the third largest (60.6%) in the world, after Georgia's (64.9%) and Bolivia's (62.3%). The rapid growth of the informal sector in the country can be traced back to the adoption and implementation of the Economic Structural Adjustment Program (ESAP) from 1990 to 1995 (Chirisa, 2009). While the informal sector already existed in the country before this period, it was small in size. The implementation of ESAP and the consequent shrinking of the economy increased unemployment as industries shut down. By 2008, industry capacity utilization had declined by close to 90% (Ndiweni & Verhoeven, 2013). While the informal sector employed only 10% of the labour force in Zimbabwe in 1980 (Labour & Economic Development Research Institute of Zimbabwe, 2017), by 2007–2008 the figure was estimated at between 80 and 94% (Dekker, 2009), as huge numbers of people entered the informal sector to earn their livelihoods. The Growth and Equity through Micro-enterprise Investments and Institutions (GEMINI) Study, 1991–1993, showed that micro and small enterprises increased in the country as the economy deindustrialized (Daniels, 1994).

In the post-2000 period, the economy further contracted due to the implementation of the fast-track land reform program. Close to two decades of political disagreements in the country also negatively affected the formal economy and have given further impetus to informal sector growth. A 2011 Finscope Micro, Small, and Medium Enterprise Survey estimated that there were close to 3.5 million micro, small, and medium enterprises in the country, with an estimated turnover of US$7.4 billion, employing about 5.7 million people (Njaya, 2015; Zimbabwe National Statistics Agency, 2011). It is the informal sector that has enabled the country to survive during a period of economic crisis (Ndiweni & Verhoeven, 2013). The food system accounts for a large share of informal sector activities. Its dual functions of creating income-earning opportunities and providing food to city dwellers are both critically important for Zimbabwean households during the protracted crisis. Clainos Chidoko et al. (2011) investigated the economic impact of the informal sector in the country and concluded that, despite numerous challenges (e.g., lack of capital and a shortage of accounting and managerial skills), the sector was indispensable to the economy of the country. Assessing the contribution of the informal sector to poverty reduction in Rusape Tatenda Saunyama (2013) concluded that income from informal sector trade enabled households to meet their food, medical, housing, and educational needs. Takunda Chirau (2012) researched women's livelihoods in Magaba, Harare and highlighted that although most women were operating in highly vulnerable sociopolitical contexts, they were nevertheless able to construct stable livelihoods in the informal sector.

For the informal food sector specifically, several studies have detailed the key contributions of the sector to the overall economy and to food security. O. I. Manyanhaire et al. (2007), while investigating the gendered dimensions of vending activities in Mutare, indicated that women dominated in food vending and that food vending constituted the women's largest source of income. Besides employment, the informal food sector serves to distribute food to consumers in areas not well-served by the formal food system. In most of urban Zimbabwe, the distribution of vegetables is primarily through the informal food system. According to Blessing Chitanda (2015), informal vegetable markets are embedded in the social and cultural spaces of the residents and thus serve not only the important food needs of the residents, but also enable residents to interact as they transact and bargain over quality, pricing, or other such aspects regarding the food. Hence, Tavonga Njaya (2014) argues that informal food trade is responsible for the upkeep of thousands of households that would otherwise struggle to survive in the challenging city environment.

The informal food sector includes cross-border trading. Thubelihe Jamela (2013) indicated that a significant proportion of cross-border traders in Zimbabwe were trading in food. The food items that are generally imported into Zimbabwe include cooking oil, rice, salt, sugar, powdered milk, maize-meal, and tinned foods. Although a small proportion of the foods imported by informal food traders may be sold to formal food retailers for resale, the majority of the food is generally sold on the informal food markets, contributing substantially to the food system of the urban areas. Thus, Peter Makaya and Constantine Munhande (2008), studying the contribution of informal food trading to food security in Gweru, argued that food trading is important because it increases food access. The importance of the informal food sector can therefore not be underestimated.

Despite the significant contributions detailed above, the informal food sector in Zimbabwe continues to be suppressed by governments. Periodic raids are carried out on the sector, which is projected as being unhygienic, dirty, and a constant source of disease, hence warranting destruction. While politicians at times support the existence of the sector, such support is conveniently voiced during election periods when the traders' support is needed. Thereafter, most politicians rejoin the bandwagon of those calling for the sector's destruction and advocate for the clean-up of the city to "international" standards (Potts, 2006). In the process, the crucial role that the sector plays in the economy, as well as the food system, is forgotten. Against this background, this chapter examines the role of the informal food sector in the food system of Epworth, Zimbabwe.

A Theoretical Perspective on the Informal Sector

The political tensions exposed through Zimbabwe's informal sector mirror tensions in academic discourse on the subject as a development problem. Discussions on the informal sector are generally controversial, based on the different viewpoints from which one can observe the sector (Njaya, 2015). Ralf Hussmanns (2001) outlines several of these viewpoints. First, the informal sector can be viewed positively as a provider of employment and income. Second, it can be seen negatively as an ungoverned sector operating outside the dictates of existing norms. Third, the sector can be romanticized as a breeding ground of entrepreneurship which, if left unencumbered by regulation and bureaucracy, could thrive and enhance prosperity. Fourth, it can be condemned as a vast area of illegality, backwardness, poverty, and unsanitary conditions.

Despite decades of debate, Cathy-Austin Otekhile and Oluwatoyin Matthew (2017) argue that knowledge about the structure, conduct, and performance of the informal sector remains limited, as little factual information exists, due to lack of interest from governments, academics, and public administrators. Debate on the sector has largely been polarized and two major discourses can be identified: the reformist and the Marxist perspectives.

The reformist perspective takes a positive view of the informal sector. It is rooted in the idea that the informal sector's contribution to the economy is largely positive. Drawing from the various studies commissioned by the ILO in the 1980s and 1990s, reformists believe that the informal sector has vast potential in creating employment, providing opportunities for on-the-job training, honing entrepreneurial skills of people in the sector, generating income, and spurring economic growth (United Nations, 1999). The role that the informal economy has played in sustaining livelihoods, especially in countries whose economies are in crisis, has solidified the reformists' contention that the informal sector is crucial to overall economic development (Lubell, 1991). It is not seen as a transient sector that operates in periods of economic crisis, but as one whose effects on the economy can be increased by enhancing macro-economic conditions to strengthen its viability.

Contrary to this perspective is the Marxist perspective. According to Otekhile and Matthew (2017), the Marxist perspective focuses on the structural dependency and exploitative relations between the formal and informal sectors. It sees the informal sector as marginal and petty, producing low-quality products, and supplying cheap labour that creates conditions for exploitative labour practices (Tokman, 1978). The sector is seen as problematic, as it reverses economic advancement and makes workers vulnerable. The fact that workers and employers in the sector are rarely registered makes it difficult for authorities to monitor or enforce compliance with policies (Chen, 2005). Marxist theorists thus see this sector as being "subordinate" to the formal and believe its eradication would not affect the national economy much because its contribution to the national fiscus is marginal. It is this view that some governments and local authorities use to advocate for the eradication of the informal sector.

This study is partial to the reformist perspective and argues that the informal sector in Zimbabwe is an integral part of the national economy, especially in an environment where economic growth has long stagnated and unemployment is high. It is imperative that efforts be made to better understand the sector, and the informal food sector in particular, as it plays a significant role in distributing food, "bulk-breaking," and generally increasing food access for the majority of the poor people who may otherwise remain on the fringes of the city's food system.

The Study Area

The study was carried out in Epworth, which is a peri-urban settlement located about 15 km to the eastern side of the Harare metropolitan area. The settlement was started by the Wesleyan Methodist Mission Trust in the late nineteenth century (Rakodi, 1995). Most of the initial inhabitants were farmers and were joined by church converts who sought sanctuary from the predominant traditionalist society in the surrounding area. In the 1970s, Epworth grew rapidly as refugees arrived from rural areas that had been rendered insecure by the intensification of the liberation war. At independence, the removal of influx-controls regulations resulted in more people moving into the area (Chitekwe-Biti et al., 2012). One of the major attractions of the area was its informal nature (Butcher, 1998), which allowed new migrants to settle without adherence to the regulations usually encountered in formally structured settlements. The population of the area increased from 20,000 in 1980 to 120,000 in 2009 and to 167,462 by 2012 (Zimbabwe Central Statistics Office, 1992, 2002; Zimbabwe National Statistics Agency, 2014). The Methodist Church passed ownership of Epworth to the government in 1986 and the Epworth Local Board was constituted to oversee administration.

Epworth's rapid growth in the late 1990s and the post-2000 period is attributable to economic challenges that forced large numbers of people to move away from Harare, where rentals are expensive (Marongwe et al., 2011). Epworth is divided into seven administrative wards. While the government's vision was for Epworth to eventually integrate it into Harare, this did not happen, presumably because Harare did not want to be saddled with the development challenges facing the area. Epworth has sparse infrastructure: only two main tarred roads, few service centres, low water and electricity connection, and a poorly developed sewer system. It is one of the poorest urban centers in the country, with a poverty prevalence of 64.5% (ZIMSTAT, 2013). The formal food system in the area is also poorly developed, with only a few shopping centers available. The food system of the area is thus largely informal.

Study Methodology

To investigate the importance of the informal sector to the food system of Epworth, this paper used data collected from four separate but interrelated surveys conducted in 2016 and 2017: a retail mapping census; a retail survey; a household food security survey; and a value chain analysis. The collected data enabled a better understanding of the role that the informal sector plays in the food system of Epworth. For the retail mapping census, research assistants identified and mapped all food retail outlets in Epworth. The aim was to get a picture of the various retail enterprises that operated in the area, the range of goods traded, operating times, and food distribution patterns. In total, 1,607 food retail outlets were identified and mapped. This information was necessary to understand Epworth's food system, especially which areas were better served and which ones were not, hence influencing food access.

Questionnaires were administered to a systematically selected sample of 297 retailers. Information sought included: the demographics of food traders, food sourcing, business income and expenditure, foods traded, and business challenges and opportunities. This enabled a better understanding of the central role that the informal food trade played in the overall food system of Epworth. A standardized food security questionnaire was administered to a systematically selected sample of 483 households in Epworth. Among other things, the questionnaire captured information on household demographics, poverty, income and expenditure patterns, food insecurity experiences, dietary information, and household coping mechanisms.

In addition, in-depth interviews were carried out with randomly selected households. The value chain analysis identified five food items (maize-meal, rice, *maputi*, vegetables, and offal) central to food security in Epworth. The food items were then tracked along the value chain, identifying the multiple packaging and pricing processes that occurred, as far back as the source or a place where no further information was available. This approach increased the chances of identifying more food suppliers to the area than would have been the case in a traditional "farm to fork" value chain analysis.

A two-day intensive training session was conducted for all enumerators before each survey component was carried out. This training also included piloting and refining of the research instruments. Data were collected and captured using the Open Data Kit technology, allowing for the simultaneous collection and capturing of data. While the four study components discussed above collected detailed data regarding retailing, household food security and food sources, this study extracted and used only data that is relevant to understanding the informal food sector's integral role in Epworth's food system.

Household Livelihood and Food Strategies in Epworth

The economic conditions in Zimbabwe in the past two decades have been dire, resulting in the GDP of the country contracting drastically and unemployment levels rising. Survey results indicate that only 13.3% of the sample population was working full time, while 13.4% were working part time. The greater proportion of the sample population (28.5%) reported to be self-employed, while 16.1% were looking for work, 9.4% were not looking for work, 2.4% were medically unfit to work, 0.4% were pensioners, and 16.5% categorized themselves as "housewives." It is worth noting that most of those that indicated self-employment were generally earning their livelihoods in the informal sector. The dominance of people in self-employment arises from the fact that salary-based employment in the country is limited. Munsaka (2013) estimated that Zimbabwe's unemployment level was about 95% by 2012. Thus, the majority of people ended up in self-employment in the informal sector. As one household respondent indicated, survival in Epworth largely hinged on eking out a living in the informal sector: "Our survival is hinged on selling our wares at Mbare Market. My husband goes and sells at market. On some occasions he gets some money, but on other occasions the money is not enough. For food to be considered adequate, we have to spend $100 a month, given the large size of family present" (Interview 7, June 25, 2016, Epworth).

The informal sector therefore formed the backbone of the household economy, and households were using income from this sector to feed themselves. Skilled individuals also worked in the informal sector when they could not find employment in the formal sector, as in the case of one woman who said: "My husband is a driver, with all the papers necessary to be employed as one, but the problem is that finding jobs nowadays is very difficult" (Interview 1, June 25, 2016, Epworth). While the income that some households were getting from their economic activities in the informal sector were not able to meet all their basic needs, it was nevertheless the only reliable source of income in an environment where the formal system had virtually collapsed.

A stable and adequate income is necessary for sustenance in the city because households pay for food and other expenses such as rent, electricity, water, medical fees, and transport, as well as educational costs. Out of 483 surveyed households, the majority (133) indicated that their major source of income was casual work. While a significant number of households (114) reported that they were getting income from formal wage work, many more earned income from informal sources, including informal wage work (118), informal business-sale of goods (71), informal business-not household produced (14), informal business-rent (13), and other informal business (10). While the total monthly income reported by surveyed households ranged from no income at all to a maximum of US$3,500, the mean monthly household income from all sources was very low at US$180.90. This was at a time when the monthly poverty datum line stood at between US$430 and US$574 for an average household of five persons (Zaba, 2017). Hence, the majority of the households were earning inadequate incomes. Sources with the highest average income were informal business selling something the household produced (US$415), followed by formal wage work (US$250), informal business selling something the household did not produce (US$193), grants (US$139), informal wage work (US$136), casual work (US$110), and informal rental income (US$108). These average income levels demonstrate that informal business and employment can be as lucrative as formal sector employment.

Consumer theory posits that there is a rational way in which limited income is allocated to different goods and services expenses and that such income allocation is done in order to maximize utility (Mafuru & Marsh, 2003). In Epworth, most households allocated their expenditure toward food and groceries (441), fuel (345), rent (240), telecommunications (230), and education (215). This was expected: in conditions of economic crisis, most households will end up focusing their expenditure on food rather than non-essential expenses, hence the limited expenditure on items not central to household survival. The total monthly household expenditure in Epworth was US$143. While savings had the highest average monthly expenditure of US$68, this should be treated with caution as there were few households saving on a monthly basis. The major expenses for most of the households were on education (US$57) and on food and groceries (US$54).

The informal sector was not only a major source of income for households in Epworth, but the households' main source of food as well. The major food sources for households in Epworth in the year preceding the research were tuck-shop/house-shop in Epworth (78%), hammer-mill in Epworth (58%), and vendors/hawkers in Epworth (42%). While other households indicated sourcing food from wholesalers (34%), informal food sources were more often patronized daily by household members. The other dominant food source indicated by the households was that of butcheries in Epworth (53%), which households patronized weekly. It is worth noting that while butcheries predominantly sell fresh meat, they were also selling other products such as matemba/*kapenta* (dried fish). Although 33% of the households indicated buying from supermarkets and 18% from wholesalers, these purchases were usually monthly. The trend that emerges here is that informal food sources were patronized more frequently and closer to home than the formal food sources that were generally located further from Epworth. Informal food sources thus played a critical role in Epworth's food system, as most households could access food closer to home. Only 33% of the surveyed households reported making purchases from supermarkets. Residents indicated that the supermarkets were too far away (56%), did not provide credit (54%), were too expensive (46%), were only for the rich (35%), and did not sell the food they needed (17%). The households' partiality toward informal food sources appears to be based on convenience and price.

Food Retailers in Epworth

Informal food retailers are critical to the food system of Epworth. According to survey results, the dominant type of food retailers in the area were tuck shops (19.5%), followed by market stand/stall vendor (16.3%), house-shop (7.7%), and table at intersection (7.1%) (Fig. 5.1). The dominance of informal food sources derived from the fact that there were few shopping centres in the area, as well as negligible trading infrastructure. Thus, there were only a few wholesalers (0.6%) and superettes (0.2%) or large/independent supermarkets (0.1%). The few wholesalers that were operating in Epworth were relatively expensive compared to the prices paid at wholesalers in Harare. Most residents therefore preferred to purchase from the streets, where the prices were comparatively lower.

An analysis of the spatial location of food retailers in Epworth illustrates some general aspects that are key to Epworth's food system. The first is that informal food retail outlets are spread out across all the wards, including within residential areas. Such location is important for the area's food system: it provides the most convenient location for the poor to access food, given that they have limited time and access to transportation to travel far for purchases. The location allows consumers to access food outside of traditional business hours, as some of these stores open

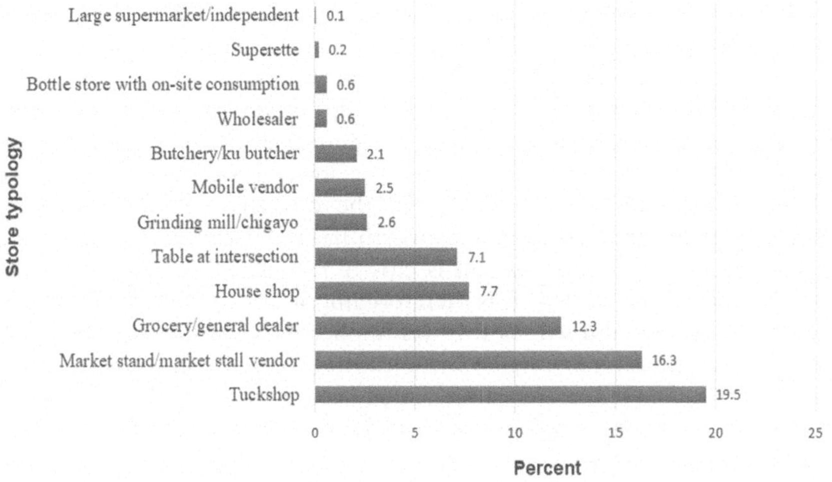

Fig. 5.1 Store typology in Epworth

early and close late. The second aspect relates to the predominance of street vendors, who play an important role in supplying food in Epworth. In a context where the formal food system had generally collapsed, street traders were expanding their stock to include such products as raw meat, traditionally sold in butcheries in the formal sector. Other foods found on the streets included the staple dish of *sadza* (meat and cooked vegetable), as well as other foods one would expect to find in restaurants. The food was often broken down into smaller, more affordable portions (bulk-breaking), commonly referred to as *tsaona*. Almost all food items could be found in the informal sector in Epworth. While the informal sector predominates in the area, the formal sector is not excluded. Rather, these two sectors exist side-by-side in complementarity to deliver food to residents. This symbiotic relationship is clearly seen in the meat trade, where street vendors often buy in bulk from butcheries and break the bulk to sell on the street.

While the formal food system is usually credited with supplying a wide variety of food to the market, thereby satisfying customer needs, an analysis of the food sold by retailers in Epworth showed that the informal sector is equally indispensable. According to survey results, informal retail outlets were selling the following key products: supplier-packed products (43%), vegetables (42%), non-food products (42%), self-packed items/bulk-broken items (37%), fresh food (34%), refrigerated food items (27%), and mixed foods (25%). Fewer retailers, however, were trading in traditional food items (18%) or in cooked/prepared foods (13%). The lower proportion of retailers dealing in traditional foods is indicative of the westernized diets in urban areas. The lower proportion of retailers selling refrigerated foods results from the challenge of low access to electricity in the area, as most of Epworth has no electricity.

The viability of a food system is closely related to the way in which food is sourced and traded. It is therefore important to understand where the key foods are sourced, as this has implications for the availability of the food as well as its pricing, which ultimately affects consumers' access. Jeremy Swift and Kate Hamilton (2001) argue that retail distribution systems are generally tilted in favour of high-income areas where infrastructure and marketing systems are well-established and more functional than in low-income areas. The reverse value chain analysis carried out in Epworth enabled a better understanding of where food is sourced and the pricing dynamics. The five food items studied were maize-meal, rice, maputi, vegetables, and offal. Their value chains are discussed below in turn.

Maize is Zimbabwe's staple crop, and it is used to make the traditional dish known as *sadza*. The value chain for maize-meal showed that much of the maize retailed in Epworth was imported from South Africa and Zambia. Although the Grain Marketing Board (GMB) was also importing maize, its role was minimal. Much of the maize-meal was imported by cross-border traders for resale to wholesalers. The majority were small-scale traders who brought in small quantities that could be imported duty-free as part of their travelers' rebate. Big importers generally shied away from maize importation, owing to high import duties. The contribution of cross-border traders to maize importation was immense, underlying the key role that the informal food sector plays in the food system of the country. Small traders in Epworth also bought maize or maize-meal from wholesalers in Harare and repackaged it into smaller quantities for resale to consumers who could not afford to purchase in large quantities. While big businesses sold maize-meal in 10-kg or 20-kg packages, the street vendors repackaged and sold maize-meal in packages of 1 or 2 kgs, pricing it at US$0.50 and US$1, respectively—an amount that consumers could afford. A number of hammer-mills were dotted around Epworth, supplying residents with maize-meal.

The rice value chain supply into Epworth was generally long, with rice coming from South Africa, Mozambique, China, Vietnam, and Singapore. While the length of the rice value chain reaffirms the embeddedness of Epworth's food system into the global and regional food system, it also shows the crucial role that the informal sector plays in importing the product. The major players in the importation of rice were wholesalers and the government, as well as cross-border traders, indicating the multiple entry points for rice into the country's food system. The role of the informal food sector in the rice value chain was significant: while the GMB imported rice in 50-kg bags, they also repackaged rice into smaller 20-kg, 10-kg, 5-kg, and even 2-kg packets for resale to wholesalers. In Epworth, however, traders in the informal sector bought rice in larger packaging at the GMB depot in Msasa, less than five kilometres away. They repackaged it into even smaller quantities (1-kg, 500-g, 250-g, and 100-g packets) that most residents could afford.

Vegetables are integral to the diets and nutritional security of residents of Epworth. While some households were growing vegetables in Epworth, the greater proportion of the traded vegetables was bought from Mbare Market in Harare. The vegetables at Mbare Market are supplied by farmers from nearby communities such as Mazowe, Dema, Marondera, and Mutoko. The vegetable trade was dominated by informal sector players. At Mbare Market, for example, the pricing was determined by middlemen (*makoronyera*) who controlled access for farmers and buyers. While traders would purchase a bundle of vegetables at the market for US$2, they would later break it down into smaller bundles for resale to consumers for US$0.10 for five to seven vegetable leaves. Although the street traders almost doubled what residents around the Mbare Market would pay, they enabled Epworth residents to access vegetables that would be significantly more expensive should a resident have to make a return trip to the market by taxi, which would cost an average of US$4. Selling vegetables in Epworth in pushcarts, wheelbarrows or on bicycles, vegetable traders in the informal market thus increase access of households to this most precious part of the local diet.

In Epworth, offal meat was available from both the formal and informal food economy. Few butcheries in Epworth were trading in offal. Rather, street traders and mobile vendors dominated, moving around with buckets of meat. The main source of offal meat was butcheries in Harare. In Harare, this meat was acquired from local abattoirs. While meat is part of the traditional dish in the country, harsh economic conditions militated against its consumption by households. The formal food system was unable to supply the offals needed in Epworth and thus the informal sector became the substitute.

Maputi is a type of popped corn that is consumed as a snack. Packaged in 50- to 80-g packets, it was generally sold along the roadside and at intersections by both stationary and mobile vendors. While it was also sold in formal shops, it was more abundant in the informal food sector. Traditionally, most of the maputi sold in Epworth used to be processed in factories in the Harare neighborhoods of Mbare, at Msasa Industrial Park, and in Chitungwiza. However, it was now being locally produced in Epworth by numerous small-scale manufacturers that have emerged in the past few years owing to the easy availability of the production technology. At about US$0.10 per packet, maputi is an affordable snack whose price has remained relatively unchanged even during the country's economic crisis. Its consumption by the majority of the people in the area, and its production by local manufacturers in the informal sector, means that it has assumed an important role in the area's food system.

Conclusion

The informal sector is an important component of the food system in Zimbabwe's urban areas. As survey results have shown, this sector is responsible for ensuring that food is available to residents of Epworth in quantities that are affordable to consumers. Besides supplying food, the informal food sector in Epworth is also providing employment that is critical for the survival of many households, especially in a context where the economy was depressed and unemployment very high. The analysis of the food sources also revealed some key facts. First, that Epworth's food system is indeed linked to the global food system, owing to some of the food that is sourced internationally. Second, even though the value chain for some of the food items consumed in the area (such as maize and rice) spans international boundaries, these value chains are embedded in the informal sector at the local scale. Third, formal and informal food systems interplay in Epworth to supply the area with the food that residents need.

Given these findings, this study makes a number of recommendations relating to the informal food sector. First, the local authority must repeal all laws that criminalize the sector, owing to its crucial role in the area's food system. Second, the local authority must improve the infrastructure that is available to businesses in the sector so that they can operate in an orderly and sustainable manner. In Epworth, as is the case in other urban areas in Zimbabwe, food vending is no longer just a temporary activity that households engage in during times of crisis. Rather, it is an enduring activity that is an integral part of the food system of the area and the economy at large.

Acknowledgements The author acknowledges the support of the Consuming Urban Poverty project (formally named Governing Food Systems for Alleviating Poverty in Secondary Cities in Africa), through which data used in this chapter were collected.

References

Butcher, C. (1998). *Epworth: A Review of the Densification Policy, Report Prepared for the Epworth Local Board*, Harare, Zimbabwe.

Chen, M. (2005). *Rethinking the informal economy: Linkages with the formal economy and the formal regulatory environment* (Working Paper No 146). UN University World Institute for Development Economic Research.

Chidoko, C., Makuyana, G., Matungamire, P., & Bemani, J. (2011). Impact of the informal sector on the current Zimbabwean economic environment. *International Journal of Economics and Research, 2*(6), 26–28.

Chirau, T. (2012). *Understanding livelihood strategies of urban women traders: A case of Magaba, Harare in Zimbabwe* (Master's thesis), Department of Sociology, Rhodes University.

Chirisa, I. (2009). The geography of informal sector operations (ISOs): A perspective of urban Zimbabwe. *Journal of Geography and Regional Planning, 2*(4), 66–79.

Chitanda, B. (2015). *A sociological analysis of vegetable markets in Dzivarasekwa 3, Harare* (Master's thesis), Department of International Development Studies, Wageningen University.

Chitekwe-Biti, B., Mudimu, P., Nyama, G. M., & Jera, T. (2012). Developing an Informal Settlement Upgrading Protocol in Zimbabwe—the Epworth Story. *Environment & Urbanization, 24*(1), 131–148.

Daniels, L. (1994). *Changes in the small-scale enterprises sector from 1991 to 1993: Results of a second nationwide survey in Zimbabwe* (GEMINI Technical Report No. 71). USAID and GEMINI.

Dekker, M. (2009). *Livelihoods and economic crisis: The case of smallholder farmers in Zimbabwe (1999–2008)*. University of Oxford.

Hussmanns, R. (2001, October 16–18). *Informal sector and informal employment: Elements of a conceptual framework*. Presented at the ILO/WIEGO Workshop on Informal Employment Statistics in Latin America, Santiago, Chile.

International Conference of Labour Statisticians (ICLS). (2013). *Report III: Report of the conference, 19th International Conference of Labour Statisticians*. International Labour Organization.

ILO. (2018). *Women and Men in the Informal Economy: A Statistical Picture*, Third Edition, International Labour Office, Geneva.

Jamela, T. (2013). *Experiences and coping strategies of informal women cross-border traders in unstable political and economic conditions: The case of Bulawayo traders* (Master's thesis), Department of Anthropology and Development Studies, University of Johannesburg.

Jutting, J. P., & de Laiglesia, J. R. (2009). Employment, poverty reduction and development: What's new? In J. P. Jutting & J. R. de Laiglesia (Eds.), *Is informal normal? Towards more and better jobs in developing countries* (pp. 17–26). OECD.

Labour and Economic Development Research Institute of Zimbabwe. (2017). *Situational analysis of women in the informal sector in Zimbabwe*. Government of Zimbabwe and International Labour Organization.

Lubell, H. (1991). *The informal sector in the 1980s and 1990s*. OECD.

Mafuru, J., & Marsh, T. (2003, July 13–16). *An analysis of household food expenditure systems in Tanzania*. Paper presented at the 2003 annual meeting of the Western Agricultural Economics Association.

Makaya, P., & Munhande, C. (2008). The contribution of informal food trading to urban food security in Zimbabwe: The case of Gweru 2000–2007. *Journal of Sustainable Development in Africa, 10*(2), 312–332.

Manyanhaire, O. I., Murenje, T., Chibisa, P., Munasirei, D., & Svotwa, E. (2007). Investigating the gendered dimensions in vending activities in the City of Mutare, Zimbabwe. *Journal of Sustainable Development in Africa, 9*(4), 169–186.

Marongwe, N., Mukoto, S., & Chatiza, K. (2011). *Scoping Study: Governance of Urban Land Markets in Zimbabwe*. Urban LandMark, Harare.

Medina, L., & Schneider, F. (2018). *Shadow economies around the world: What did we learn over the last 20 years?* (IMF Working Paper No. 18/17). International Monetary Fund.

Munsaka, T. (2013). *Human Resources management: A Critical Analysis of Zimbabwe's Unemployment Crisis from 1990–2012*, GRIN Verlag, Munich.

Ndiweni, E., & Verhoeven, H. (2013). The rise of informal entrepreneurs in Zimbabwe: Evidence of economic growth or failure of economic policies? *African Journal of Accounting, Auditing and Finance, 2*(3), 260–276.

Njaya, T. (2015). Informal sector, Panacea to the high unemployment in Zimbabwe? Case of informal sector enterprises in Harare metropolitan. *International Journal of Research in Humanities and Social Studies, 2*(2), 97–106.

Otekhile, C., & Matthew, O. (2017, March 9–10). *An explorative study of the contribution of the informal sector to economic activities in Lagos, Nigeria.* Paper presented at the 20th International Scientific Conference, "enterprise and competitive environment". Brno, Czech Republic.

Paradza, G. (2009). The retail and service informal sector challenge to urban design in Zimbabwe. In A. Brown & C. Davidson (Eds.), *Responsive design and plan implementation.* Department of City and Regional Planning, Cardiff University.

Potts, D. (2006). "Restoring order?" Operation Murambatsvina and the urban crisis in Zimbabwe. *Journal of Southern African Studies, 32*, 273–291.

Rakodi, C. (1995). *Harare: Inheriting a Settler Colonial City—Change or Continuity?* John Wiley and Sons, New York.

Saunyama, T. (2013). *The contribution of informal sector trade to poverty reduction in Rusape, Zimbabwe* (Master's thesis). Department of Social Work and Criminology, University of Pretoria.

Steel, W., & Snodgrass, D. (2008). *Raising productivity and reducing risks of household enterprises.* Diagnostic Methodology Framework, World Bank.

Swift, J., & Hamilton, K. (2001). Household food and livelihood security. In S. Devereux & S. Maxwell (Eds.), *Food security in Sub-Saharan Africa.* ITDG.

Tokman, V. E. (1978). An exploration into the nature of informal–formal sector relationships. *World Development, 6*(9), 1065–1075.

United Nations. (1999). *World Survey on the Role of Women in Development: Globalization, Gender and Work.*

Vanek, J., Alter Chen, M., Carré, F., Heintz, J., & Hussmanns, R. (2014). *Statistics on the informal economy: Definitions, regional estimates & challenges* (WIEGO Working Paper (Statistics) No. 2).

Yelwa, M., & Adam, A. J. (2017). Informality and economic growth in Nigeria: 1980–2014. *Journal of Economics and Public Finance, 3*(3), 405.

Zaba, F. (2017, June 16). Soaring health costs: Poor caught in death trap. *The Independent.*

Zimbabwe Central Statistics Office. (1992). Census national report, Harare. CSO, 2001. *Income, consumption and expenditure survey* 2001.

Zimbabwe Central Statistics Office. (2002). *Census 2002; Zimbabwe preliminary report.*

Zimbabwe National Statistics Agency. (2011). *FinScope consumer survey Zimbabwe.* Zimbabwe National Statistics Agency.

Zimbabwe National Statistics Agency. (2013). Poverty and poverty datum line analysis in Zimbabwe 2011/12, ZimStat. Available at: https://www.zimstat.co.zw/sites/default/files/img/publications/Finance/Poverty2011.pdf [Accessed 20 May 2017].

Zimbabwe National Statistics Agency. (2014). *Zimbabwe population census 2012*. Population Census Office.

CHAPTER 6

The Enabling Environment for Informal Food Traders in Nigeria's Secondary Cities

Danielle Resnick, Bhavna Sivasubramanian, Idiong Christopher Idiong, Michael Akindele Ojo, and Likita Tanko

Introduction

Rapid urbanization undeniably has altered food systems. The informal food retail sector represents a critical source of food security for the urban poor, for whom food access is heavily conditioned on income, housing, transport, and time (Tinker, 1997). In addition, the sector is a major

D. Resnick (✉)
Brookings Institution, Washington, DC, USA
e-mail: dresnick@brookings.edu

B. Sivasubramanian
Department of City and Regional Planning, Cornell University, Ithaca, NY, USA

I. C. Idiong
University of Calabar, Calabar, Nigeria

M. A. Ojo · L. Tanko
Federal Technical University of Minna, Minna, Nigeria

L. Riley and J. Crush (eds.), *Transforming Urban Food Systems in Secondary Cities in Africa*,
https://doi.org/10.1007/978-3-030-93072-1_6

source of employment for the poor in many cities of the developing world. This is particularly true in sub-Saharan Africa, where a significant share of the population depends on the informal economy for their livelihoods and most informal workers are concentrated in food retail.

In Nigeria, one of Africa's most populous and fastest urbanizing countries, informal food retail plays a central role. The informal economy is two-thirds the size of Nigeria's formal economy and constitutes about 80% of Nigeria's workforce (Neuwirth, 2012). Within the informal economy, retail represents the employment sector of around 40% of the population (Nigerian Bureau of Statistics, 2014). Food retail is a major sub-sector within informal trade. In fact, despite the growing presence of supermarkets in Nigeria, traditional open-air markets constitute the main source of food for two-thirds of the country's population (Nzeka, 2011). Informal food retail is also a major source of employment for women, who control a large share of market activity and commodity trading in Nigeria (Mangdon & Chintem, 2014). By providing consumers with access to processed foods, the informal economy also has the potential to foster stronger linkages for Nigeria's agricultural value chains, such as in poultry (Liverpool-Tasie et al., 2016).

Yet, despite the importance of informal retail to food security and employment needs in Nigeria and elsewhere, very little is known about the livelihoods of such vendors and the governance constraints they encounter. Much of the existing literature on Nigerian food traders focuses on Lagos, but with Lagos being a megacity with a population of approximately 15 million, such studies may not be reflective of the experiences of traders in urban centres in the country's secondary cities. Furthermore, a substantial share of existing literature focuses on the harassment encountered by traders by the local government areas (LGA) and state governments (Onodugo et al., 2016), with minimal attention given to the broader enabling environment in which these traders operate.

Consequently, this chapter addresses these gaps by combining a survey of 1,097 food traders in the Nigerian cities of Calabar and Minna, with key informant interviews with local and state policymakers as well as civil society organizations. We find some key differences across the two cities in terms of the demographic profiles of traders and the service delivery constraints they encounter within their trading locations, suggesting that policy responses to this constituency need to be properly nuanced, even at the subnational level. Furthermore, we observe that contrary to findings from other Nigerian cities, especially Lagos, the levels of harassment encountered by traders in these secondary cities is not especially high. However, there is a large gap between the de jure roles of local bureaucrats in regulating traders and the de facto oversight of traders on the ground. Many traders are unaware of government policies relevant to their occupation or have had little interaction with government authorities on issues of food safety.

To elaborate on these findings in more detail, the following section reviews the state of existing research on informal food traders in Nigeria. The subsequent section discusses the comparative context between Calabar and Minna by providing an institutional map of how informal food retail should be governed in the two cities. This is followed by a discussion of the survey methodology and a description of the findings about the demographic profiles of the traders, the service delivery environment in which they operate, and their degree of engagement with government officials.

Informal Food Traders in Nigeria

Informal food traders include those who are stationary within open-air markets as well as peripatetic hawkers who may operate from various locations over the course of a day or week. Within the food sector, traders may sell fresh foods such as vegetables, fruits, uncooked meats and fish, prepared foods (including cooked meals), and manufactured goods that are typically packaged in plastic or aluminum for retail. Such trading has a long tradition in Nigeria, but was exacerbated by structural adjustment in the 1980s when the contraction of formal employment forced many to resort to informal vending (Onodugo et al., 2016). As in many other developing countries, traders in Nigeria operate in high-risk and unfavourable conditions, in which they are frequently exposed to a hazardous working environment (Abe, 2012).

Much of the existing literature on traders in Nigeria is concentrated on two issues. First, food safety is a major concern, especially for those involved in fresh and prepared foods. Although street foods serve as a crucial link in meeting urban food demands, they pose numerous health dangers as proper sanitary measures are not commonly practiced, frequently due to lack of appropriate infrastructure. A number of studies have found that food vendors lack necessary food safety and hygiene knowledge, as well as sufficient public services, such as clean water and toilets, in order to pursue safe handling practices (Chukuezi, 2010; Dipeolu et al., 2007). For instance, A. Omemu and S. Aderoju (2008) find that only 12% of food vendors in the city of Abeokuta had formal training in food preparation, while only 31% had the medical health certificate indicating their completion of the recommended physical and medical examinations. Volume and price were valued over freshness and cleanliness of produce. Furthermore, certain known safety practices could not be executed due to a lack of basic services, including water and toilets, at the vending locations.

Second, government harassment features prominently in both academic scholarship and media reports in Nigeria. Such harassment includes bullying, beating, seizure of wares, forceful extortions, and occasional lock-ups in police cells (Olurinola et al., 2014). Such violent acts exacerbate the sufferings of street vendors, pushing them deeper into the cycle of poverty (Abe, 2012).

For more than a decade, and under successive governors, Lagos in particular has been known for its "zero tolerance" policy toward vendors. In 2003, the Lagos State Street Trading and Illegal Market Prohibition Law was passed to prohibit street trading and involves a ₦90,000 fine or a six-month jail sentence (At the time of writing, US$1 = 363 Nigerian naira.). Under former governor Babatunde Fashola, a series of beautification projects began in 2007 that resulted in the demolition of markets throughout the city. Street traders became especially vulnerable to the Kick Against Indiscipline brigades, which are an environmental law enforcement unit within the Lagos State government that seized the goods of traders deemed to be violating environmental laws (Nwanna, 2018). Another former governor, Akinwunmi Ambode, retained the same policy under the guise of Operation Clean Lagos. In 2015, he stated, "The next hawker could be a robber or a terrorist. You are now well-advised. If we all cooperate and decide not to buy, gradually and collectively, the hawkers will not come to the highways and streets anymore" (Vanguard Nigeria, 2015).

A study in Lagos found that 29% of the surveyed vendors had been arrested for street trading within the previous six months, while 44% reported less severe harassment measures, including frequent seizure and confiscation of merchandise (Basinski, 2009). Another study in Lagos corroborates harassment by government officials, showing that 32.2% of their sample had had their goods confiscated, 18.8% had paid a bribe to retrieve the said goods, 16.1% had had goods destroyed, 9.6% had been arrested, and 21.6% had been evicted from their trading spot (Lawanson, 2014).

While such harassment is sometimes justified in terms of addressing traffic congestion and pollution, some more recent forced removals in Nigeria have been motivated by the country's industrial development policy, which focuses on a cluster strategy. Specifically, some informal businesses were reorganized into "enterprise zones" to ensure a clustering of the same types of trading within a specific geographic location for the purposes of economies of scale in delivering infrastructure (Iwuagwu, 2011).

Notwithstanding a few studies in Enugu City (Onodugo et al., 2016) and in Ogun State (Omemu & Aderoju, 2008), a majority of the insights about informal food trade have been concentrated on Nigeria's primate city, Lagos. There are a few reasons to expand the focus to secondary cities in Nigeria. First, with a population of more than 15 million, Lagos has a high level of inequality that can precipitate conflicts between poor, informal traders, and more affluent residents who want to live in a world-class city with clean streets and reduced congestion. Elsewhere in Africa, the numerical size of wealthier citizens in capital and primate cities, and their resultant electoral significance, has been found to influence government reactions to informal traders (see Resnick, 2019). These dynamics may be less prominent in smaller cities with less sizeable middle classes.

Second, secondary cities are increasingly recognized as requiring greater policy attention, as they are often driving much of the developing world's rapid urbanization. The population of secondary cities can range from a few hundred thousand to several million people (Roberts, 2014). Some secondary cities take the form of satellite city clusters that surround larger urban metropolitan areas, while others assume the role of key industrial centres or development growth poles. Regardless of size, secondary cities are an important catalyst in enabling local production, transportation and transfer of goods, people, and ideas between the subnational, metropolitan, national, regional, and global systems of cities. Yet, insufficient data on the economic and political conditions of these cities inhibits the capacity to support urban development and encourage employment (Roberts, 2014).

Third, given Nigeria's federal system, oversight of informal trade is devolved primarily to the state and LGA levels. The LGA is constitutionally mandated to oversee the markets as well as to collect taxes from traders (Grossman, 2016). This means that traders' experiences, and government reactions to traders, should vary across cities in the country.

Comparative Context: Calabar and Minna

This chapter focuses on two secondary cities of approximately 500,000 people: Calabar and Minna. These two cities were chosen due to the variation they offer in terms of geopolitical location, ethnic and religious composition, and regulatory framework regarding street vending. Minna, located in the North-Central state of Niger and within the Middle Belt region, consists of three major ethnic groups, namely Hausa, Nupe, and Gwari. Calabar sits in the South-South state of Cross River and is home to the Efik people. While a majority of Minna's residents are Muslim, Calabar's residents are predominantly Christian. Such traditions have implications for food safety practices as well as the gender dimensions of informal employment. In addition, given the geographically biased nature of Nigeria's political system, the two cities have favoured different political parties. At the time of fieldwork, Minna residents heavily voted for the now-ruling All Progressives Congress (APC) in the last elections and there is an APC governor for Niger State, while Calabar was a stronghold for the People's Democratic Party. Such political differences may translate into different policy stances of the state government toward traders. In Niger State, there has been repeated demolition of vendors' stalls in Tunga and Central markets, while in Cross River State, the governor sent a Hawkers' Rights Bill to the House of Assembly in 2016 to legalize hawking in the state. The described regional differences allow us to observe the presence of subnational variation across informal food retailers and understand the causes at play.

To better understand the enabling environment for informal food trade in the two cities, semi-structured interviews were conducted with almost two dozen key informants at the LGA and state levels between April and May 2018. Collectively, they provide an overview of the complex governance landscape, from both an institutional and regulatory perspective, in which informal food traders are expected to operate. In Calabar, two main pieces of legislation govern food vending. These are the Cross River State Public Health Laws Cap 16 Vol. V of 2004, which stipulates how foods should be prepared for safety reasons, and the Hawkers' Rights Protection Bill of 2016. The latter bill was introduced by Governor Benedict Ayade and passed by the Cross River State House of Assembly in

2016 and signed into law. The bill legalizes street hawking as a livelihood option for persons living in the state and protects hawkers against harassment by government officials while carrying out their daily vending activities. According to one politician involved in pushing for the law, "The vendors are very much aware of the Bill given the wide publicity by government following its passage" (Interview with the Honorable State House Assembly Member, Yala, Cross River).

Figure 6.1 illustrates the complex array of actors involved in overseeing informal food traders. The city of Calabar consists of two LGAs, which are the Calabar Municipal Council and Calabar South. The LGAs technically are responsible for providing adequate infrastructure within the markets, which is why revenue collection is so critical. The revenue department in each LGA plays a key role in collecting levies from traders in the market and on the streets. The vendors in the market and on the streets pay ₦150 and ₦250 daily, respectively. Those in the markets pay a lower daily rate since they must also pay rent on their stalls. The revenue department also issues an operational permit to food vendors and requires them to

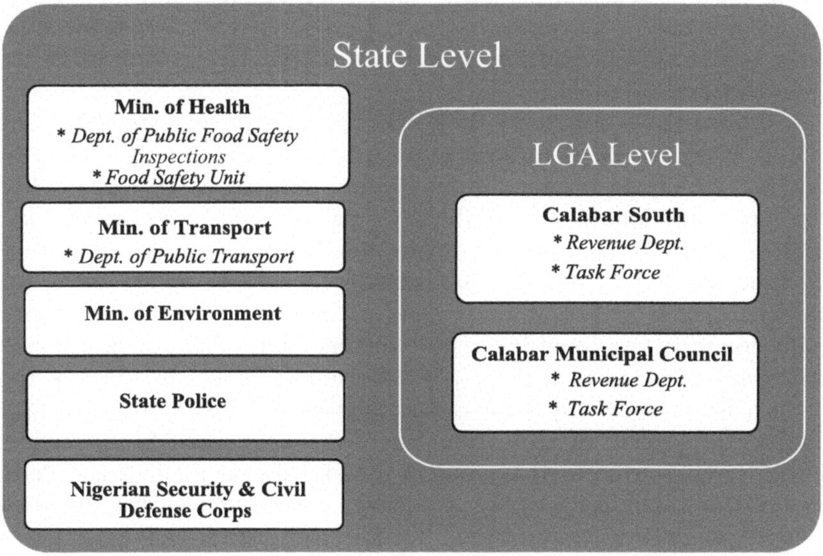

Fig. 6.1 Institutional actors involved in Informal Food Trade Regulation, Calabar

submit a medical certificate to assess their health status (Interview with the revenue officer, Calabar). The Calabar Task Force consists of temporary employees of the LGA who monitor and enforce vendors' adherence to LGA regulations, often with the help of the police.

At the level of Cross River State, five actors are relevant. The Ministry of Health, through the Public Food Safety Inspection Unit, is mandated to certify the safety of foods sold to the consumers by food vendors. Inspections ideally are carried out monthly to ensure that food vendors operate under hygienic conditions and that the vendors possess their required medical certificate. The Food Safety Unit within the Ministry of Health provides traders with hygiene training, increasing awareness on food handling from production to point of delivery to consumers (Interview with the environmental health officer, Ministry of Health, Calabar). The Ministry of Transport through the Department of Public Transportation collaborates with the Ministry of Environment and law enforcement agencies like the police and the Nigerian Security and Civil Defence Corps (NSCDC) in the monitoring of all street activities (including trading and public transportation) to forestall breakdown of law and order. They ensure that the roads in market areas are decongested, and they apprehend vendors who go against the regulations regarding street hawking in the city (Interview with the road traffic officer, Department of Public Transportation, Malabar).

In Minna, laws over food safety have been in place since 1976, but were amended in 2013 to ensure clean, healthy, and safe processing of food and packaging for consumption. Street vending is not banned, but there are restrictions in some of the city's main streets. As in Calabar, food vendors are likewise required to obtain a food vending license before they sell food. According to the LGA councils for Minna, which are Bosso and Chanchaga, food vendors are regularly monitored, and the food they sell is routinely inspected by health workers from the LGA health care department. In addition, there are also sanitary offices in some of the major markets in the city, such as Kure market and Kasuwan Gwari. Like Calabar, the revenue department of the LGA is responsible for the collection of revenue from food vendors in the markets on a daily basis. Petty traders with a stall in the market pay ₦30 for a daily ticket, hawkers with no stall pay ₦20 a day, and food vendors with established shops pay a monthly rate that ranges from ₦500 to ₦800 (Interview with the revenue officer, Chanchaga LGA).

Figure 6.2 shows that seven key organizations are relevant at the state level. The state Ministry of Agricultural and Rural Development has no formal role but does apparently retain a repository of data on formal/informal traders/vendors, such as names and locations of business outfits, types of food businesses, and scale of operation (Interview with the Ministry of Agriculture and Rural Development, Planning, and Statistics Directorate). The Ministry of Livestock, Meat, and Fish monitors the sale of meat and fish in the open market and ensures products sold are safe for human consumption. Along with the Ministry of Health, they inspect animals before slaughter and inspect meat and milk before they are taken out of the abattoir and off the farm (Interview with the Ministry of Livestock, Meat and Fish, Directorate of Veterinary Public Health). The Niger State Urban Development Board is responsible for the siting of markets and maps out where traders are allowed or prohibited.

The Ministry of Environment's oversight of informal vendors is primarily through a parastatal that it houses, known as the Niger State

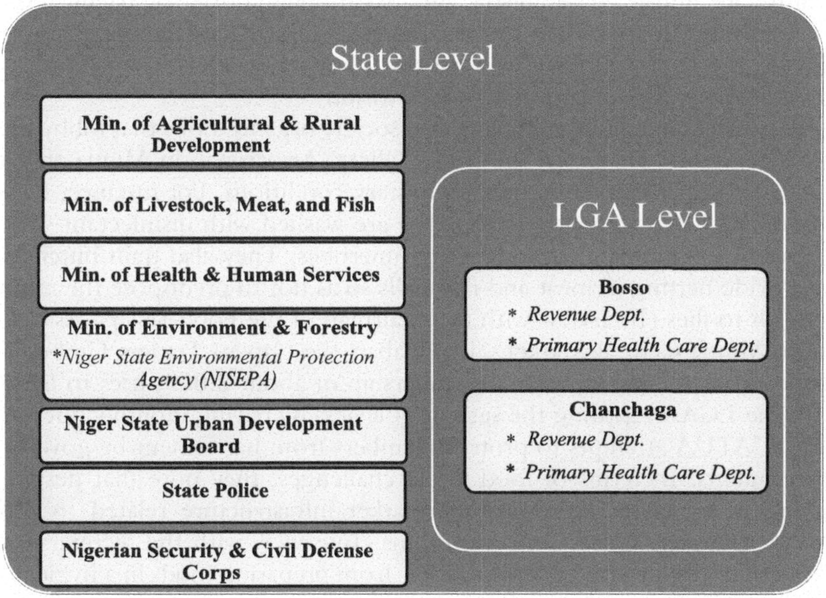

Fig. 6.2 Institutional actors involved in Informal Food Trade Regulation, Minna

Environmental Protection Agency (NISEPA). NISEPA primarily has an oversight and enforcement role focused on bakeries, hoteliers, corporate food vendors, and informal traders.

NISEPA also issues a special license to sell food in the market, which needs to be registered with the Environmental Health Officers Registration Council. With NISEPA's mobile courts, which are simply NISEPA's offices rather than a formal court of law, traders who have violated environmental safety regulations or refused to pay their LGA fees can be arrested by the police and tried with assistance of the NSCDC. In addition, NISEPA works with the LGA's Sanitary Vanguards on enforcing basic sanitary rules within places where food is traded. NISEPA's functions are limited, however, by inadequate logistics and funding as well as a lack of staff (Interview with the general manager of NISEPA).

The Ministry of Health and Human Services, through its Department of Food and Nutrition, also regulates food vending activities by routine inspections and claims to offer periodic training of food vendors on food safety requirements. In doing so, it interacts closely with the LGA primary health care units. This ministry also has mobile courts for trying food vendors who do not have a special license to sell food in the markets and on the streets (Interview with the Ministry of Health and Human Services, Department of Food and Nutrition).

In neither cities are there many civil society organizations that lobby for vendors' rights. However, the Food Sellers' Association in Minna trains butchers on personal hygiene and sanitary conditions. For instance, they ensure that slabs where meat is placed are washed with disinfectant such as Dettol to eliminate pathogens and microbes. They also train butchers to provide netting in meat and fish stalls so as not to predispose the meat and fish to flies (Interview with the chairman of the Food Sellers' Association, Minna Central Market). In Calabar, the Capital Traders Umbrella Association (CATUA), with a membership of about 3,000, tries to liaise with the LGAs regarding the setting of levies. By paying monthly dues of ₦50, CATUA attempts to protect members from harassment by government agents. In terms of food safety challenges, they note that despite paying heavy taxes, there is poor market infrastructure related to the quality of water, toilets, and waste bins (Interview with the secretary of CATUA). This in turn prevents traders from preparing foods in a hygienic manner.

Survey Approach

To determine how well traders are aware of the governance landscape that oversees their activities and how well-written policies actually are implemented, as well as to analyze their livelihoods in greater depth, a survey of 1,097 informal food sellers in nine markets within each city was conducted, for a total of 530 respondents in Calabar and 567 respondents in Minna. The interviews took place from mid-May to early June 2018. Figure 6.3 presents the location and names of the sampled markets in each city. Markets were selected because they are typically where a large concentration of informal food vendors are found. To ensure a representative sample, nine markets were selected in each city.

During the survey, we stratified between those vendors located within markets and those on the periphery of the markets, trading on the pavement or roads. This is because those outside the market theoretically can be more vulnerable to harassment, while those inside the market can provide a better assessment of market infrastructure. As such, having a relatively equal combination of two groups allows for a more comprehensive understanding of the institutional and governance constraints they encounter. To be eligible to participate in the survey, the respondents needed to be at least 18 years of age and be a food seller, vending either fresh, prepared, or packaged food.

Survey modules explored the demographic characteristics of the vendors, including educational attainment and migratory history, measures of food security, experiences with local government authorities, and services offered within the market. Due to the density of respondents, surveying informal traders proved much more complex than a typical household survey because traders can be organized in multiple rows along streets or alleyways within the markets. Therefore, a random walk methodology using a zigzagged pattern and targeting every third food trader was employed to both remain systematic and maintain a degree of randomness as to not inappropriately target specific traders. In more densely populated markets, every fifth food trader was targeted.

Profile of Informal Food Vendors

Table 6.1 presents an overview of informal food vendors in the survey locations in Nigeria. In Calabar, there were almost equivalent shares of traders who permanently traded from the same shop or stall (i.e., "fixed")

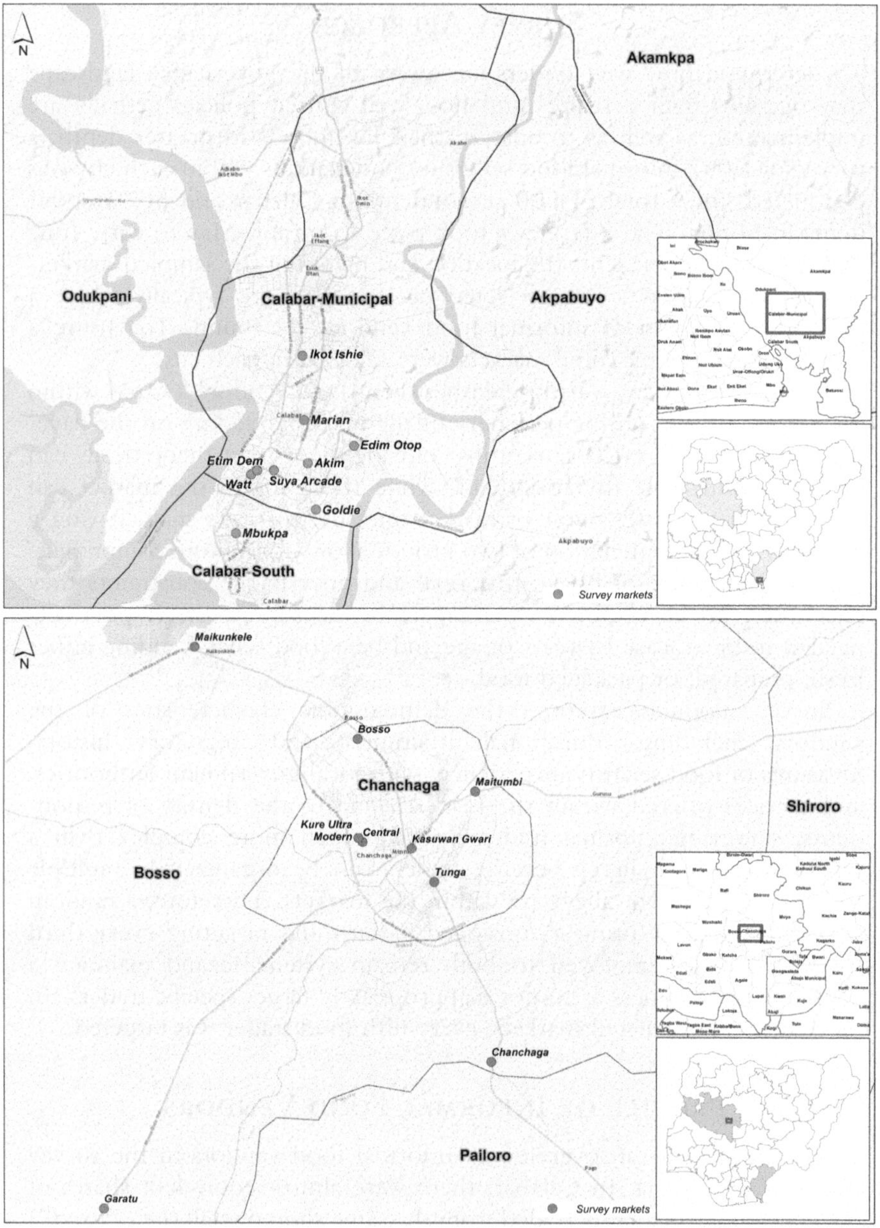

Fig. 6.3 Map of survey locations in Calabar and Minna

Table 6.1 Livelihood profile of informal food vendors

Variable	*Calabar*	*Minna*
Respondent location	50.8	61.4
Inside the market		
Outside the market on pavement/street	49.2	38.6
Nature of vending	53.0	50.8
Fixed		
Semi-static	43.8	28.9
Itinerant/peddler	3.2	20.3
Food category	54.8	52.7
Fresh foods		
Prepared foods	11.1	3.3
Processed foods	34.1	44.0
Holds a second job	3.2	2.5
Self-employed	95.7	90.7
Average number of days per week you are working in the market	6.3	6.5
Average number of hours per day working in the market	10.7	10.1
Average number of minutes to travel one way to place of trade	20.3	13.1
Approximately how much do you sell per day	1.1	0.4
Less than ₦1000		
Between ₦1000 and ₦3000	9.3	6.9
Between ₦3000 and ₦5000	21.3	31.8
Between ₦5000 and ₦7000	22.5	21.9
Between ₦7000 and ₦9000	14.3	13.4
More than ₦9000	30.6	21.7
Average monthly taxes	₦4468.7	₦688.5

N = 530 in Calabar and 567 in Minna

and those who traded from the same table or place on the pavement every day but had to remove their wares every evening (i.e., "semi-static").

More than half of the sampled vendors were selling fresh foods, considerably fewer traders sold prepared foods, and a sizeable portion vended packaged foods. Since food vendors could sell foods that belonged to multiple categories, the total percentages aggregate to more than 100%. Further breakdown of the food categories reveals that fresh vegetables were the top vended fresh food, while roots and tubers were also a commonly sold commodity. In terms of prepared foods, the top items sold were roasted yams, soup, and prepared rice, while bottled oils and canned goods constituted the top packaged foods.

Almost the entire sample is self-employed, rather than employees working for someone else or unpaid family workers, and food vending is their primary occupation. Based on the average number of minutes to their trading location, most are working relatively close to where they live. Nonetheless, they are working, on average, more than six days a week and very long hours. In both cities, more than half have a daily turnover that equals or is less than ₦7000 (US$19 at the time of fieldwork). Despite being informal vendors, they need to pay a variety of taxes and fees to multiple actors, including the LGA revenue authorities, the market chairperson, and product associations (e.g., tomato and onion).

Notably, the gender distribution also varies across the two cities. Table 6.2 shows that females represent 64.7% of food traders in Calabar and only 35.1% in Minna. The lower share in Minna may be strongly related to religious differences in the two states, which have implications for female employment. Indeed, while most traders overall in Minna are from the traditionally Muslim Hausa ethnic group, most of the female traders in Minna are from the Gwarri, Igbo, or Yoruba ethnic groups. While young people in Africa are often believed to be disproportionately employed in the informal economy (Banks, 2016; Filmer & Fox, 2014), this needs to be nuanced with respect to food traders. Those who are aged 18–24 comprise a relatively small share of food traders compared with those who are in the 25–34 cohort. The latter represent more than one-third of the traders in each city, as do those in the 35–44 range.

Importantly, the data reveal that food vendors in secondary cities of Nigeria are relatively well-educated, validating previous studies showing that lack of employment, rather than lack of employable skill, shapes informal vending (Adedeji et al., 2014; Olurinola et al., 2014; Onodugo et al., 2016). In other words, informal food vending is no more an activity for illiterates than it is for college graduates. There are, though, some notable differences across the two cities, with Calabar displaying higher levels of education than Minna. In both cities, today's traders have, on average, higher levels of education than that of their primary caretaker when they were children. Indeed, even in Minna, many more traders have completed their secondary education than the previous generation. This supports findings elsewhere in Africa that a major generational shift in Africa has been the expansion of education (see Resnick & Thurlow, 2015).

A number of scholars argue that a significant increase in informal vending is due to rural-urban migration (Bhowmik & Saha, 2012;

Table 6.2 Demographic profile of informal food vendors

Variable	*Calabar*	*Minna*
Ethnic community, cultural group, or tribe		
Efik	35.5	0.0
Gwari	0.0	14.8
Hausa	3.2	58.2
Ibibio	23.1	0.0
Igbo	22.5	8.5
Ijaw	0.9	0.0
Nupe	0.0	2.7
Yoruba	1.9	6.5
Others	13.8	9.3
Female	64.7	35.1
Age		
18–24	9.2	6.0
25–34	34.2	34.7
35–44	34.5	36.0
45–54	15.8	18.0
55–64	5.5	4.9
65 or older	0.8	0.4

(continued)

Table 6.2 (continued)

Variable	*Calabar*	*Minna*
Education		
No school	2.4	14.8
Some primary	2.1	7.2
Primary completed	11.3	21.3
Some secondary	10.2	5.3
Secondary completed	46.8	37.6
Post-secondary qualification	27.2	13.8
Education of caretaker		
No school	13.2	38.8
Some primary	10.4	13.2
Primary completed	22.3	16.1
Some secondary	11.7	4.1
Secondary completed	26.8	15.5
Post-secondary qualification	15.7	11.1
Migrant to LGA	54.9	32.3
From different LGA[a]	11.5	3.5
From different state/country[a]	43.4	28.7
Average number of years in LGA	10.3	14.2
Parent was a trader	80.8	77.3
Member of any association/organization related to job	2.1	4.6
Public participation within the last 12 months	35.3	50.8
Attended a community meeting		
Attended a meeting or rally of a political party	20.9	20.6
Participated in a protest or demonstration	15.5	16.1

[a]The observations for this variable are 285 and 180 for Calabar and Minna, respectively, since it only refers to those who are migrants to the LGA

$N = 530$ in Calabar and 567 in Minna

Bromley, 2000; Skinner, 2008). Our data reflect the prevalence of migrants within the food vending sector as well, especially in Calabar, where 55% of the sample had moved to the LGA in which they currently reside.

Of this migrant population, 21% moved from a different LGA within the Cross River State and the remaining 79% migrated from either a different state within Nigeria or a neighbouring country. A lower proportion of Minna's food traders are migrants, with 32.3% reporting that they were not born within that LGA. Of this migrant population among Minna food vendors, 10.9% migrated from a different LGA within Niger State, while 89.1% moved from a different state in Nigeria.

Evidently, there is far more migration occurring across state boundaries than across LGAs of the same state. A closer examination of the migration patterns reveals from where most traders are moving. As Calabar is located in the South-South geopolitical zone in Nigeria, it is unsurprising that 80% of the surveyed migrants moved from either the South-South region or the neighbouring South-East zone. As for Minna, located in the North-Central zone, 44.2% of its food vendor migrants reported having moved from the bordering North-West zone, while 19.0% migrated within the same geopolitical zone of North-Central. Twenty percent moved from the further away South-East region. Overall, migration to Calabar tends to occur from geographically adjacent areas, while migration to Minna appears to take a larger cross-country movement. Notably, in both cities, women constitute a larger share of migrants, with their primary reason for moving being either for marriage or to join their families, while men noted that their main motivation was to find work.

In both cities, traders spend an extended period of time in one location. In Calabar, the average number of years in the current LGA was 10.3 years, while a typical food vendor in Minna had been living in the current LGA for an average of 14.2 years. These data demonstrate a sense of permanency within their present living situation and, as a result, their occupation, emphasizing the importance of trading for their livelihoods. Data on length of time trading reveal that 49.5% of traders in Calabar and 35% of traders in Minna have been vending in their current location for at least six years. Food vending therefore is not a temporary job, but rather for most it is the sole means of income. This notion is further highlighted by the strikingly small number of traders who hold a second job—3.2% in Calabar and 2.5% in Minna.

Ray Bromley (2000) notes that street vending provides pathways of upward social mobility through entrepreneurial channels, in particular for marginalized groups that have limited wealth and power. Similarly, Margaret S. McMillan, Dani Rodrik, and Claudia Sepúlveda (2017) argue that as structural transformation unfolds, individuals shift from agriculture to modern economic activities, including informal trade, where incomes are marginally higher. To examine whether informal vending is either a stepping-stone or the status quo, respondents were asked if at least one of their parents had also been a trader. Table 6.2 shows that 80.8% and 77.3% of traders in Calabar and Minna, respectively, had a parent who is or was a trader. On the one hand, this suggests that vending is more of a family tradition for most food traders rather than an active step toward a new livelihood. On the other hand, of the smaller share of vendors whose parents were not traders, the most common employment for mothers and fathers included farming or they had been unemployed. As such, this suggests that for a segment of these vendors informal trade signifies a step toward potentially higher incomes than their previous generation had attained.

As it has been found more generally in Africa regarding informal sector workers (Lindell, 2010), these vendors rarely participate in associations representing their interests because of the potential costs involved, as well as the difficulty of an association being able to speak on behalf of a highly heterogeneous constituency. Nonetheless, they do engage in some forms of political participation, with half the Minna sample claiming they attended a community meeting in the last year and 20% in each city claiming to have engaged in a political party meeting or rally.

Studies have demonstrated that even with the recent expansion of supermarkets, the urban poor continue to obtain their food through informal means (Abrahams, 2006; Crush & Frayne, 2011). While this survey did not explicitly look at urban consumers, informal traders are themselves a major share of the urban poor. Table 6.3 strongly supports trends found elsewhere, emphasizing the importance of informal food retail on urban food security. Respondents were asked to specify where they regularly source eight different groupings of food that are viewed by major international organizations as critical to a healthy diet and for dietary diversity (see FAO, 2016). Except for milk and other milk products, such as cheese and yogurt, supermarkets play a negligible role as a food source for these traders.

Table 6.3 Share of informal traders who obtain food from listed sources

Food Groups	*Calabar*			*Minna*		
	Street vendor/ market trader	*Small shop/ kiosk*	*Supermarket*	*Street vendor/ market trader*	*Small shop/ kiosk*	*Supermarket*
Fresh fruits	99.4	69.1	8.1	99.3	79.9	8.1
Fresh vegetables	99.1	70.4	4.2	99.1	84.5	6.9
Meat, fish, other animal products	98.3	68.9	4.0	99.1	86.2	6.5
Eggs	93.6	74.3	8.7	83.4	89.6	5.8
Grains and flours	97.7	70.2	8.3	91.2	88.2	7.6
Milk and other milk products	91.5	71.5	15.3	85.0	90.3	12.3
Roots, tubers, and plantains	97.9	70.2	5.1	93.8	88.9	7.2
Prepared snacks and meals	95.3	63.4	6.6	99.5	67.7	6.2

N = 530 in Calabar and 567 in Minna
NB Values do not total to 100% since respondents may purchase food from more than one source

Enabling Environment for Informal Vendors

As noted earlier, informal traders often face a complex array of actors regulating their activities, which can create opaque lines of accountability and opportunities for extortion and harassment. As such, respondents were asked about their experiences with three types of harassment, including seizure of goods by authorities, forced removals, and arrests. In addition, traders were given the opportunity to share any other types of harassment they might have encountered from government authorities. Table 6.4 shows that in contrast to Lagos, where documented experiences of harassment and even brutality are quite high, this has not been a major problem in these two secondary cities. Eighteen percent of food traders in Calabar have experienced some type of harassment, while

Table 6.4 Experiences with forms of harassment by government authorities (%)

Types of harassment	*Calabar*	*Minna*
Ever harassed at all	18.3	3.2
Goods seized	13.0	2.1
Forcibly relocated	8.7	1.4
Arrested	0.8	0
Other	2.3	0
Never harassed at all	81.7	96.8

N = 530 in Calabar and 567 in Minna

NB Percentages do not sum exactly to the overall harassment percentage since some respondents experienced multiple types of harassment

the corresponding figure in Minna is only three percent. Despite being relatively low percentages overall, the difference in harassment between the two cities is statistically significant. Importantly, of the 115 traders who were harassed across the two cities, 70% were women and most of them reported experiencing a seizure of their goods. This is a troubling dynamic, given the importance of informal trade to their livelihoods.

Figure 6.4 highlights that government harassment, through either forced relocations or confiscations, is not perceived to be the biggest challenge faced by traders in either city. Instead, a lack of sufficient

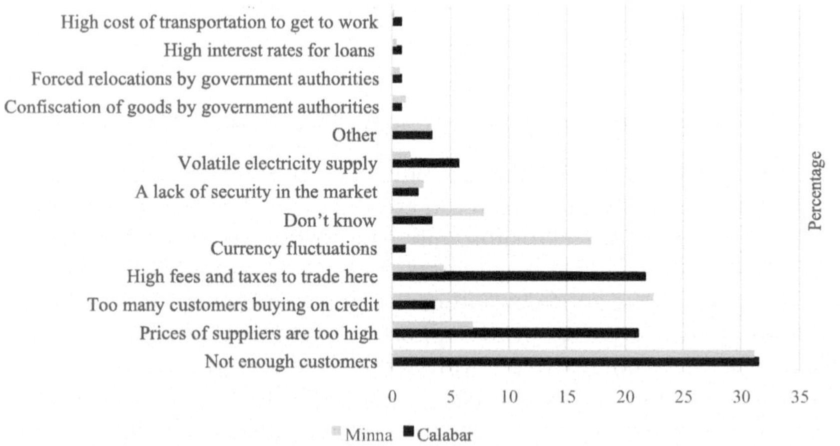

N=530 in Calabar and 567 in Minna.

Fig. 6.4 Top challenges faced by traders

customers is the main concern of traders since that obviously affects their profitability. Beyond that, there are some stark inter-city differences. In Calabar, the other top concerns include high prices by suppliers, high fees, and taxes by authorities who oversee the markets and volatile electricity supply. In Minna, currency fluctuations and the tendency of customers to buy on credit are perceived as the major challenges. These divergences can be attributed to economic, geographical, and cultural differences. For instance, it is not surprising that fees and taxes are a source of complaint in Calabar since, as noted earlier, they are between five and six times higher than they are in Minna. On average, Table 6.1 shows that respondents reported paying a little over ₦4,400 per month in Calabar to trade compared with ₦687 in Minna. Moreover, volatile electricity supply is less pronounced in Minna because it is near the Kainji Dam, which provides the city with an almost constant electricity supply.

Customers buying on credit is most likely more problematic in Minna because this is a common practice within the Hausa community (Aker et al., 2014; Lyon & Porter, 2009), which, as shown earlier, tends to dominate informal trading in Minna. While harassment is a relatively small problem for traders compared with other anxieties, how do they engage with the broader governance landscape they encounter related to food licensing and safety? As noted earlier, there are a variety of actors at the state and LGA levels that are responsible for regulating informal food trade in these two cities. However, in contrast to the official statements by state and LGA officials in the two cities, less than one percent of the survey sample in each city claimed that they need a license to sell food. Moreover, on average, 78% of the sample in each city noted that they had never had an inspection by a health officer of their food handling or food quality in at least the six months prior to the implementation of the survey. This was even true for markets that have sanitation officers housed in the market, such as Kure market and Kasuwan Gwari, as well as for a major meat market in Calabar known as the Suya Arcade. As another example of the gaps between policymakers' beliefs and informal traders' knowledge, 78% of those in Calabar had never even heard of the Hawkers' Bill.

The traders were also asked which government officials they come into contact with the most over the course of a month in the context of their work as a food trader. In Minna, half the sample claimed that they did not encounter any authority. Otherwise, the LGA revenue officer was the main authority identified by most of the respondents in each city

Table 6.5 Government official that traders engage with most often in the course of a month (%)

Official	*Calabar*	*Minna*
Local government market managers	18.7	21.2
LGA health officers	6.4	4.4
LGA revenue officers	33.8	22.8
State-level authorities	2.1	0.4
Task Force	23.2	0.9
Police	1.9	1.0
None	13.0	49.4
Other	0.9	0.0

N = 530 in Calabar and 567 in Minna

(Table 6.5). Neither state-level authorities nor LGA health officers were encountered frequently by these food traders. Importantly, in Calabar, 23% identified engaging often with the municipal task force.

These patterns have important linkages to the broader institutional environment. For instance, when comparing Figs. 6.1 and 6.2, a key difference is the presence of a task force within the relevant LGAs in Calabar and its absence in Minna. Traders in Calabar that reported experiencing some type of harassment often attributed it to the task force locking their shops until they paid the required taxes and fees to operate. Since no task force exists in Minna, engagement with that authority is not reported there. Furthermore, more respondents in Calabar (85%) than Minna (69%) reported being asked to pay taxes and fees to the LGA, which suggests that reach of an LGA revenue department may be enhanced by an enforcement actor, i.e., the task force.

Of the relatively small number of respondents who vend prepared foods (Table 6.1), 30% noted that they lack clean water and soap within the markets to wash their hands and utensils. This situation exists even though 81% of prepared-food vendors in Calabar and 95% in Minna make the meals and snacks they sell within the marketplaces. The latter point suggests that for food safety concerns to be addressed, market infrastructure requires as much attention as regulation and oversight.

Indeed, Table 6.6, which only focused on traders inside markets who would be more knowledgeable about market services, reveals that access to clean running water and health facilities are major deficiencies in the markets of both cities. Toilet access is a particular concern in Minna, while insufficient drainage, which can contribute to cholera and

Table 6.6 Share of Respondents Reporting Access to Key Services in the Market (%)

Service	*Calabar*	*Minna*
Trash collection	43.1	44.5
Toilets	62.1	36.2
Electricity	40.2	66.1
Clean, running water	16.7	8.9
Safe storage facilities for merchandise	28.3	37.6
Shelter during bad weather	28.3	55.2
Fire extinguishers	2.2	1.1
Security	74.3	82.5
Health facilities	1.9	2.9
Proper drainage	15.2	53.7

N = 269 in Calabar and 348 in Minna

other foodborne diseases during the rainy season, is more problematic in Calabar.

In addition, more than half the traders in each city lack access to trash collection, which can attract pests and vermin with negative impacts on human health. In many cases, having physical access does not provide a comprehensive perspective since those utilizing the services often must pay a repeat user fee to either a market official or a private citizen who is renting out a toilet or collecting garbage as a means of earning additional money.

Overall, then, while traders do not face high levels of government harassment, their daily efforts to earn a living as traders can be thwarted by the substandard service environment in which they operate. This presumably is even more pronounced for those operating on the streets and pavements that may not even be entitled to use the services that do exist within the markets. Since access to services was limited to those who worked inside a market, the number of observations is smaller than for the entire sample of informal traders.

Summary and Conclusions

Despite the recognized importance of traders to urban food security in Africa and elsewhere (Resnick, 2017; Skinner, 2016), very little is known about this relatively heterogeneous constituency. The challenge is exacerbated by the predominance of empirical work that is concentrated in just one city, even though the demographic trends and governance structures that characterize cities can vary quite substantially. By focusing on two similarly sized but culturally and geographically distinct secondary cities in one of the world's fastest urbanizing countries, this chapter used an original survey of informal food vendors to understand their livelihood profiles, demographic composition, work challenges, and public opinions. This was combined with semi-structured interviews with knowledgeable policymakers involved in regulatory and oversight activities relevant to informal food trade. The joint integration of perspectives from both government and vendors allows for identifying where gaps exist to improve the livelihoods of the traders.

In doing so, the analysis presented here reveals two key findings. First, the enabling environment encountered by vendors in Calabar and Minna varies substantially from the findings in other Nigerian cities, especially Lagos, where the government is well-known for its draconian treatment of informal vendors. A number of reasons have been offered for harsh crackdowns on informal vendors in capital or primate cities, including aspirations to create world-class cities that attract tourists and business (McDonald, 2008) and electoral incentives to co-opt middle-class voters (Holland, 2016). However, these circumstances may be less relevant in Africa's smaller secondary cities, where a negligible middle-class and less dense urbanization reduce the need for crackdowns against informal vendors. As noted by one respondent in Minna, who explains that there is no ban on street vending and hawking in the city, "Economic realities are part of the reason why there was no explicit ban" (Interview with the revenue officer, Chanchaga LGA, April 2018). In Nigeria in particular, the ability of the governor to dissolve opposition-led LGAs also reduces the likelihood that crackdowns on vendors are tied to inter-governmental power struggles, as has been found in some African capital cities (see Resnick, 2019).

Instead of harsh repression of their activities, these food traders are operating more in an environment of benign neglect by the government. Despite the range of hygiene training, food licensing and oversight activities that state and LGA officials claim to execute over traders, many of those in our sample had not been exposed to these activities, except for revenue collection. Notwithstanding the collection of taxes and fees, important investments in services essential for hygienic food preparation have not sufficiently materialized. In Nigeria and elsewhere, informal food traders are frequently blamed for outbreaks of foodborne disease. Establishing laws and regulations for governing their behaviours is necessary but not sufficient to improve food safety. Regular sensitization to food sourcing and preparation issues is critical, as is the investment of revenues collected from traders into the relevant water, sanitation, and waste collection infrastructure that will enable them to implement these practices. This is especially important given recent analysis that suggests that food safety becomes most problematic in rapidly urbanizing countries, due to longer supply chains coupled with growing incomes to consume foods outside of the home (Jaffee et al., 2019).

This benign neglect seems closely tied to both low capacity and a high degree of opacity in the governance of informal vending. Low levels of staff and resources were directly reported through interviews by the government actors in charge of overseeing informal food trade. Insufficient compliance with revenue payments exacerbates the situation as there are not enough resources to investment back into the infrastructure of the markets (Interviews with the revenue officers in Chanchaga LGA and Calabar Municipality, April 2018). Figures 6.1 and 6.2 further show that there are multiple institutions involved in informal food trade; as has been found elsewhere (Adamtey, 2015; Roever, 2005), this undoubtedly stretches scarce resources and undermines accountability for oversight. Moreover, in both cities, traders report very little engagement with any of the state-level authorities charged with overseeing the sector and instead interact much more with their respective LGAs. This points to the importance of strengthening and consolidating the functions of the LGA into potentially one unit focused explicitly on informal traders and that collectively concentrates on collecting revenue from, and enforcing environmental and health regulations among, this constituency.

Secondly, while sharing many similarities, the profile of vendors in Calabar and Minna is quite diverse in notable ways, including their gender and religious composition, experience with government harassment, the priority challenges to their business profitability, and the profile of services to which they have the greatest access. Recognizing this variation is critical to ensure that policy responses for food safety, gender inclusion, taxation, market siting, and other key issues are properly nuanced to actually resonate with traders. Therefore, in the clear absence of strong associations that voice the concerns of informal traders in these two cities, LGAs need to invest in mechanisms of outreach and engagement, such as monthly town hall meetings or regular visits to market chairpersons, which enable the communication of the diverse priorities of traders into the local government policy arena.

Acknowledgements The authors are grateful for funding from the USAID Nigerian Agricultural Productivity Project and from the CGIAR Research Program on Policies, Institutions and Markets, led by the International Food Policy Research Institute. They also thank the CLEEN Foundation and NoiPolls Nigeria for their implementation of the surveys with informal traders in Calabar and Minna. Mekamu Kedir provided excellent research assistance by developing maps of the survey locations. All errors remain those of the authors alone. A modified version of this chapter was published in 2019 in *Urban Forum* 30: 385–405.

References

Abe, T. (2012). Bracing the odds in the face of double tragedy: The dilemma of street trading in Ibadan Metropolis of Nigeria. *Journal of Sustainable Development in Africa, 14*(8), 104–118.

Abrahams, C. (2006). Globally useful conceptions of alternative food networks in the developing south: The case of Johannesburg's urban food supply system. *Institute of Geography Online Paper Series, 1*, 1–39.

Adamtey, N. (2015). *Informal economy budget analysis: Accra metropolis*. WIEGO.

Adedeji, J., Fadamiro, J., & Adeoye, A. (2014). Spatial implications of street trading in Osogbo traditional city Centre, Nigeria. *Architecture Research, 4*(1A), 34–44.

Aker, J., Klein, M., O'Connell, S., & Yang, M. (2014). Borders, ethnicity and trade. *Journal of Development Economics, 107*, 1–16.

Banks, N. (2016). Youth poverty, employment and livelihoods: Social and economic implications of living with insecurity in Arusha, Tanzania. *Environment and Urbanization, 28*(2), 437–454.

Basinski, S. (2009). *All fingers are not equal: A report on street vendors in Lagos, Nigeria*. CLEEN Foundation.

Bhowmik, S., & Saha, S. (2012). *Street vending in ten cities in India* (Report written for National Association of Street Vendors of India). Tata Institute of Social Sciences.

Bromley, R. (2000). Street vending and public policy: A global review. *International Journal of Sociology and Social Policy, 20*(1–2), 1–28.

Chukuezi, C. (2010). Food safety and hygienic practices of street food vendors in Owerri, Nigeria. *Studies in Sociology of Science, 1*(1), 50–57.

Crush, J., & Frayne, B. (2011). Supermarket expansion and the informal food economy in southern African cities: Implications for urban food security. *Journal of Southern African Studies, 37*(4), 781–807.

Dipeolu, A., Akinbode, S., & Okuneye, P. (2007). Income generating potentials of street food vending businesses in Ogun state, Nigeria. *ASSET Series C, 2*(1), 180–189.

Food and Agriculture Organization (FAO). (2016). *Minimum dietary diversity for women: A guide to measurement.* Food and Agriculture Organization.

Filmer, D., & Fox, L. (2014). *Youth employment in Sub-Saharan Africa.* World Bank.

Grossman, S. (2016). *The politics of order in informal markets: Evidence from Lagos.* PhD dissertation, Harvard University.

Holland, A. (2016). Forbearance. *American Political Science Review, 110*(2), 232–246.

Iwuagwu, O. (2011). The cluster concept: Will Nigeria's new industrial development strategy jumpstart the country's industrial takeoff? *Afro Asian Journal of Social Sciences, 2*(2), 1–24.

Jaffee, S., Henson, S., Unnevehre, L., Grace, D., & Cassou, E. (2019). *The safe food imperative: Accelerating progress in low- and middle-income countries.* World Bank.

Lawanson, T. (2014). Illegal urban entrepreneurship? The case of street vendors in Lagos, Nigeria. *Journal of Architecture and Environment, 13*(1), 33–48.

Lindell, I. (2010). *Africa's informal workers: Collective agency, alliances and transnational organizing in urban Africa.* Zed Books.

Liverpool-Tasie, S., Omonona, B., Sanou, A., Ogunleye, W., Padilla, S., & Reardon, T. (2016). *Growth and transformation of chicken and eggs value chains in Nigeria* (Feed the future innovation lab for Food Security (FSP) Research Paper No. 22). Michigan State University.

Lyon, F., & Porter, G. (2009). Market institutions, trust and norms: Exploring moral economies in Nigerian food systems. *Cambridge Journal of Economics, 33*(5), 903–920.

Mangdon, T., & Chintem, W. (2014). Hygiene and sanitary practices of street food vendors in southern Kaduna, Nigeria. *International Journal of Health and Medical Information, 3*(2), 49–57.

McDonald, D. (2008). *World city syndrome: Neoliberalism and inequality in Cape Town.* Routledge.

McMillan, M., Rodrik, D., & Sepúlveda, C. (2017). *Structural change, fundamentals, and growth: A framework and case studies.* International Food Policy Research Institute.

Neuwirth, R. (2012). *Stealth of nations: The global rise of the informal economy.* Random House.

Nigerian Bureau of Statistics. (2014). *Nigerian labor force survey*. Central Bank of Nigeria.

Nwanna, C. (2018). Right to the city: Lagos, an emerging Revanchist City in Nigeria? *Advances in Social Sciences Research Journal, 5*(4), 59–69.

Nzeka, U. (2011). *Steady growth of Nigeria's retail food sector.* Global Agricultural Information Network, United States Department of Agriculture.

Olurinola, I., Fadayomi, T., Amoo, E., & Ola-David, O. (2014). Occupational health and safety among street traders in Nigeria. *International Journal of Economics and Finance, 6*(4), 59–68.

Omemu, A., & Aderoju, S. (2008). Food safety knowledge and practices of street food vendors in the City of Abeokuta, Nigeria. *Food Control, 19*(4), 396–402.

Onodugo, V., Ezeadichie, N., Onwuneme, C., & Anosike, A. (2016). The dilemma of managing the challenges of street vending in public spaces: The case of Enugu City, Nigeria. *Cities, 59*, 95–101.

Resnick, D. (2017). *Governance: Informal food markets in Africa's cities in 2017* (Global food policy Report). International Food Policy Research Institute.

Resnick, D. (2019). The politics of crackdowns on Africa's informal traders. *Comparative Politics, 52*(1), 21–51.

Resnick, D., & Thurlow, J. (Eds.). (2015). *African youth and the persistence of marginalization*. Routledge.

Roberts, B. (2014). *Managing systems of secondary cities: Policy responses in international development*. Cities Alliance.

Roever, S. (2005). *Negotiating formality: Informal sector, market, and state in Peru*. PhD dissertation, University of California.

Skinner, C. (2008). *Street trade in Africa: A review* (School of Development Studies Working Paper No. 51). University of Kwazulu-Natal.

Skinner, C. (2016). *Informal food retail in Africa: A review of evidence* (Consuming urban poverty Working Paper No. 2). University of Cape Town.

Tinker, I. (1997). *Street foods: Urban food and employment in developing countries*. Oxford University Press.

Vanguard Nigeria. (2015, November 6). *Ambode reads riot act to traffic offenders, okada riders, street hawkers*. Vanguard Nigeria. www.vanguardngr.com/2015/11/ambode-reads-riot-act-to-traffic-offenders-okada-riders-street-hawkers/.

CHAPTER 7

Secondary Cities and Urban Agriculture in Sub-Saharan Africa

Daniel Tevera

Introduction

Since the beginning of the twenty-first century, urban agriculture (UA) in sub-Saharan African cities has been an expanding feature of the urban landscape and economy, mainly because of increasing food insecurity, poverty, and unemployment in many parts of the continent. Informal food enterprises, including street food, have evolved to be important sources of urban employment and food. Previous food studies in cities that dominate their national urban systems, such as Harare (Drakakis-Smith et al., 1995; Mbiba, 1995), Lusaka (Hampwaye et al., 2007), Kampala (Maxwell et al., 1998), Maputo (Paganini et al., 2018), and Cape Town (Battersby & Marshak, 2013), have provided interesting insights into the drivers of UA and the spatial extent of the activity in

D. Tevera (✉)
Department of Geography, Environmental Studies and Tourism,
University of the Western Cape, Cape Town, South Africa
e-mail: dtevera@uwc.ac.za

L. Riley and J. Crush (eds.), *Transforming Urban Food Systems in Secondary Cities in Africa*,
https://doi.org/10.1007/978-3-030-93072-1_7

several national cities across sub-Saharan Africa. Studies that have been done in the past decade have shed light on food production issues in secondary cities and towns. However, most of the studies have focused on single cities and have not generated comparative cross-country discussions on UA in different size and function urban centres. In recent years, secondary cities in Africa have been receiving increased attention, which is largely attributable to their vast numbers (about 7,500 across the continent) and their rapid growth since the new millennium (Tacoli & Agergaard, 2017). Secondary cities and towns are closely connected to the apex and usually primate cities (their main source of consumer and capital goods) and to their rural hinterlands (their primary source of cultural foods and various environmental resources, such as wild fruits and edible insects).

The Food and Agriculture Organization of the United Nations (FAO) defines UA as any agricultural activity that entails the raising, growing, or processing (and distribution) of agricultural produce in cities and towns (FAO, 2001). In sub-Saharan Africa, UA takes place across a whole range of spaces, including backyard home gardens (especially in middle-income suburbs), community gardens (at sites such as school grounds, church premises, and community centres) and open space cultivation on vacant lots and road verges. UA in African cities has generated considerable debate regarding the impact of its contribution to household food and nutrition security (Battersby & Haysom, 2016; Binns & Lynch, 1998; Crush et al., 2011, 2018; Drescher, 1996; Ferreira et al., 2018; Hampwaye, 2013; Riley et al., 2018). Its contribution to urban employment and incomes has also been hotly contested, largely because of the non-availability of reliable data on the activity, which takes place behind high security walls (in the case of urban home gardens in many Zimbabwean urban centres) and well beyond the watchful gaze of officials.

The benefits of UA have been highlighted in the literature and they include the improvement of environmental quality (through green spaces), improvement of household food security, and the production of cost-effective food supplies in urban spaces (Modibedi et al., 2021; Simatele & Binns, 2008; Simatele et al., 2012). Several scholars have also acknowledged the importance of UA in Africa and its ubiquity in post-independence African cities. For example, Thabo Modibedi, Mosima Masekoameng, and Matome Maake (2021) have argued that urban community gardens have contributed to the mitigation of food insecurity in low-income urban areas in the Gauteng Province, South Africa's commercial and industrial hub.

However, studies done by the African Food Security Urban Network (AFSUN) across 11 cities in Southern Africa reveal that the alleged contribution of UA to the nutritional status of urban households is exaggerated, if one considers the low levels of participation in many low-income areas (Crush et al., 2018). Similarly, studies by Jane Battersby and Gareth Haysom (2016), Bruce Frayne, Cameron McCordic, and Helena Shilomboleni (2014), and Madhav Badami and Navin Ramankutty (2015) support the view that the contribution of UA in Southern Africa to the nutritional status of urban households is exaggerated.

This chapter is based on a desktop study of three secondary urban centres that perform different functions and have experienced different growth trajectories during the past two decades, but have similar UA experiences. Nakuru is a commercial and transport city in Kenya, Ndola is a mining and industrial city in Zambia, and Karoi is an agricultural town in Zimbabwe. This chapter seeks to pursue a discussion that allows a re-imagining of UA in sub-Saharan Africa's secondary urban centres, especially the gendered food production landscapes that have become theaters of colliding visions of urban farmers and local authorities.

Secondary Cities and the Urban Food Question

According to UN-DESA (2019), the urban population in Africa will exceed 1.3 billion by 2050 and a significant portion of this population will be residing in secondary cities with less than 500,000 residents. Many of these secondary cities are experiencing rapid growth that is attributable to rural–urban migration. These secondary cities experienced considerable disruptions to their local food systems following various government-enforced economic lockdowns implemented in 2020 and 2021 to slow the spread of the COVID-19 virus (FAO, 2020). The disruption to food systems by COVID-19 highlights the need to reconnect local production and consumption. One way to achieve this might be through deliberate promotion of commercial agricultural activities in the peri-urban zones of these cities so as to make them better positioned to withstand future shocks.

The levels of urban planning and management vary considerably from city to city, as do the urban planning policies that aim to control UA. The limited levels of urban land use planning are generally a reflection of the underdeveloped institutional capacities to plan and enforce regulations. According to David Satterthwaite (2017), a majority of local authorities that manage small urban centres in sub-Saharan Africa are constrained by having insufficient capacity or funding to fulfill their mandates. Also, urban planning in many secondary urban centres has largely been top-down, with limited input from stakeholders.

Urban transformations in secondary cities in sub-Saharan Africa have generated renewed interest in food systems resilience in smaller towns. Two main questions are pertinent. First, will the 7,500 secondary cities that account for 48% of Africa's urban centres be able to achieve Sustainable Development Goal 2 on hunger eradication by 2030? Second, is UA making a meaningful contribution to the urban food system, especially food production? While urban development in sub-Saharan Africa is characterized by rapid population growth and increasing urban food insecurity, secondary urban centres such as Nakuru (Kenya), Ndola (Zambia), and Karoi (Zimbabwe) have recorded high growth rates over the past 20 years, and this has had consequential effects on urban food security livelihoods.

Nakuru is a secondary city whose population growth during the past two decades has been in excess of job creation capacity. This has given rise to urban informality characterized by numerous street trading activities and informal settlements. Between 1969 and 2020, the population of Nakuru increased from 47,000 to 383,000. The high rate of population growth has been accompanied by rising poverty and an uptake of UA (Foeken & Owuor, 2008; Owuor, 2006). In Nakuru, UA activities in the peri-urban zone have contributed to the city's local food system, as some of the produce finds its way to local vegetable markets (Foeken & Owuor, 2008; Owuor, 2006). However, it is unlikely that the contribution of UA to local food production can be maintained because the urban development projects that have been implemented in recent years in the peri-urban zone (Willkomm et al., 2021) and in open spaces will reduce the space available for UA. Ndola is an intermediate city with a population of about 500,000, including a large labour force that is employed in copper mining and allied industries. The city experienced a long period of steady growth from the 1960s to the early 1990s. However, following the decline of copper mining and de-urbanization in the copper belt during the 1990s, the rate of growth has declined. The location of Ndola in the country's copper mining region has resulted in the growth of close ties with the global economy and also with a huge rural hinterland that has provided the mines with cheap semi-skilled labour. Karoi, on the other hand, is a small agricultural town of fewer than 35,000 people, situated in the heartland of one of Zimbabwe's most productive farming regions. However, with the repeal of influx control measures following the country's attainment of majority rule in 1980, Karoi experienced relatively rapid growth during the 1980s and 1990s.

This chapter examines city and food production issues in three secondary urban centres that have varied UA experiences. The studies on Karoi (Gondo et al., 2017), Nakuru (Foeken & Owuor, 2008; Owuor, 2006; Willkomm et al., 2021), and Ndola (Mwitwa, 2008; Smart et al., 2015; Voleníková & Opršal, 2016) are significant because they emphasize the role of UA in various African secondary towns of different sizes. Maximilian Willkomm, Alexander Follmann, and Peter Dannenberg (2021) have argued that close attention needs to be paid to how rapid secondary urbanization is presenting food insecurity challenges through the displacement of urban food producers on one hand and the disruption of local food systems (especially in peri-urban zones) on the other hand. This state of affairs could be a reflection of an inability by both local urban authorities and national governments to structure urban space, through spatial planning, in ways that can accommodate UA and also nurture food systems in urban systems. One way to address this spatial challenge is by mainstreaming African urban food systems into the much broader urban policy and planning programs, because the two are interconnected. However, this requires the development of city strategic agendas on UA that involve local authorities, urban farmers, and other stakeholders, including community-based organizations (de Zeeuw, 2010).

Synthesis of Major UA Themes

Table 7.1 highlights several common elements of the UA experiences of Karoi, Nakuru, and Ndola, despite their considerable differences in terms of size, economic base, and UA policies. The table presents key aspects of UA in these urban centres, captured in the following broad themes: Theme 1 highlights the main factors that motivate households to engage in UA; Theme 2 presents the socio-demographic profiles of the urban farmers in terms of age, gender, educational attainment, and income class; Theme 3 highlights the dominant UA activities; Theme 4 identifies the bottlenecks that are blocking the uptake of UA; and Theme 5 provides an

Table 7.1 Key aspects of urban agriculture in Karoi (Zimbabwe), Nakuru (Kenya), and Ndola (Zambia)

	Karoi (Zimbabwe)	*Nakuru (Kenya)*	*Ndola (Zambia)*
Theme 1: Motivation for engaging in UA	– Food and nutrition benefits – Income generation – Community building	– Food and nutrition benefits – Income generation – Survival strategy	– Food and nutrition benefits – Income generation – Community building
Theme 2: Urban farmer profile	– Most farmers are female – Most male and female farmers are 26–35 years old – Low and middle income – Most have secondary school education – Most are regularly employed in formal/informal economy	– Most farmers are female – Most male and female farmers are 40+ years old – Low and middle income – Most have secondary school education – Most are regularly employed in formal/informal economy	– Most farmers are female – Most male and female farmers are 40+ years old – Low and middle income – Most have secondary school education – Most are regularly employed in formal/informal economy
Theme 3: Dominant activities	– Open space land is acquired through squatting – Mostly crop cultivation (e.g., leafy vegetables in home gardens and maize in open spaces) mostly on 20–49 m^2	– Crop cultivation (mainly maize) and livestock keeping (mostly chickens, goats, sheep, cattle, and pigs)	– Crop cultivation and livestock keeping (mostly chickens in the large plot areas)
Theme 4: Challenges	– Land permits are expensive – Favoritism determines land allocation – Field is too far – Land and water shortages – Periodic destruction of crops by urban local authorities	-Lack of suitable land – Difficult to get land for cultivation – Pests and insects are a problem	– Land and water shortages – Difficult to get suitable land for cultivation – Pests and insects are a problem

	Karoi (Zimbabwe)	*Nakuru (Kenya)*	*Ndola (Zambia)*
Theme 5: Policy frameworks	– Bylaws don't allow livestock keeping –Open space UA is technically illegal although there is less harassment of urban farmers – National Environmental Policy prohibits stream bank cultivation –UA not integrated into urban planning	– Cultivation of selected crops in open spaces is permitted – Municipal bylaws prohibit river bank and road verge cultivation –UA is not integrated into urban planning	– No legal land tenure in open spaces – The policy environment for UA activities is generally prohibitive – UA not integrated into urban planning – City supports sustainable UA

Sources Willkomm et al. (2021); Foeken and Owuor (2008); Gondo et al. (2017); Mwitwa (2008); Owuor (2006); Smart et al. (2015); Voleníková and Opršal (2016)

overview of the policy environments in the three urban centres and their effects on UA.

Theme 1: Drivers of UA

UA is driven by a variety of factors, including urban food insecurity, urban poverty, entrepreneurial aspirations, desire for ethnic foods, and shocks to the urban food systems that are caused by economic, environmental (e.g., drought), and health crises (e.g., government-induced COVID-19 lockdowns). Table 7.1 highlights the main reasons why both poor households and relatively well-resourced commercial farmers have engaged in UA in the three secondary urban centres. Lack of access to food, because it is either too expensive or unavailable for purchase, is one of the main reasons why many low-income households have resorted to UA under fairly challenging conditions (Foeken & Owuor, 2008; Gondo et al., 2017; Voleníková & Opršal, 2016; Willkomm et al., 2021). Urban food production is largely a response by the urban poor to inadequate, unreliable, and irregular access to food supplies (Foeken & Owuor, 2008; Modibedi et al., 2021; Zezza & Tasciotti, 2010). The need to produce fresh and culturally preferable food has also compelled some middle-income households to grow the food they consume. The persistence of UA in African cities is attributed to factors such as expensive urban food, urban food deserts, high unemployment levels, and food rights initiatives.

Theme 2: Profile of Urban Farmers

Table 7.1 shows that in Nakuru, Ndola, and Karoi, the major groups involved in UA are women, middle-aged residents (over 40 years old), people with regular incomes, and fairly educated urban dwellers with high school education (Foeken & Owuor, 2008; Gondo et al., 2017; Voleníková & Opršal, 2016). The limited involvement of young people in the activity implies reduced opportunities for potential transfer of UA skills and knowledge to the younger generation. Regarding the role of women involved in UA, it is essential to think beyond the numbers and to consider women's involvement in terms of their contribution to the gendering of UA (Hovorka & Lee-Smith, 2006; Hovorka et al., 2009; Ishani, 2009). It is clear from all three urban centres that since women are actively engaged in UA activities they are contributing to urban food and nutrition security in ways that are often missed by urban food studies that focus on where households buy their food. However, further studies in Nakuru, Ndola, and Karoi could try to establish whether households that are engaged in UA are more food secure than households that do not participate in the activity. An earlier study by Alice Hovorka (2004) also argues for looking at UA through an alternative entrepreneurship prism.

The various studies conducted in the three secondary urban centres help to dispel the myth that people engaged in UA in Africa are predominantly poor and recent arrivals from rural areas who have run out of livelihood options. A study by David Mkwambisi, Evan Fraser, and Andy Dougill (2011) on UA in Lilongwe and Blantyre (Malawi) also found that UA is not a preserve of the poor. The study also noted that urban food production in the two cities in Malawi is dominated by high-income households because, unlike low-income households, this group of urban farmers has the capacity to cultivate larger pieces of land and to purchase the required agricultural inputs, such as seeds, fertilizers, and pesticides. They also hire farm labour from low-income urban neighbourhoods. While this creates employment for a group of people who would have been unemployed, it also tends to create a group of urban farm labourers and to reproduce social inequalities. Urban farmers are commonly portrayed through a poverty lens as resource-poor urbanites who engage in undercapitalized micro-farming activities as a livelihood strategy, but the experiences of Ndola and Nakuru suggest that while poverty is the driving force for many low-income urban farmers, many households have engaged in the activity for entrepreneurial purposes. It is also worth noting that the most poverty-stricken households do not own spaces they can use for cultivation, and neither can they afford to rent. In fact, the poor tend to be transient and do not have the cash and social capital to enable them to access urban land for cultivation. In view of this, the open spaces in Karoi, Nakuru, and Ndola should not be perceived as sites of survivalist pursuits of the poor; instead they should be viewed as spaces of resistance to top-down urban planning approaches by both resource-rich and resource-poor households.

Theme 3: Dominant Activities

Cultivation of maize and leafy vegetables is practiced by urban farmers in the three secondary urban centres, with leafy vegetables generally grown in home gardens and maize in open spaces during the rain season. Urban farmers in Nakuru and Ndola with relatively large plots are able to keep chickens, but in Karoi the municipal bylaws make it illegal to keep livestock in urban areas.

Theme 4: Challenges Facing Urban Farmers

Urban farmers in the three secondary urban centres face several common challenges, such as limited access to suitable land for cultivation, land tenure insecurity, and theft of produce before harvesting. Karoi's urban farmers are also confronted with the issue of expensive permits that are beyond their reach. In Nakuru and Ndola, pests and insects are a common problem. Limited access to water and inability to obtain adequate resource inputs are key challenges faced by urban farmers in Nakuru (Foeken & Owuor, 2008; Willkomm et al., 2021).

Theme 5: Policy Frameworks

UA has long been a feature of sub-Saharan African cities, despite earlier efforts by both municipal authorities and national governments to supress the activity (Davies et al., 2020; Drakakis-Smith et al., 1995; Mbiba, 1995; Schmidt, 2012; Simatele & Binns, 2008; Tevera, 1996). The policy environment for UA activities in Karoi is generally prohibitive, despite the relaxation of controls over the past decade. In Nakuru and Ndola, official responses to UA have been generally supportive of the activity. However, a closer examination reveals that there is need for a policy shift in the three urban centres for UA to contribute in a meaningful and sustainable way to local food production. The findings also reveal the struggles engaged in by those in the urban informal food sector as they seek to negotiate a web of restrictive municipal policies that are generally anti-informal sector. Urban planning in the three urban centres has paid minimal attention to food issues and, as a result, UA continues to remain outside national urban development and planning frameworks. On the other hand, in both Ndola and Karoi a general lack of coherent UA policies and planning has resulted in inconsistent policy reactions toward UA activities on unauthorized open land. In Karoi, as in other Zimbabwean cities, the central government has generally been unenthusiastic about supporting UA. The destruction by council workers of crops grown on open spaces, although less frequent during the past five years, is justified by officials on the basis that the activity is not authorized and that the land is earmarked for urban development, such as housing construction. Such actions account for the tension between urban developers and urban farmers in some African cities. In order to overcome this tension, it is essential to encourage the development of multi-stakeholder UA platforms, involving municipal councils and urban farmer groups, that can promote the development of ecologically sustainable and socially friendly local food production systems.

Conclusion

This chapter has contributed to the discussion of UA practices in Karoi, Nakuru, and Ndola and has reflected on the changing UA policy environment in the three urban centres. Close attention needs to be paid to how rapid secondary urbanization is presenting food insecurity challenges through the displacement of urban food producers on the one hand and the disruption of local food systems on the other hand. One way to address this spatial challenge is by mainstreaming UA into the much broader urban policy and planning programs, as the two are interconnected. However, many local governments across sub-Saharan Africa fail to include UA in urban planning policies. The experiences of Ndola, Nakuru, and Karoi reveal how the three urban councils have responded to urban food challenges by being (or becoming) more tolerant of UA activities in open spaces. The big challenge for the local urban authorities managing secondary cities and towns in sub-Saharan Africa is to find ways of creating space for the urban poor to improve their nutritional status through pro-poor planning processes that allow the urban poor to produce some of their food. Lessons learned from elsewhere in sub-Saharan Africa reveal that city councils and national governments need to support livelihood strategies pursued by the poor, such as UA, in order to help them to be more food secure (Tevera, 2009). The experiences of Karoi, Nakuru, and Ndola provoke a rethink in the understanding of African realities regarding UA in secondary urban centres.

Acknowledgements This chapter is based on a keynote address that was presented at the international workshop on *Transforming Cities and Urban Agriculture in Africa*, supported by the Balsillie School of International Affairs and the Social Sciences and Humanities Research Council of Canada, May 14, 2021.

References

Badami, M., & Ramankutty, N. (2015) Urban agriculture and food security: A critique based on an assessment of urban land constraints, *Global Food Security,4*, 8–15.

Battersby, J., & Haysom, G. (2016) Why Urban Agriculture Isn't a Panacea for Africa's Food Crisis, *The Conversation,* 14 April.

Battersby, J., & Marshak, M. (2013). Growing communities: Integrating the social and economic benefits of urban agriculture in Cape Town. *Urban Forum, 24*(4), 447–461.

Binns, T., & Lynch, K. (1998). Feeding Africa's growing cities into the 21st century: The potential of urban agriculture. *Journal of International Development, 10*(6), 777–793.

Crush, J., Hovorka, A., & Tevera, D. (2011) Food security in Southern African cities: The place of urban agriculture. *Progress in Development Studies, 11*(4), 285–305.

Crush, J., Hovorka, A., & Tevera, D. (2018). Farming the city: The broken promise of urban agriculture. In B. Frayne, J. Crush, & C. McCordic (Eds.), *Food and nutrition security in Southern African Cities* (pp. 101–117). Routledge.

Davies, J., Hannah, C., Guido, Z., Zimmer, A., McCann, L., Battersby, J., & Evans, T. (2021). Barriers to urban agriculture in Sub-Saharan Africa, *Food Policy, 103.*

de Zeeuw, H. (2010). Cities farming for the future: Multi-stakeholder policy formulation and action planning on urban agriculture in developing countries. *Acta Horticulaturae, 881*, 97–109.

Drakakis-Smith, D., Bowyer-Bower, T., & Tevera, D. (1995). Urban poverty and urban agriculture: An overview of the linkages in Harare. *Habitat International, 19*(2), 183–193.

Drescher, A. (1996) Urban micro farming in central Southern Africa: A case study of Lusaka, Zambia, *African Urban Quarterly, 11*(2 and 3), 229–249.

Food and Agriculture Organization of the United Nations (FAO). (2001). *A briefing guide for the successful implementation of urban and Peri-Urban agriculture in developing countries and countries of transition.* FAO.

Food and Agriculture Organization of the United Nations (FAO). (2020). *Cities and local governments at the forefront in building inclusive and resilient food systems: Key results from the FAO survey 'Urban Food Systems and COVID-19.'* FAO.

Ferreira, A., Guilherme, R., Ferreira, C., & Oliveira, M. (2018). Urban agriculture: A tool towards more resilient urban communities? *Current Opinion Environmental Science Health, 5*, 93–97.

Foeken, D., & Owuor, S. (2008). Farming as a livelihood source for the urban poor of Nakuru, Kenya. *Geoforum, 39*(6), 1978–1990.

Frayne, B., McCordic, C., & Shilomboleni, H. (2014). Growing out of poverty: Does urban agriculture contribute to household food security in Southern African cities? *Urban Forum, 25*(2), 177–189.

Gondo, R., Madigele, P., Mogomotsi, G., Tokwe, T., Chakuya, J., & Chirefu, H. (2017). Sustainability of urban agriculture under economic and political instability in Karoi, Zimbabwe. *Global Journal of Advanced Research, 4*(2), 52–62.

Hampwaye, G., Nel, E., & Rogerson, C. (2007). Urban agriculture as local initiative in Lusaka, Zambia. *Environment and Planning C: Government and Policy, 25*, 553–572.

Hampwaye, G. (2013). Benefits of urban agriculture: Reality or illusion? *Geoforum, 49*, R7–R8.

Hovorka, A. (2004). Entrepreneurial opportunities in Botswana: (Re)shaping urban agriculture discourse. *Journal of Contemporary African Studies, 22*(3), 367–388.

Hovorka, A., & Lee-Smith, D. (2006) Gendering the urban agriculture agenda. In R. Van Veenhuizen (Ed), *Cities farming for the future: Urban agriculture for green and productive cities* (pp.125–137). IDRC.

Hovorka, A., de Zeeuw, H., & Njenga, M. (Eds.). (2009). *Women feeding cities: Mainstreaming gender in urban agriculture and food security*. Action Publishing.

Ishani, Z. (2009) Key gender issues in Urban livestock keeping and food security in Kisumu, Keny. In A. Hovorka, H. de Zeeuw & M. Njenga (Eds), *Women feeding cities: Mainstreaming gender in urban agriculture and food security* (pp. 105–122). Action Publishing.

Maxwell, D., Levin, C., & Csete, J. (1998). 'Does urban agriculture help prevent malnutrition? *Evidence from Kampala', Food Policy, 23*, 411–424.

Mbiba, B. (1995). *Urban agriculture in Zimbabwe: Implications for Urban management and poverty*. Avebury.

Mkwambisi, D., Fraser, E., & Dougill, A. (2011). Urban agriculture and poverty reduction: Evaluating how food production in cities contributes to food security, employment and income in Malawi. *Journal of International Development, 23*(2), 181–203.

Modibedi, T., Masekoameng, M., & Maake, M. (2021). The contribution of urban community gardens to food availability in Emfuleni Local Municipality, Gauteng Province. *Urban Ecosystems, 24*, 301–309.

Mwitwa J. (2008). *Farming system analysis of Ndola Urban and Peri-Urban agriculture*, MDP-ESA and RUAF Foundation.

Owuor, S. (2006). *Bridging the Urban–Rural divide: Multi-Spatial Llivelihoods in Nakuru Town* (Research Report No. 81). Kenya. African Studies Centre.

Paganini, N., Lemke, S., & Raimundo, I. (2018). The potential of urban agriculture towards a more sustainable urban food system in food-insecure neighbourhoods in Cape Town and Maputo. *Food Economy, 20*(3), 399–421.

Puppim de Oliveira, J., & Ahmed, A. (2021). Governance of urban agriculture in African cities: Gaps and opportunities for innovation in Accra, Ghana. *Journal of Cleaner Production, 312*, 127730.

Riley, L., Chilanga, E., Zuze, L., & Joynt, A. (2018). *Food security in Africa's secondary cities: No. 1. Mzuzu, Malawi* (Food Security Series No. 27). African Food Security Urban Network (AFSUN).

Satterthwaite, D. (2017). The impact of urban development on risk in Sub-Saharan Africa's cities with a focus on small and intermediate urban centres. *International Journal of Disaster Risk Reduction, 26*, 16–23.

Schmidt, S. (2012). Getting the policy right: Urban agriculture in Dar es Salaam, Tanzania. *International Development Planning Review, 34*(2), 129–145.

Simatele, D., & Binns, T. (2008). Motivation and marginalization in African urban agriculture: The case of Lusaka, Zambia. *Urban Forum, 19*, 1–21.

Simatele, D., Binns, T., & Simatele, M. (2012). Urban livelihoods under a changing climate: Perspectives on urban agriculture and planning in Lusaka, Zambia. *Journal of Human Development and Capabilities, 13*(2), 269–293.

Smart, J., Nel, E., & Binns, T. (2015). Economic crisis and food security in Africa: Exploring the significance of urban agriculture in Zambia's Copperbelt province. *Geoforum, 65*, 37–45.

Tacoli, C., & Agergaard, J. (2017). *Urbanisation, rural transformations and food systems: The role of small towns*. IIED.

Tevera, D. (1996). Urban agriculture in Africa: A comparative analysis of findings from Zimbabwe, Kenya and Zambia. *African Urban Quarterly, 11*(2–3), 181–187.

Tevera, D. (2009). Management and governance issues in the urban development process in Sub-Saharan Africa. In D. Seck & D. Busari (Eds.), *Growth and development in Africa* (pp. 365–388). Africa World Press.

UN-Department of Economic and Social Affairs (UN-DESA). (2019). *World Population Prospects 2019.*

Voleníková, L., & Opršal, Z. (2016). The role of urban agriculture in household well-being: A case study of community-based urban agriculture in Ndola, Zambia. *Development, Environment and Foresight*, *2*(2), 80–90.

Willkomm, M., Follmann, A., & Dannenberg, P. (2021). Between replacement and intensification: Spatiotemporal dynamics of different land use types of urban and peri-urban agriculture under rapid urban growth in Nakuru, Kenya. *Professional Geographer, 73*(2), 186–199.

Zezza, A., & Tasciotti, L. (2010). Urban agriculture, poverty, and food security: Empirical evidence from a sample of developing countries. *Food Policy, 35*(1), 265–273.

PART II

Food Security, Poverty & Livelihoods

CHAPTER 8

Spatial Characteristics of Urban Food Systems and Food Retailers in Smaller Urban Areas

Jordan Blekking, Stacey Giroux, Kurt Waldman, and Tom Evans

Introduction

In urban sub-Saharan Africa (SSA), the food retailer environment is made up of different food retailers offering both fresh and processed products in different quantities and available at different prices (Wanyama et al.,

J. Blekking (✉) · K. Waldman
Department of Geography, Indiana University, Bloomington, IN, USA
e-mail: jblekkin@iu.edu

S. Giroux
Ostrom Workshop, Indiana University, Bloomington, IN, USA

T. Evans
School of Geography, Development and Environment, University of Arizona, Tucson, AZ, USA

L. Riley and J. Crush (eds.), *Transforming Urban Food Systems in Secondary Cities in Africa*,
https://doi.org/10.1007/978-3-030-93072-1_8

2019). Access to food of different quantities and prices is essential for low-income urban consumers because they often shop at multiple retailers to meet their food needs (Battersby & Haysom, 2018). Three important food retailers that low-income households rely on are street vendors, small grocers and traditional public markets. Street vendors typically provide smaller quantities of food at more affordable prices and typically sell fresh fruits and vegetables. Small grocers may sell fresh produce, but they typically sell processed and non-perishable foods such as cooking oil, packaged foods, soft drinks and maize-meal. Traditional markets provide fresh foods and meat, as well as processed foods. Traditional markets sell directly to consumers and are used as wholesalers by street vendors and small grocers.

The food retailer environment in SSA is rapidly transforming due to a number of large-scale processes, including urbanization throughout the region, the growth of international supermarket chains and the ongoing transitions in dietary preferences and habits (Haysom & Tawodzera, 2018; Peyton et al., 2015; Reardon & Zilberman, 2018; Seto & Ramankutty, 2016). These large-scale changes have implications for food access and food security at the city, household and individual levels. As secondary and tertiary urban areas (as opposed to primary urban areas, such as national capitals) expand in population and size, it is important for these urban food systems to be able to equitably meet the food access needs of all urban consumers, regardless of their socio-economic status or their location within the city. It is necessary to understand which retailers operate in a city and its residential areas, how consumers interact with the different retailer types and how food access varies both financially and physically with respect to those food retailers.

A considerable amount of research highlights the relationship between food access in urban SSA and household economic characteristics, such as employment, income and food sourcing (Blekking et al., 2020; Davies et al., 2020; Demmler et al., 2017; Tuholske et al., 2020), but less research addresses the spatial aspects of the relationship between consumers and food retailers. To understand the spatial characteristics of urban food systems and food retailers, we use primary data collected in Monze and Mumbwa, Zambia. There are three key pieces of information underpinning this study: the location of household within each respective city; the frequency with which households purchase food from street vendors, small grocers and traditional markets during a 14-day period; and the Household Food Insecurity Access Scale (HFIAS) score for each household. We then use spatial analysis to determine hotspots of retailer use in order to understand how space relates to food purchasing behaviour. Once hotspots are identified, we compare household characteristics of those within and outside of the hotspots in order to understand the relationship between food purchasing behaviour and the spatial characteristics of the urban food retailer environment.

Background

In general, greater rates of urbanization lead to higher rates of development, but this has not occurred in SSA (Ravallion, 2007). This is cause for concern, considering urbanization rates have steadily increased over the past 20 years to the point that SSA is currently the site of the most rapid rates of urbanization in the world (UN-DESA, 2018). At the same time, urban poverty has increased, with the increase in the share of poor households in urban areas sometimes outpacing the rate of overall population increase (Tefft et al., 2017). Because urbanization in SSA has not led to expected improvements in development, and ultimately livelihoods, ensuring equitable food access in terms of economic and physical access is critical to ensuring urban food security. Unfortunately, it is hard for Southern African governments to effectively plan for equitable food systems because there are many different actors with competing interests (AGRA, 2020). The little planning that has occurred has been largely targeted towards modernizing urban areas via development initiatives that favour larger retailers, such as international supermarket chains (Battersby, 2017). However, in their effort to push for more modern cities, private firms and the government are not transforming local urban food systems sustainably and equitably by including smaller, local food retailers.

Urbanization in SSA and other regions of the Global South tends to manifest as spatial extensification, rather than intensification via greater population densities (Nagendra et al., 2018). Extensification often leads to rapid urban expansion with uneven or inconsistent levels of infrastructure and development, characterized by urban sprawl outward from the city centre with patches of undeveloped land between developed areas (Cobbinah & Amoako, 2012). Recently settled areas are not necessarily low-income areas, but can also feature middle- and upper-income areas. Heterogeneous spatial expansion of urban areas in SSA can outpace the ability of local governments to adequately provide required services such as water, sanitation and roads (Cohen, 2006). Because of the limited services and relatively low population density, few bricks-and-mortar food retailers (such as supermarkets, grocers, butchers or traditional markets) exist in newly developed, unplanned settlements and smaller food retailers (such as street vendors), often fill this spatial void. Street vendors are well-suited for this role because they have the ability to pick up and move in order to operate where consumers live or travel, and they provide food in the quantities and at the prices consumers require (Battersby & Watson, 2018).

In Zambia, urban agriculture is not substantially associated with household food security (Davies et al., 2020). Instead, urban consumers must purchase food from food retailers (Crush & Frayne, 2010). In general, low-income households spend a larger share of their income on food (Cohen & Garrett, 2010), so decreased incomes or increased food prices can present challenges to household food access (Caesar & Crush, 2016). Even holding income constant, households must also balance the costs associated with other living expenses, such as transportation, housing and energy (Pothukuchi & Kaufman, 1999). To stretch their purchasing power, urban households often visit multiple food retailers to meet their food needs (Battersby & Watson, 2018). Considering the relationship between food access and economic characteristics is important and well understood, but it does not tend to take into account the association between food access and space.

Both food access and purchasing power can exhibit spatial patterns, like the spatial patterns associated with poverty (Giroux et al., 2020). How food retailers are distributed throughout an urban area can create disparities in household food access to the point that food security is reduced for some. In the Global North, this type of spatial disparity in food access is often referred to in terms of food deserts. However, applying this term and its framing to the food retailer environment in the Global South is less straightforward (Battersby and Crush, 2014). The confluence of legacies of colonialism, global capitalism, rapid urbanization and lack of local food planning and policy means that the spatial dynamics of food access unfolds in ways that are not well understood. Spatial features of these food systems are of paramount importance and by mapping and analysing the diversity of food retailers that exist and the ways households rely on them, we begin to identify the roles of different food retailer types in ensuring equitable food access across space, while also taking household economic factors into consideration.

Data and Methods

We analyse household-level data collected from two smaller urban areas in Zambia: Mumbwa and Monze. As of 2019, Zambia's population was about 17.4 million (Bank of Zambia, 2019). As in other countries in Southern Africa, smaller urban areas in Zambia are undergoing urbanization at a rate of approximately 2.5% annually (Zimmer et al., 2020). Both Mumbwa and Monze serve as district capitals, have populations below 50,000 people, and their designated boundaries are less than 30 square kilometres. Both urban areas feature numerous food retailers that range in market share and placement throughout the city, such as traditional markets, wholesalers, small grocers, butchers and street vendors. Monze has three traditional markets, while Mumbwa has one. Notably absent from Mumbwa and Monze are supermarket chains common in the region such as Shoprite or Pick n Pay. Since the Mumbwa household survey was conducted in 2019, the supermarket chain Choppies has started to operate there. Mumbwa is located approximately two hours west of Lusaka, the national capital, and Monze is approximately three hours south of Lusaka. Both cities are located along paved roads and serve as important economic hubs for farmers in the surrounding area.

In Monze, we collected data from 219 households in June 2018. In Mumbwa, we collected data from 135 households in June 2019. In both cities, we selected residential areas for sampling, in consultation with local government officials. We used a stratified area random sampling technique to sample households from within low- and low-middle-income residential areas in both cities (Fig. 8.1) (Montello & Sutton, 2006). The northwest portion of Mumbwa was not sampled because that area is primarily higher-income residential housing. In Monze, the northeast portion of the city contains higher-income households and was also not sampled. Sampled residential areas represent both formally designated and informally established residential areas. We sampled a minimum of 30 households in each selected low- and low-middle-income residential area in order to ensure geographic distribution across areas of interest throughout the city and to ensure sufficient observations for purposes of statistical analysis.

One household member over the age of 18 and with knowledge of the household and its members responded to the survey. The survey included modules regarding household member demographics, employment characteristics and food purchasing and consumption behaviour, including the

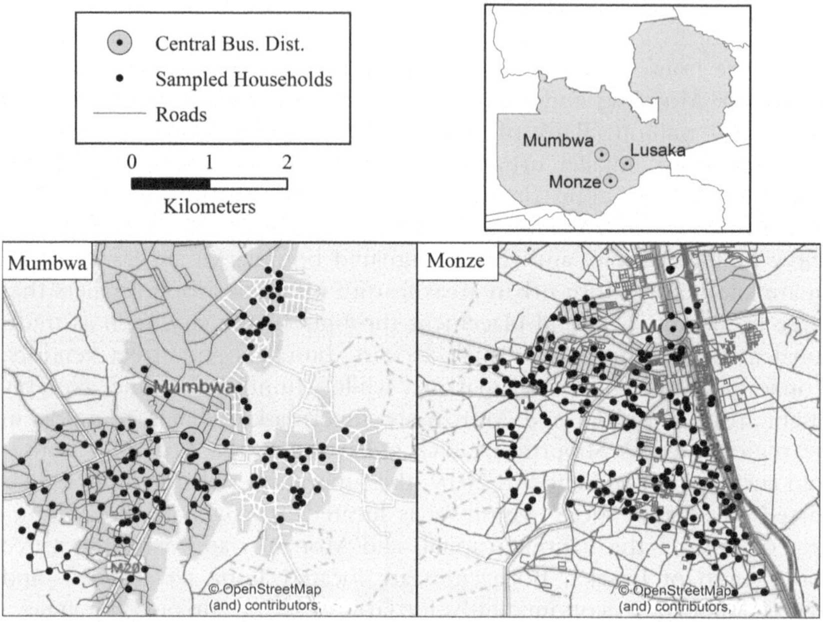

Fig. 8.1 Maps of Mumbwa and Monze, Zambia (*NB* To maintain confidentiality, all household locations were skewed using the Gaussian Displacement process. This process manipulates the latitude and longitude slightly in a random direction and at a random distance, following a Gaussian distribution)

frequency with which household representatives reported purchasing food from street vendors, small grocers and traditional public markets over the prior 14-day period. We define street vendors as individuals selling food along the side of the road. Small grocers are defined as individually owned stores selling an array of products, including food. Traditional markets can either be managed formally by the city council or informally by a cooperative or community group. In our analysis, we consider small grocers inside the demarcated boundary of a traditional market to be part of the market, not separate grocers.

Spatial analysis is commonly used in public health research in SSA (Chimoyi & Musenge, 2014; Sedda et al., 2018), but less often in food systems and food security research. However, the application of spatial analysis to these types of research can provide a nuanced approach to understanding how household-level outcomes vary across space. In this research, we use hotspot analysis (Getis-Ord Gi* statistic) to discern whether there are areas of the respective cities where households rely more heavily on particular retailers. This analytical approach identifies statistically significant clusters, called hotspots, of high values for the variable of interest. In this case, hotspots delineate areas where households with high values (more frequent use of particular food retailers) are surrounded by other households with similarly high values. When household values are substantially different from those of the global population, a statistically significant value is observed. It is important to note that hotspot analysis uses values from the individual and neighbouring households to calculate the Getis-Ord Gi* statistic. Only when the statistic is calculated and found to be statistically significant are households identified as hotspots. Some households may be identified as hotspots when their neighbours are not and will therefore appear to be spatially independent.

A spatially independent household indicates an area in space where high values are occurring, even if neighbouring households with high values are not indicated as hotspots. For example, a household with a value of 14 for the number of times it has visited a traditional market in the past 14 days, surrounded by other households with the same or similarly high values will be indicated as a hotspot, as will the other households with similarly high values. However, a value of 14 surrounded by low values will be represented as an independent hotspot, while the same value surrounded by a mixture of high and low values will likely not be indicated as a hotspot. We use Fixed Distance Band to conceptualize the spatial relationship. A Fixed Distance Band creates a weight matrix that restricts the use of observations outside of a set distance buffer. Using this method, neighbouring observations within a fixed buffer of 500m receive a weight of 1, while observations beyond the fixed buffer receive a weight of 0. We use hotspots of greater than 95% statistically significant confidence to illustrate our findings more clearly.

We then compare characteristics of the households inside and outside the hotspots. We pay particular attention to differences in household food access, as measured using a condensed version of the HFIAS (rHFIAS). This version of rHFIAS was amended by local collaborators to represent a more culturally appropriate metric. The condensed version retained the four-week recall period suggested by Jennifer Coates et al. (2007) and measured the frequency of occurrences for each question. Possible responses were "never", "rarely", "sometimes" or "often". Because the questions pertain to negative outcomes (i.e. lack of food access), lower values represent better food access, while higher values represent poor food access. For this condensed version of rHFIAS, values can range from 0 (better food access) to 15 (worse food access).

Results

In Mumbwa, households are smaller, they visit food retailers less often, and the rHFIAS score is higher, indicating poorer food access, despite spending more per capita on food for the household (Table 8.1). The higher household food cost per capita per month in Mumbwa is due, in part, to the two surveys being conducted in separate years. In 2018, the average exchange rate was 10.45 ZMK to 1 USD, while one year later, in 2019, the kwacha had devalued to a rate of 12.91 ZMK to 1 USD.

Table 8.1 Descriptive statistics for household food purchasing behaviour in Mumbwa and Monze, Zambia

	Mumbwa Households N = *135*	*Monze Households* N = *219*
Variable	Mean (SD)	Mean (SD)
rHFIAS (0 = best food access, 15 = worst)	4.87 (3.77)	3.85 (2.74)
Household members	5.52 (2.43)	5.95 (3.19)
Food cost per capita per month (ZMK)	202.01 (166.30)	120.53 (112.10)
Household has water inside the home	13.97%	22.37%
Distance to nearest traditional market (minutes walking)	25.77 (15.98)	22.23 (16.45)
Number of times in the past 14 days the household purchased food from a roadside food retailer	5.83 (5.76)	7.40 (5.51)
Number of times in the past 14 days the household purchased food from a traditional market	5.49 (5.08)	6.38 (4.96)
Number of times in the past 14 days the household purchased food from a small grocery store	5.05 (4.81)	6.10 (4.07)

Households in Mumbwa are, on average, a 26-min walk from the centrally located traditional market, while in Monze households are approximately a 22-min walk from one of the three traditional markets.

Street vendor hotspots are in the top row of panels, traditional market hotspots are identified in the middle row of panels and small grocer hotspots in the bottom row (Fig. 8.2). All panels in the left column represent Monze households and hotspots, while all panels in the right column represent the same for Mumbwa. In Monze, we find two hotspots of street vendor patronage: the largest cluster, to the northwest of the city centre, is made up of 19 households; the second cluster, to the southeast, is made up of five households (Panel A). Three main household clusters and one spatially independent household are identified as hotspots for purchases from traditional markets in Monze (Panel B). The large cluster located towards the southern part of the city is located near the Mungungu market, while the two clusters to the north are located near the city's other two traditional markets. Our spatial analysis indicates three clusters of households identified as hotspots for making frequent purchases from small grocers (Panel C). The two hotspots located south of the spatially independent household are near the Mungungu market, while the cluster located in the west is near the settled periphery of the city (Fig. 8.2).

In Mumbwa, we find two areas of households identified as street vendor hotspots. The largest of the two areas is made up of 10 households, while the smaller area to the east is made up of two households (Panel D). Spatial analysis identifies two areas of hotspots for households making frequent purchases from traditional markets, plus one spatially independent household in Mumbwa (Panel E). The two main clusters are located near the only traditional market in the city. Lastly, we find one small grocer hotspot of households located in the east and two spatially independent households: one located north of the town centre and one located to the southwest of the town centre (Panel F).

Our findings suggest that Monze and Mumbwa exhibit similar spatial patterns regarding street vendors. In both cities, households located further from the town centre purchase more frequently from street vendors. Households living near the city centre in both urban areas use street vendors or small grocers less frequently (Panels B and E), which suggests that traditional markets play a stronger role in providing food for centrally located households. It is important to note that Monze has three traditional markets, while Mumbwa has only one. As a result, there are

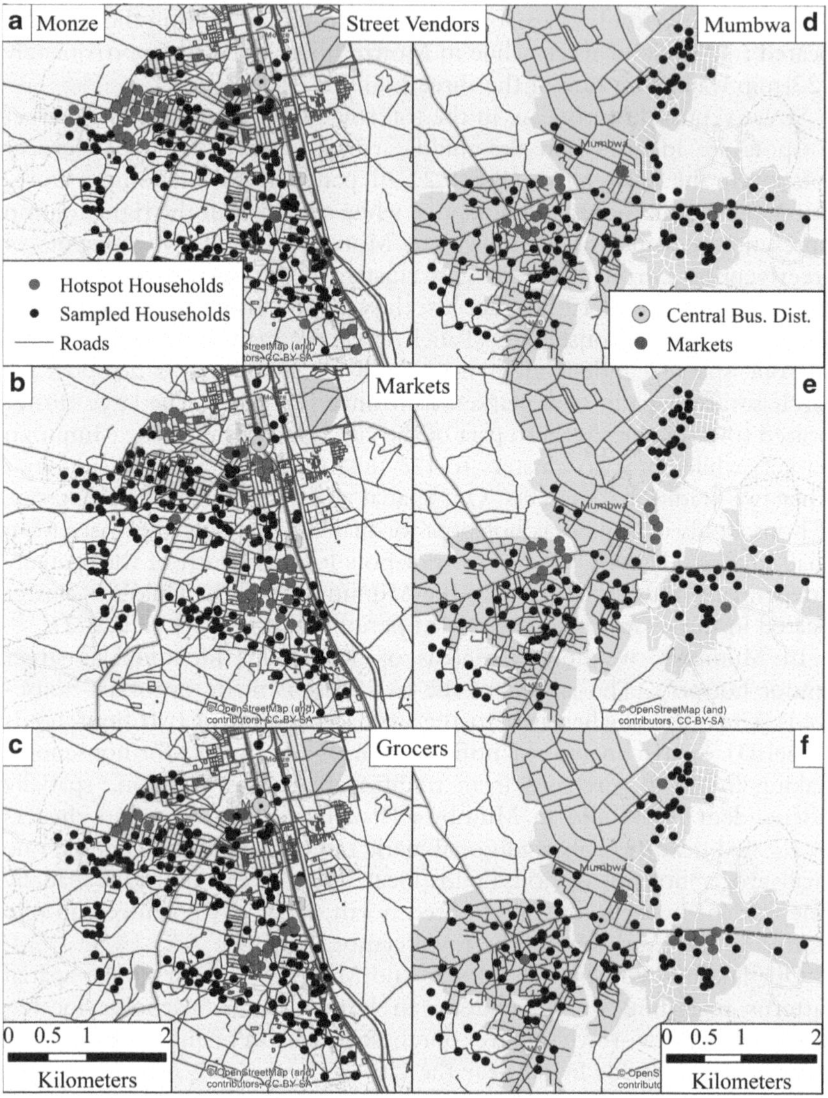

Fig. 8.2 Sampled and hotspot household locations within the two study cities (*NB* Hotspots by retailer type are identified in red and indicate >95% confidence of belonging to a hotspot)

multiple traditional market-related hotspots spread across Monze, while in Mumbwa the hotspots are located near the single market.

We find that households in street vendor hotspots in both Mumbwa and Monze visit street vendors more than 10 times on average in a 14-day period (Table 8.2). A noteworthy difference between the two groups of households is that despite the high frequency of visits to street vendors by households in both urban areas, Mumbwa households in the street vendor hotspots have worse food access (6.75) as compared to households in street vendor hotspots in Monze (3.88). In Monze, households in street vendor hotspots have nearly twice the assets that their outside hotspot counterparts do. In Mumbwa, the asset index is only slightly different between households inside and outside the street vendor hotspot. Mumbwa households in traditional market hotspots have similarly worse food access (6.31) compared to market hotspot households in Monze (3.97). Mumbwa households in market hotspots have marginally greater assets than those outside the hotspots. For both Mumbwa and Monze traditional market hotspot households, they are about half the distance to the nearest traditional market than outside hotspot households. This suggests that these households are centrally located in their respective urban area, near to where traditional markets are also located, and that this proximity improves their food access. Furthermore, these results indicate that financial accessibility is not necessarily the limiting factor for why households visit these food retailers, and that households that buy food from these retailers are not necessarily food insecure.

Discussion

Given that most urban areas in SSA are growing in terms of population and spatial extent, there exists a need to consider the spatial process of urban growth in the context of food retailers. In Zambia, efforts to modernize the urban areas have led to policies that favour large retailers such as supermarkets (Abrahams, 2010); however, this does not appear to support sustainability and equitability of food access for low-income urban households. To transform food systems in smaller urban areas in SSA, policymakers must take a proactive approach to integrating street vendors and other small food retailers in parallel to supermarkets (Skinner, 2018). Consider that in Mumbwa a new supermarket opened since this study was conducted. Local news outlets report that the impetus for these openings comes from the favourable investment policies set forth by the

Table 8.2 Comparison of household characteristics for household hotspots of purchasing from street vendors, traditional markets and small grocers during a 14-day period

Variable	*Street vendors*				*Traditional markets*				*Small grocers*			
	Mumbwa		*Monze*		*Mumbwa*		*Monze*		*Mumbwa*		*Monze*	
	Hotspots (N = 12)	*Outside Hotspots (N = 123)*	*Hotspots (N = 25)*	*Outside Hotspots (N = 194)*	*Hotspots (N = 13)*	*Outside Hotspots (N = 122)*	*Hotspots (N = 33)*	*Outside Hotspots (N = 186)*	*Hotspots (N = 8)*	*Outside Hotspots (N = 127)*	*Hotspots (N = 18)*	*Outside Hotspots (N = 201)*
	Mean (SD)	*Mean (SD)*	*Mean (SD)*	*Mean (SD)*	*Mean (SD)*	*Mean (SD)*	*Mean (SD)*	*Mean (SD)*	*Mean (SD)*	*Mean (SD)*	*Mean (SD)*	*Mean (SD)*
HFIAS (0 = best food access, 15 = worst)	6.75 (3.41)	4.69 (3.77)	3.88 (1.94)	3.85 (2.84)	6.31 (3.66)	4.72 (3.76)	3.97 (2.16)	3.83 (2.84)	3.25 (1.83)	4.97 (3.84)	4.00 (1.85)	3.84 (2.81)
Household Members	5.83 (3.16)	5.49 (2.37)	6.12 (2.17)	5.93 (3.30)	5.92 (2.69)	5.48 (2.41)	6.52 (4.47)	5.85 (2.91)	6.00 (3.02)	5.49 (2.40)	7.94 (5.25)	5.78 (2.89)
Food Cost per Capita per Month (ZMK)	175.73 (153.93)	204.57 (167.82)	126.46 (139.32)	119.76 (108.51)	182.43 (160.40)	204.09 (167.42)	149.63 (152.71)	115.37 (102.93)	174.63 (140.43)	203.73 (168.12)	92.24 (53.63)	123.06 (115.65)
Asset index (unitless)	1.67 (1.39)	1.49 (1.54)	2.29 (2.27)	1.26 (1.32)	1.60 (1.31)	1.49 (1.55)	1.33 (1.44)	1.39 (1.50)	1.85 (1.50)	1.48 (1.53)	1.44 (1.73)	1.37 (1.47)
Distance to nearest traditional market (minutes walking)	21.08 (8.28)	26.23 (16.48)	39.00 (15.55)	20.07 (15.32)	17.69 (9.04)	26.63 (16.33)	12.97 (9.31)	23.88 (16.91)	18.13 (7.99)	26.25 (16.24)	16.94 (14.38)	22.71 (16.58)
Roadside food retailer purchases(frequency)	10.42 (5.23)	5.38 (5.63)	13.32 (2.10)	6.64 (5.34)	6.46 (6.72)	5.76 (5.68)	6.64 (6.09)	7.54 (5.40)	9.13 (5.96)	5.62 (5.71)	10.94 (5.21)	7.08 (5.43)
Traditional market purchases (frequency)	9.50 (6.05)	5.10 (4.83)	5.96 (4.40)	6.43 (5.04)	10.54 (5.06)	4.96 (4.80)	11.73 (3.73)	5.43 (4.54)	5.38 (5.68)	5.50 (5.07)	11.56 (3.91)	5.92 (4.78)
Small grocer purchase(frequency)	7.92 (5.71)	4.77 (4.65)	7.76 (3.64)	5.88 (4.08)	5.07 (4.82)	5.05 (4.83)	7.55 (4.44)	5.84 (3.95)	9.13 (5.46)	4.80 (4.67)	10.28 (4.87)	5.72 (3.78)

Zambian government, which encourages "chain store[s] to rapidly expand [their] business prospects across the country" (Jere, 2020). The same enthusiasm to integrate street vendors and other small, local retailers into food systems is absent in Zambia. By exclusively considering the role of large food retailers, officials and investors are ensuring that food access is not equal, in particular for low-income households. Because urban consumers require many types of food retailers to meet their food needs (Wanyama et al., 2019), a one-size-fits-all approach to transforming food systems in smaller urban areas will likely have limited success in ensuring food access as smaller cities grow. To ensure a strong, equitable environment for a variety of retailer types, policy must focus on including smaller, local food retailers.

The spatial clustering of households that purchase food from street vendors reflects aspects of the spatial distribution of poverty and limited food access, and reaffirms the importance of street vendors as nodes within the overall food system. Street vendors not only facilitate food access for poorer households, but they also play an important role in ensuring equitable physical access across an urban area (Battersby & Watson, 2018); however, our analyses show that the association among poorer households, spatial location of vendors and households, and food access can be mixed. Consider that households in the street vendor hotspots in Mumbwa are, on average, a 20-min walk from the nearest traditional market, and spend substantially less per month on food than non-hotspot households. In Monze, households within the street vendor hotspot are, on average, a 39-min walk from the nearest traditional market and spend marginally more on food per month as compared to households outside these hotspots. In both instances, households rely substantially on street vendors, but associated household characteristics vary between the study sites. Assuming either household economic characteristics or spatial proximity to retailers are the limiting factor to food access does not consider other structural factors related to food access, including other non-food costs, employment opportunities, gendered aspects or even environmental challenges.

Unfortunately, street vendors are typically left out of urban planning initiatives in favour of large, modern food retailers such as supermarkets. Rather than viewing street vendors as assets within the food retail environment, the Zambian government often targets street vendors through policies and initiatives that target their removal. This is a short-sighted approach that only harms urban household food access (Giroux et al.,

2020). An alternative approach would be for national and local governments to treat street vendors and other small, local retailers as crucial actors within the local food retailer environment. Evidence for this as a useful approach exists. In Liberia, for example, street vendors and the Monrovia City Corporation have agreed to a legal understanding that normalizes street vending and allows street vendors to operate in designated areas without harassment (Reed & Bird, 2019). Nothing like this currently exists in Zambia, but it does provide an example of an approach to consider moving forward.

Transforming the food systems of smaller urban areas in SSA also extends to re-examining the role of traditional markets. We find the relationship between households and traditional markets is the result of differences in both the number and placement of traditional markets. Traditional markets provide food directly to households and also act as wholesalers to street vendors (Battersby, 2019; Giroux et al., 2020). In this capacity, traditional markets play a central role in the food distribution network, both in terms of bringing food from farmers into the city and then in terms of distributing food to street vendors and consumers. Because traditional markets have a central position within the food distribution system of smaller urban areas, including them in plans to transform smaller urban areas is critical and can have implications for food access among the poorest households through street vendors.

Conclusion

Mumbwa and Monze, like other urban areas included in this book, represent important sites for evaluating how food systems of smaller urban areas are transforming. Consumer retail dynamics and food access are changing as a result of rapid population growth and spatial extensification. While street vendors are critical for low-income households, characterizing food access or food security in terms of unidimensional patterns of household purchasing misrepresents the actual food purchasing behaviour of these households, which is related not only to economic resources but also to spatial location. Developing urban policy that promotes a mix of retailers, thus improving economic and physical access to food across socio-economic class, will become increasingly critical as secondary and tertiary urban areas continue to grow.

References

Abrahams, C. (2010). Transforming the region: Supermarkets and the local food economy. *African Affairs, 109*(434), 115–134.

AGRA. (2020). *Feeding Africa's cities: Opportunities, challenges, and policies for linking African farmers with growing urban food markets* (AGRA Issue No. 8). Alliance for a Green Revolution in Africa.

Bank of Zambia. (2019). *National Statistics*. Bank of Zambia. https://www.boz.zm/statistics.htm.

Battersby, J. (2017). Food system transformation in the absence of food system planning: The case of supermarket and shopping mall retail expansion in Cape Town, South Africa. *Built Environment, 43*(3), 417–430.

Battersby, J. (2019). The food desert as a concept and policy tool in African cities: an opportunity and a risk. *Sustainability: Science Practice and Policy, 11*(2), 458.

Battersby, J., & Crush, J. (2014). Africa's urban food deserts. *Urban Forum, 25*(2), 143–151. https://doi.org/10.1007/s12132-014-9225-5

Battersby, J., & Haysom, G. (2018). Linking urban food security, urban food systems, poverty, and urbanisation. In J. Battersby & V. Watson (Eds.), *Urban food systems, governance, and poverty in African Cities* (pp. 56–67). Routledge.

Battersby, J., & Watson, V. (2018). Addressing food security in African cities. *Nature Sustainability, 1*(4), 153–155.

Blekking, J., Waldman, K., Tuholske, C., & Evans, T. (2020). Formal/informal employment and urban food security in Sub-Saharan Africa. *Applied Geography, 114*, 102131.

Caesar, M., & Crush, J. (2016). *Urban food deserts and climate change in Africa's hungry cities* (HCP Discussion Paper No. 3). Hungry Cities Partnership.

Chimoyi, L., & Musenge, E. (2014). Spatial analysis of factors associated with HIV infection among young people in Uganda, 2011. *BMC Public Health, 14*, 555.

Coates, J., Swindale, A., & Bilinsky, P. (2007). *Household Food Insecurity Access Scale (HFIAS) for measurement of household food access: Indicator Guide (Version 3)*. FHI 360/FANTA.

Cobbinah, P., & Amoako, C. (2012). Urban sprawl and the loss of peri-urban land in Kumasi, Ghana. *International Journal of Social and Human Sciences, 6*(388), e397.

Cohen, B. (2006). Urbanization in developing countries: Current trends, future projections, and key challenges for sustainability. *Technology in Society, 28*(1–2), 63–80.

Cohen, M., & Garrett, J. (2010). The food price crisis and urban food (in) security. *Environment and Urbanization, 22*(2), 467–482.

Crush, J., & Frayne, B. (2010). *The invisible crisis: Urban food security in Southern Africa* (Urban Food Security Series No. 1). African Food Security Urban Network (AFSUN).

Davies, J., Hannah, C., Guido, Z., Zimmer, A., McCann, L., Battersby, J., & Evans, T. (2020). Barriers to urban agriculture in Sub-Saharan Africa. *Food Policy, 1,* 101999.

Demmler, K., Klasen, S., Nzuma, J., & Qaim, M. (2017). Supermarket purchase contributes to nutrition-related non-communicable diseases in urban Kenya. *PloS One, 12*(9), e0185148.

Giroux, S., Blekking, J., Waldman, K., Resnick, D., & Fobi, D. (2020). Informal vendors and food systems planning in an emerging African city, *Food Policy, 1,* 101997.

Haysom, G., & Tawodzera, G. (2018). 'Measurement drives diagnosis and response': Gaps in transferring food security assessment to the urban scale. *Food Policy, 74,* 117–125.

Jere, J. (2020, June 26). *Choppies to open five stores*. Zambia National Broadcasting Corporation.

Montello, D., & Sutton, P. (2006). *An introduction to scientific research methods in geography*. SAGE Publications.

Nagendra, H., Bai, X., Brondizio, E., & Lwasa, S. (2018). The urban south and the predicament of global sustainability. *Nature Sustainability, 1*(7), 341–349.

Peyton, S., Moseley, W., & Battersby, J. (2015). Implications of supermarket expansion on urban food security in Cape Town, South Africa. *African Geographical Review, 34*(1), 36–54.

Pothukuchi, K., & Kaufman, J. (1999). Placing the food system on the urban agenda: The role of municipal institutions in food systems planning. *Agriculture and Human Values, 16*(2), 213–224.

Ravallion, M. (2007). Urban poverty. *Finance and Development, 44*(3), 15–17.

Reardon, T., & Zilberman, D. (2018). Climate smart food supply chains in developing countries in an era of rapid dual change in agrifood systems and the climate. In L. Lipper, N. McCarthy, D. Zilberman, S. Asfaw, & G. Branca (Eds.), *Climate smart agriculture* (pp. 335–351). Springer.

Reed, S., & Bird, M. (2019). *Liberia's street vendors pioneer new approach with city officials*. WIEGO. Street Vendors and Public Space Report.

Sedda, L., Vilela, A., Aguiar, E., Gaspar, C., Gonçalves, A., Olmo, R., Silva, A., de Cássia da Silveira, L., Eiras, Á., Drumond, B., Kroon, E., & Marques, J. (2018). The spatial and temporal scales of local dengue virus transmission in natural settings: A retrospective analysis. *Parasites and Vectors, 11*(1), 79.

Seto, K., & Ramankutty, N. (2016). Hidden linkages between urbanization and food systems. *Science, 352*(6288), 943–945.

Skinner, C. (2018). Contributing yet excluded? Informal food retail in African cities. In J. Battersby & V. Watson (Eds.), *Urban food systems governance and poverty in African Cities* (pp. 104–115). Routledge.

Tefft, J., Jonasova, M., Adjao, R., & Morgan, A. (2017). *Food systems for an Urbanizing World: Knowledge product* (Working Paper No. 122014). World Bank Group.

Tuholske, C., Andam, K., Blekking, J., Evans, T., & Caylor, K. (2020). Comparing measures of urban food security in Accra, Ghana. *Food Security, 12*, 417–431.

UN-DESA. (2018). *World Urbanization Prospects 2018.*

Wanyama, R., Gödecke, T., Chege, C., & Qaim, M. (2019). How important are supermarkets for the diets of the urban poor in Africa? *Food Security, 11*(6), 1339–1353.

Zimmer, A., Guido, Z., Tuholske, C., Pakalniskis, A., Lopus, S., Caylor, K., & Evans, T. (2020). Dynamics of population growth in secondary cities across southern Africa. *Landscape Ecology, 35*(11), 2501–2516.

CHAPTER 9

Food Insecurity, Food Sourcing and Food Coping Strategies in the OOO Urban Corridor, Namibia

Ndeyapo Nickanor, Lawrence Kazembe, and Jonathan Crush

Introduction

The world is rapidly urbanizing, with half the global population estimated to be living in cities and towns by 2050 (UN-DESA, 2018). Namibia was already 50% urbanized in 2018, a figure projected to rise to 62% by 2030.

N. Nickanor
Faculty of Agriculture, Engineering and Natural Science, University of Namibia, Windhoek, Namibia

L. Kazembe (✉)
Department of Computing, Mathematical and Statistical Science, School of Science, Faculty of Agriculture, Engineering and Natural Science, University of Namibia, Windhoek, Namibia
e-mail: lkazembe@unam.na

J. Crush
Balsillie School of International Affairs, Wilfrid Laurier University, Waterloo, ON, Canada
e-mail: jcrush@balsillieschool.ca

University of the Western Cape, Cape Town, South Africa

L. Riley and J. Crush (eds.), *Transforming Urban Food Systems in Secondary Cities in Africa*,
https://doi.org/10.1007/978-3-030-93072-1_9

While the African continent has witnessed a rise in megacities such as Lagos and Kinshasa, with populations in excess of 10 million, much of the growth is taking place further down the urban hierarchy. The challenges that follow rapid urbanization in large cities, including rising poverty and food insecurity, and the mushrooming of marginalized people on the urban fringes, are also being replicated in smaller towns and cities (Irham, 2012; Knorr et al., 2018). Lack of access to food presents a particular challenge for sustainability in secondary cities. Hunger and poverty are intrinsically linked, with the urban poor spending a large proportion of their income on food (Chung & Myers, 1999; UNCTAD, 2008). Food insecurity is also closely tied to health and nutrition (Hassan, 2017; Sassi, 2018), quality of life (Casey et al., 2005; Moafi et al., 2018; Sharkey et al., 2011) and general well-being (Frongillo et al., 2017; Jaron & Galal, 2009). Various coping strategies are commonly used as fallback mechanisms for dealing with insufficient food supplies (Farzana et al., 2017; Maxwell et al., 1999; Shariff & Khor, 2008). Long-term coping strategies include selling assets to meet food demands, which may leave the household even more vulnerable to future shocks (Heltberg et al., 2012; Miller et al., 2010).

Common measures of food insecurity include the Household Food Insecurity Access Scale (HFIAS), the Household Food Insecurity Access Prevalence (HFIAP), the Household Dietary Diversity Score (HDDS) and the Months of Adequate Household Food Provisioning indicator (Sassi, 2018; Swindale & Bilinsky, 2006). While the HFIAS and HFIAP are the most widely used indicators in the literature, dietary diversity is critically important because of its linkages to nutrition insecurity (Hassan, 2017). The association between food insecurity, dietary diversity and food sources has been increasingly examined. Various studies have focused on the impact of formal food sources, especially the effect of supermarket expansion on food insecurity (Crush & Frayne, 2011; Heltberg et al., 2012; Peyton et al., 2015; Shariff & Khor, 2008). However, much less attention has been paid to the relationship between food and nutrition security, dietary diversity and food sources, especially in secondary cities. Similarly, the role of informal community social support systems is not highlighted in the mainstream literature on food security (Nyikahadzoi et al., 2013). These support systems include begging, food sharing, borrowing from neighbours and food provisioning at schools or communal kitchens. Another emerging food security strategy is reliance on rural-to-urban food remittances, sent by either family or friends (Crush & Caesar, 2018; Frayne, 2010).

In this chapter, we unpack the food sources and coping strategies adopted by marginalized and low-income groups in three interlinked secondary cities in Namibia. The goal of the study was to assess the relationship between food insecurity, household characteristics and food sources in the northern urban corridor of Oshakati-Ongwediva-Ondangwa (OOO). This chapter also considers the use of household coping strategies in response to food insecurity in OOO. The first section of this chapter provides geographical context, information about the household survey methodology, and the descriptive and statistical modeling techniques used. The section also identifies and explains the food security, food sourcing and socio-economic variables and measures selected for analysis. The following section of this chapter presents the results of the statistical analysis, identifying which independent food sourcing and household variables are most closely associated with the odds of being food insecure, having low dietary diversity and needing to employ various coping strategies. This chapter concludes with reflections on the significance of the analysis for understanding why, despite considerable food system transformation, these levels remain extremely high.

Methodology

Study Context

The corridor towns of OOO are located in the northern part of Namibia (see Nickanor et al. [2019b] for further details). The combined population in 2020 was 114,472, increasing from 35,705 at independence in 1991. At the time of the 2011 Population and Housing Census, 21% of the houses were informal or semi-permanent. Rural-to-urban migration is the main contributing factor to growth, with 60% of the population born in rural areas (Nickanor et al., 2019a, b). The proportion of the population that is severely poor increased from 4.4% in 2009–2010 to 4.8% in 2015–2016 (Nickanor et al., 2019a, b). Administratively, all three towns are located within Namibia's Oshana Region, with a total 2011 population of 176,674, of whom 45% were urban-based (NSA, 2014). Oshana Region is divided into 11 constituencies and includes the four corridor constituencies of Ondangwa, Ongwediva, Oshakati East and Oshakati West. Much of the surrounding countryside is communal land, where the population lives in scattered villages and primarily engages in the communal cultivation of staple crops such as pearl millet (*mahangu*), livestock-rearing and the harvesting of wild foods.

Selection of Participants

The research on which this chapter is based involved a representative household food security survey in the OOO Corridor in May 2017, using a two-stage stratified sampling design. At the first stage, 35 primary sampling units covering the entire corridor were selected. At the second stage, a fixed number of 26 households was selected in each of the sampling units. A total of 910 households were targeted for interview, and 853 household heads were actually interviewed, giving a response rate of 94%. The wide-ranging questionnaire from the African Food Security Urban Network—Food, Urbanization, Environment and Livelihoods project (AFSUN-FUEL)—was used and collected comprehensive data on household membership and characteristics, food security, food sources, dietary diversity and coping strategies. (For further details of the sampling strategy, see Nickanor et al. [2019b].)

Variables

A number of dependent variables of interest were chosen from the data set for further analysis. The first is food insecurity (defined as the prevalence of food inaccessibility in the four weeks prior to the survey). Three measures of food insecurity were calculated. First, for each household, the HFIAS and HFIAP were derived from a set of 9 questions that capture different aspects and the increasing severity of food insecurity. Here, we use a binary indicator (0 = food secure; 1 = food insecure) derived from the scores generated by the 9 questions. Second, dietary diversity was measured using the HDDS, which is based on the number of food groups consumed within the household in the previous 24h (Swindale & Bilinsky, 2006). The HDDS scale runs from 0 to 12, and a score is calculated for each household. An increase in the average number of different food groups consumed provides a quantifiable measure of improved household dietary diversity. The HDDS scores were binned into two categories: more food diverse (6–12 on the HDDS) and less food diverse (1–5 on the HDDS). The third measure used was the reduced coping strategies index (CSI), which deploys 12 questions related to food unavailability. Respondents were asked the number of days per week that they applied each coping strategy and a mean score was obtained for each strategy. An overall average toward a value of 7 suggests the family was more food insecure, while a mean toward 0 indicates less food insecurity. A coping index was computed using the principal component technique and split into three categories: 1 = food secure; 2 = moderately food insecure; 3 = severely food insecure.

The primary explanatory variables used in the analysis include sources of food and the socio-economic characteristics of the respondent households, including sources of income, income level (as a categorical variable), informality (household in informal or formal housing), subjective and objective poverty measures (defined using the lived poverty index (LPI) and the income-based marginal poverty line, respectively), household size, type of household (female-centered, male-centered, nuclear, extended) and education level of the head of household. The LPI measures the frequency with which people experience shortages of basic necessities, including lack of water, cooking fuel, electricity and medical care (Mattes et al., 2016). LPI scores range from 0 to 4, with values close to 0 indicating availability of basic items, while high values suggest lack of one or more basic necessities. Types of household were binned into two categories: female-headed (all female-centred households) and male-headed (the other three types). Food sources were grouped into food categories: (a) rural-to-urban food transfers (from relatives or friends); (b) communal food sources, including shared meals with neighbours and/or other households in community, food provided by neighbours and/or other households in the community, community food kitchens (e.g., soup kitchen), borrowing food from others, food provided at work, food provided to children at school/creche (educare) and begging; (c) informal food sources (tuck shops and street vendors) and (d) formal food sources (supermarkets, small shops such as grocers, cafés and butcheries, fast food takeaways, restaurants and open markets). We generated a score for each household based on the number of sources used to access food in the previous month.

Analysis

Descriptive summaries using cross-tabulation of the three food insecurity outcomes and the four types of explanatory variables were first generated, and a chi-square test of association carried out. Second, since both food insecurity prevalence and household dietary diversity were binned into two categories, a binary logistic regression was fitted. Third, for the coping strategy outcome, the generated variable had more than two categories so multinomial logistic regression was used. In all three outcome models, we adjusted for socio-economic and food source variables. Significant associations between the response variables and the explanatory variables were assessed at $p < 0.005$.

Results

Background Characteristics of the Sample

Of the sample of 853 households, two-fifths (40%) were female-centred (headed by a woman without a partner or spouse), while 19% were male-centred (headed by a man without a partner or spouse). Extended families comprising a couple with dependents, including children and other relatives, constituted 21% of the sample, and nuclear households of parents and their offspring accounted for only 16%. The average household size was 4.4 (standard deviation of 3.1). Half the surveyed households had members below the age of 25 years, and two-thirds of all household members were of working age (16 to 60 years).

Food Insecurity Prevalence and Household Dietary Diversity

Overall food insecurity prevalence was at 77% of households, while 65% reported a low dietary diversity score of five or fewer food groups consumed. Table 9.1 cross-tabulates the two food security indicators of food insecurity prevalence and greater dietary diversity with the four sets of independent variables. The table shows a positive relationship between food insecurity prevalence, food sources and socio-economic factors. More than 80% of households receiving rural-to-urban food transfers and sourcing food both communally and informally were in the food insecure prevalence category. Similarly, more female-headed households, informal housing households, marginally poor and low lived poverty households

Table 9.1 Cross-tabulation of food insecurity prevalence and household dietary diversity

Variable	*Categories*	*Food insecurity prevalence* *% (n)*	*Household dietary diversity* *% (n)*
Food sources			
Rural–urban transfer	1=Yes	80.7 (377)	38.8 (178)
	0=No	72.6 (277)	30.4 (112)
	χ^2_{df}(*p*-value)	**7.78 (p = 0.003)**	**6.25 (p = 0.012)**
Communal food sources	Yes	85.9 (73)	34.1 (28)
	No	76.2 (582)	34.9 (262)
	χ^2_{df}(*p*-value)	**4.09 (p = 0.043)**	**0.18 (p = 0.89)**
Informal food sources	Yes	84.8 (196)	39.2 (89)
	No	74.3 (459)	33.2 (201)
	χ^2_{df}(*p*-value)	**10.7 (p = 0.001)**	**2.65 (p = 0.061)**
Formal food sources	Yes	76.9 (644)	35.2 (290)
	No	91.7 (11)	0 (0)
	χ^2_{df}(*p*-value)	**1.46 (p = 0.023)**	**5.95 (p = 0.015)**
Socio-economic Factors			
Female-headed household	Female	81.4 (276)	32.2 (107)
	Male	75.2 (233)	40.7 (124)
	χ^2_{df}(*p*-value)	**3.74 (p = 0.033)**	**4.88 (p = 0.017)**
Housing type	Formal	73.4 (398)	42.3 (225)
	Informal	84.0 (252)	21.8 (64)
	χ^2_{df}(*p*-value)	**12.25 (p < 0.001)**	**35.1 (p < 0.001)**
Marginal poverty*	Poor	85.9 (55)	11.3 (7)
	Otherwise	78.6 (485)	36.0 (219)
	χ^2_{df}(*p*-value)	**1.89 (p < 0.001)**	**15.33 (p < 0.001)**

(continued)

Table 9.1 (continued)

Variable	*Categories*	*Food insecurity prevalence* % (n)	*Household dietary diversity* % (n)
Lived poverty index	< 1.00	64.8 (273)	45.9 (188)
	1.01–2.00	93.4 (155)	34.8 (57)
	2.01–3.00	100 (60)	21.7 (13)
	3.01–4.00	100 (16)	0 (0)
	χ^2_{df} (*p*-value)	**80.3 ($p < 0.001$)**	**26.9 ($p < 0.001$)**
Net income	< = N$1100.00	89.2 (141)	17.4 (77)
	1101.00–2100.00	84.7 (100)	19.7 (23)
	2101.00–4200.00	85.0 (113)	23.4 (31)
	4201.00–12,000.00	76.5 (114)	42.9 (63)
	N$ 12,001.00 +	58.5 (72)	67.2 (82)
	χ^2_{df} (*p*-value)	**47.24 ($p < 0.001$)**	**101.23 ($p < 0.001$)**

*Below the national lower-bound or severe poverty line (N$520.80 = US$40.1)

were food insecure than their male-headed, formal and less poor counterparts. As lived poverty increased in severity, so did food insecurity prevalence. There was a very clear relationship with income quintiles as food insecurity prevalence declined from 89% for households in the lowest quintile to 59% for households in the highest.

With regard to greater dietary diversity, there were again significant relationships, but this time between dietary diversity and food transfers from the rural areas and formal food sources. The reverse would also be true, with households that do not receive food transfers and that source food from communal and informal sources having low levels of dietary diversity. In addition, male-headed households, residents of formal housing and those households with better marginal poverty and lived poverty scores had more dietary diversity. Conversely, female-headed households, marginally poor households and those in informal settlements had low dietary diversity. As Table 9.1 shows, dietary diversity improved with income, with 17% of the lowest income quintile households in the more diverse category, compared with 67% of households in the highest income quintile.

Table 9.2 presents the results from the multiple binary regression on food insecurity prevalence as odds ratios (OR). An OR > 1 suggests increased odds of food insecurity, while an OR < 1 indicates a decreased likelihood of food insecurity. The odds of being food insecure are three times as high for households receiving rural-to-urban food transfers (OR = 3.16, 95% confidence interval (CI): 1.69–5.90) and sourcing food from informal sources (OR = 3.15, 95% CI: 1.25–7.98). They were also slightly higher for households relying on communal food sources (OR = 1.29, 95% CI: 0.24–6.94). Income level and poverty (severe and lived) were significantly associated with food insecurity. Households living in severe poverty were four times as likely to be food insecure than other households (OR = 4.02, 95% CI: 0.70–23.07), as were households in the lowest income quintile compared with households in the upper quintile (OR = 3.81, 95% CI: 3.81, CI: 0.89–16.27). In general, the odds of being food insecure declined with income. Informal housing residents also had increased odds of being food insecure. Female-headed households had marginally reduced risk of being food insecure compared to male-headed households. An important additional finding was that households receiving social grants were less likely to be food insecure (OR = 0.52, 95% CI: 0.28–0.88).

Table 9.2 Results from multiple logistic regression on food insecurity prevalence

Variable	*Categories*	*Odds Ratio (OR)*	*95% CI for OR*		*p-value*
Food Sources					
Rural–urban transfer	1=Yes	3.16	1.69	5.90	$p < 0.001$
	0=No	1.00			
Communal food sources	Yes	1.29	0.24	6.94	0.767
	No	1.00			
Informal food sources	Yes	3.15	1.25	7.98	0.015
	No	1.00			
Formal food sources	Yes	0.85	0.57	1.33	0.274
	No	1.00			
Socio-economic factors					
Female-headed household	Female	1.38	0.86	2.22	0.18
	Male	1.00			
Housing type	Formal	1.00			
	Informal	1.61	0.63	4.09	0.32
Marginal poverty	Poor	4.02	0.70	23.07	0.12
	Otherwise	1.00			
Social grants	Yes	0.52	0.28	0.88	0.041
	No	1.00			
Net income	< N$1,100	3.81	0.89	16.27	0.071
	1,101–2,100	2.09	0.70	6.18	0.182
	2,101–4,200	2.94	1.14	7.56	0.026
	4,201–12,000	2.44	1.09	5.43	0.029
	N$12,001+	1.00			
Lived poverty index		5.89	3.07	11.89	$p < 0.001$
Household size		1.18	1.04	1.27	0.203

Table 9.3 shows the results of the same multiple logistic regression with reference to dietary diversity and confirms some of the findings observed using chi-square tests in Table 9.1. For example, patronage of formal food sources was associated with increased odds of greater dietary diversity (OR = 1.66, 95% CI: 1.18–2.59, $p = 0.015$). Similarly, access to informal food sources was associated with a threefold increase in dietary

Table 9.3 Results for multiple logistic regression on high household dietary diversity

Variable	*Categories*	*Odds Ratio (OR)*	*95% CI for OR*		*p-value*
Food sources					
Rural–urban transfer	1 =Yes	1.14	0.58	1.47	0.45
	0=No	1.00			
Communal food sources	Yes	1.15	0.25	5.73	0.66
	No	1.00			
Informal food sources	Yes	3.14	1.29	7.36	0.015
	No	1.00			
Formal food sources	Yes	1.66	1.18	2.59	0.021
	No	1.00			
Socio-economic factors					
Female-headed household	Female	1.29	0.48	1.47	0.26
	Male	1.00			
Housing type	Formal	1.00			
	Informal	1.19	0.46	2.32	0.411
Marginal poverty	Poor	1.31	0.54	3.17	0.55
	Otherwise	1.00			
Social grants	Yes	1.36	0.49	3.73	0.54
	No	1.00			
Net income	<=N$1,100	0.25	0.11	0.54	0.031
	1,101–2,100	0.22	0.10	0.45	0.021
	2,101–4,200	0.23	0.12	0.48	0.075
	4,201–12,000	0.64	0.29	1.07	0.12
	N$12,001+	1.00			
Lived poverty index		0.78	0.44	0.98	0.05
Household size		0.79	0.67	0.94	0.024

diversity (OR = 3.14, 95% CI: 1.29–7.36, $p = 0.021$). However, use of communal food sources and rural-to-urban transfers did not significantly increase or decrease the odds of greater dietary diversity. Of the socio-economic variables, the most important factor increasing the chances of greater dietary diversity was net income; however, there was no significant association with whether the household was female-headed (OR = 1.29, 95% CI: 0.48–1.47), nor whether it received social grants (OR = 1.36, 95% CI: 0.49–3.73), although there were increased odds of higher dietary diversity. The odds of more dietary diversity were reduced with increased lived poverty (OR = 0.78, 95% CI: 0.44–0.94) and increased household size (OR = 0.79, 95% CI: 0.67–0.94). While social grants were associated with less risk of food insecurity, they were not positively associated with more dietary diversity, which suggests that they may increase the quantity of basic staples consumed but do not lead to a more diverse diet.

Coping Strategies

The next set of tables examines the relationship between food-related coping strategies and the food source and socio-economic variables. We first calculated how many of the strategies each household had used in the previous seven days and a mean value for all households using each strategy. The closer the mean to 0, the lower the use of that strategy. An increase in the mean value indicates more frequent use of that strategy. Table 9.4 shows that the most important coping strategies overall were relying on less preferred and less expensive foods (2.67), reducing the number of meals eaten in a day (1.34) and limiting portion size at mealtimes (0.98). Although these mean values seem quite low, the standard deviation suggests that significant numbers of households were using these strategies. For example, 70% of households relied on the coping strategy of reduced portions for five days in the week. Or again, about 70% of the households coped for close to four days in the week by reducing the number of meals eaten. Figure 9.1 identifies three levels of coping strategy use—low, moderate and intensive—and the variation in use of these strategies by food source. The highest level of use of coping strategies was observed among households obtaining food from community and informal food sources.

Table 9.5 explores the association between coping strategies and the food source and socio-economic variables, based on chi-square tests. As many as 70% of poor households fall into the intensive use category. In addition, 61% of households in the lowest income quintile fall into this category (compared to only nine percent in the upper quintile). Households in informal housing were also more likely to be using more coping strategies (52% versus 24% in formal housing). Roughly equal proportions of female- and male-headed households were in the intensive category, although fewer female-headed than male-headed households were using

Table 9.4 Coping strategy and mean/median number of days household applied them

Coping Strategy	*Mean*	*Median*	*Standard Deviation*	*N*
Relied on less preferred and less expensive foods	2.67	2	2.65	816
Had to borrow food, or rely on help from a friend or relative	0.52	0	1.43	770
Had to purchase food on credit	0.23	0	0.87	730
Had to gather wild food, hunt or harvest immature crops	0.11	0	0.63	684
Had to consume seed stock held for next season	0.12	0	0.75	674
Had to send household members to eat elsewhere	0.30	0	1.09	710
Had to send household members to beg	0.16	0	0.77	711
Had to limit portion size at mealtimes	0.98	0	1.97	778
Had to restrict consumption by adults in order for small children to eat	0.45	0	1.26	614
Had to feed working members of HH at the expense of non-working members	0.04	0	0.34	689
Had to reduce number of meals eaten in a day	1.34	0	2.35	785
Had to skip entire days without eating	0.55	0	1.32	776

the lowest level of coping strategy (25% versus 31%). In terms of food sources, 58% of households purchasing food from the informal sector were using coping strategies most intensely, followed by communal food sources (47%) and formal food sources (33 percent). Coping strategies were less strongly associated with rural-to-urban transfers, although 37% of households not receiving transfers fell into the intensive use category, compared to 30% of those receiving transfers, suggesting some benefit.

Table 9.6 shows the results from the multiple logistic regression analysis for the intensity of use of coping strategies. The greatest use of coping strategies is associated with informal food sources (OR = 4.61, 95% CI: 2.64–7.48), living in informal housing (OR = 2.82, 95% CI: 1.75–4.54), increased household size (OR = 1.56, 95% CI: 1.07–2.22) and lived poverty (OR = 2.74, 95% CI: 1.88–3.99). Those with monthly incomes

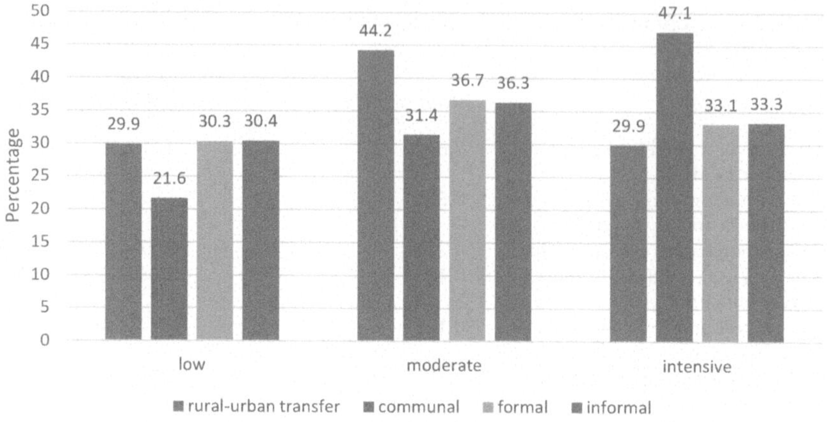

Fig. 9.1 Variation in coping strategies across food sources

of less than N$1,100 were seven times more likely to use some coping strategy (OR = 7.12, 95% CI: 2.65–20.32). Those receiving an average income of N$1,101–2,100 were five times more likely to use coping strategies (OR = 5.24, 95% CI: 1.84–14.57) than those with incomes higher than N$12,000.00. The risk of having to use coping strategies decreases consistently with increasing income. Further, households that were deemed to be marginally poor were nearly four times more likely to use coping mechanisms than those deemed otherwise (OR = 3.94, 95% CI: 1.64–9.15). In addition, households that received rural-to-urban transfers proved to be less reliant on coping strategies than those that did not (OR = 0.78, 95% CI: 0.50–1.22). Those receiving social grants were more likely to deploy coping strategies than those that were not (OR = 1.13, 95% CI: 0.62–2.41), which is consistent with Namibia's social grant policy targeting the most vulnerable households.

Conclusion

The arrival of supermarket chains in the OOO Corridor over the last decade, as part of Namibia's supermarket revolution (Nickanor et al., 2019a), has fundamentally altered the food system of these secondary cities, lengthening food supply chains well beyond the local area and improving the diversity (although not necessarily the affordability) of food

Table 9.5 Coping severity by food sources and socio-economic factors

Variable	*Categories*	*Coping strategy*			*Total*
		Less	*Moderate*	*Intensive*	*(n)*
Food sources					
Rural–urban transfer	1 =Yes	25.9	44.2	29.9	294
	0=No	35.8	26.7	37.4	243
	χ^2_{df}(*p*-value)	17.78 ($p < 0.001$)			
Communal food sources	Yes	21.6	31.4	47.1	51
	No	31.3	36.8	31.9	486
	χ^2_{df}(*p*-value)	14.99 ($p = 0.034$)			
Informal food sources	Yes	15.0	27.2	57.8	147
	No	36.3	39.7	24.1	390
	χ^2_{df}(*p*-value)	58.83 ($p < 0.001$)			
Formal food sources	Yes	30.3	36.7	33.1	532
	No	10.0	0	60.0	5
	χ^2_{df}(*p*-value)	3.06 ($p = 0.232$)			
Socio-economic factors					
Female-headed household	Yes	24.6	40.8	34.6	228
	No	30.9	32.3	36.9	217
	χ^2_{df}(*p*-value)	3.97 ($p = 0.137$)			
Housing type	Formal	35.0	41.2	23.7	354
	Informal	20.9	26.9	52.2	182
	χ^2_{df}(*p*-value)	43.91 ($p < 0.001$)			
Marginal poverty	Poor	13.3	16.7	70.0	30
	Otherwise	29.2	39.2	31.7	401
	χ^2_{df}(p-value)	18.19 ($p < 0.001$)			
Lived poverty index	<1.00	43.3	40.9	15.8	291
	1.01–2.00	87.0	46.6	44.7	103
	2.01–3.00	0	26.1	73.9	46
	3.01–4.00	0	67.0	93.3	15
	χ^2_{df}(p-value)	133.7 ($p < 0.001$)			
Net income	<=N$1,100	15.2	23.9	60.9	92
	1,101–2,100	14.6	37.8	47.6	82
	2,101–4,200	32.5	41.3	26.3	80
	4,201–12,000	36.3	39.2	25.5	102
	N$12,001+	42.7	48.0	9.3	75
	χ^2_{df}(*p*-value)	67.57 ($p < 0.001$)			

Table 9.6 Results for multiple logistic regression on coping intensity

Variable	*Categories*	*Odds Ratio (OR)*	*95% CI for OR*		p-*value*
Food Sources					
Rural–urban transfer	1=Yes	0.78	0.50	1.22	0.278
	0=No	1.00			
Communal food sources	Yes	1.17	0.49	2.65	0.719
	No	1.00			
Informal food sources	Yes	4.61	2.64	7.48	0.001
	No	1.00			
Formal food sources	Yes	0.32	0.02	3.95	0.64
	No	1.00			
Socio-economic Factors					
Female-headed household	Female	1.66	0.49	2.12	0.43
	Male	1.00			
Housing type	Formal	1.00			
	Informal	2.82	1.75	4.54	<0.001
Marginal poverty	Poor	3.94	1.64	9.15	<0.001
	Otherwise	1.00			
Social grants	Yes	1.13	0.62	2.41	0.52
	No	1.00			
Net income	<=N$1,100	7.12	2.65	20.32	<.001
	1,101–2,100	5.24	1.84	14.57	<0.001
	2,101–4,200	3.65	1.05	11.45	0.045
	4,201–12,000	1.78	0.78	4.67	0.34
	N$12,001+	1.00			
Lived poverty index		2.48	1.34	3.12	$p <$ 0.001
Household size		1.56	1.07	2.22	0.011

available in the corridor. In addition, town councils have replaced earlier informal markets with modern facilities where food vendors can rent space and take advantage of the municipal services available on site. Despite these recent changes in the food system, the informal food vending sector remains strong and vibrant and is expanding as the population grows. These informal sources include street vendors, mobile sellers, home-based enterprises and tuck shops (small shops in informal settlements). Foods from both the formal and the informal food system are further supplemented by the transfer of locally grown cereals, vegetables and wild foods from neighboring communal farming districts. Despite the abundance of food wrought by these transformations, levels of food insecurity are extremely high, dietary diversity is low and many households rely on coping strategies, such as eating fewer meals or foregoing food altogether. This conundrum—of dearth in the midst of plenty—is not confined to OOO, but it is particularly acute given the year-round availability of all varieties of food and the multiplicity of outlets. This chapter addresses the conundrum, using data from a recent AFSUN-FUEL city-wide survey of nearly 1,000 households.

Despite the proliferation of food sources and food types, the overall prevalence of food insecurity in the OOO Corridor is very high at 77% of households, with two-thirds also reporting low dietary diversity. These households are also more likely to deploy a variety of coping strategies. Over 80% of households receiving rural-to-urban food transfers and sourcing food communally and informally were food insecure. In other words, these food sources may make food more available and affordable, but they do not guarantee food security. In addition, female-headed households, those in informal housing and those living in poverty were more food insecure. The strongest relationship is between food insecurity and income: as household income decreases, food insecurity prevalence consistently increases. Greater dietary diversity is positively associated with food transfers from the rural areas and patronage of formal food sources. In addition, male-headed households, residents of formal housing and households with better poverty scores all had more dietary diversity. As with food security, dietary diversity improves with household income.

In order to further test these findings, a multiple logistic regression was applied to the data. First, the odds of being food insecure were three times higher for households receiving rural-to-urban food transfers and sourcing food informally than those that were not, confirming that access to transfers and informal vendors may mitigate, but not eliminate, food insecurity. Second, income level and poverty were significantly associated with food insecurity. Households living in severe poverty were four times as likely to be food insecure than other households, as were households in the lowest income quintile when compared with households in the upper quintile. Third, the odds of being food secure increased with household income. Fourth, patronage of formal food sources was associated with increased odds of greater dietary diversity, while the use of informal food sources and rural-to-urban transfers did not significantly affect the odds of greater dietary diversity. The likelihood of improved dietary diversity also declined with increased lived poverty, larger household size and reduced household income. Finally, while social grants were associated with lower risk of food insecurity, they were not positively associated with more dietary diversity.

The most important coping strategies relied on by food insecure and low-income households were reliance on less preferred and expensive foods, reducing the number of meals eaten in a day and limiting portion size at mealtimes. The most intensive use of coping strategies was observed among food insecure households, those sourcing from communal and informal food outlets, poor households, those in informal housing and those in the lowest income quintile. This picture is broadly consistent with the emerging scenario in the Global South more generally (Crush & Frayne, 2010). In smaller urban places, most food sources are physically accessible and even within walking distance for most residents. The central question, therefore, is why so many households are unable to take advantage of physical proximity and remain food insecure.

Although there is a growing presence of informal food sources in the OOO Corridor, these sources do not provide the necessary cushion against food insecurity and lack of dietary diversity (Nickanor et al. 2019b). The association of improved dietary diversity with formal sources of food presents an important policy message,particularly as the majority of households source some of their food from the formal market, such as supermarkets. All varieties of food—staples, cooked, fresh produce and processed foods—are obtained from supermarkets (Nickanor et al. 2019b). Access to such sources is key to maintaining nutrition security and curbing the emerging non-communicable disease epidemic (Nickanor et al., 2021).

An important characteristic of secondary urbanization in Namibia is the perpetuation of strong rural–urban linkages, including informal rural-to-urban food remittances. Over 60% of the population in the corridor's urban centres were born in the rural areas. Nickanor et al. (2019b) show that 80% of the households in the corridor own land in the rural areas, 42% were growing food in the rural areas and 55% received food from relatives in rural areas. Thus, proximity to the rural areas as well as closer rural–urban linkages than in the capital, Windhoek, are highly relevant as coping strategies related to food and nutrition insecurity. While informal food sources can play a greater role in protecting urban residents against food insecurity (Tawodzera & Crush, 2018) and formalization of urban informal food sources is desirable through greater provision of trading spaces (Bénit-Gbaffou, 2016; Kazembe et al., 2019; Morange, 2015; Skinner, 2008), the policy focus needs to be on both rural and urban areas simultaneously for secondary cities. Achieving greater food security in secondary cities in Namibia requires policies that facilitate the journey of rural produce to markets, investments in education (including building awareness among street vendors of opportunities to source food from rural areas) and infrastructure for increased production of food in the rural areas.

References

Bénit-Gbaffou, C. (2016). Do street traders have the "right to the city"? The politics of street trader organisations in inner city Johannesburg, post-operation clean sweep. *Third World Quarterly, 37*(6), 1102–1129.

Casey, P., Szeto, K., Robbins, J., Stuff, J., Connell, C., Gossett, J., & Simpson, P. (2005). Child health-related quality of life and household food security. *Archives of Pediatric and Adolescent Medicine, 159*(1), 51–56.

Chung, C., & Myers, S. (1999). Do the poor pay more for food? An analysis of grocery store availability and food price disparities. *The Journal of Consumer Affairs, 33*(2), 276–296.

Crush, J., & Caesar, M. (2018). Food remittances and food security: A review. *Migration and Development, 7*(2), 180–200.

Crush, J., & Frayne, B. (2010). *The invisible crisis: Urban food security in Southern Africa* (Urban Food Security Series No. 1). African Food Security Urban Network (AFSUN).

Crush, J., & Frayne, B. (2011). Supermarket expansion and the informal food economy in Southern African cities: Implications for urban food security. *Journal of Southern African Studies, 37*(4), 781–807.

Drysdale, R., Mosa, M., & Urmilla, B. (2019). Food Security in the district of iLembe, KwaZulu-Natal: A comparison of coping strategies between urban and rural households. *Ecology of Food and Nutrition, 58*(4), 379–396.

Farzana, F., Rahman, A., Sultana, S., Raihan, M., Haque, M., Waid, J., Choudhury, N., & Ahmed, T. (2017). Coping strategies related to food insecurity at the household level in Bangladesh. *PLoS ONE, 12*(4), e0171411.

Frongillo, E., Nguyen, H., Smith, M., & Coleman-Jensen, A. (2017). Food insecurity is associated with subjective well-being among individuals from 138 countries in the 2014 Gallup World Poll. *The Journal of Nutrition, 147*(4), 680–687.

Frayne, B. (2010). Pathways of food: Mobility and food transfers in Southern African cities. *International Development Planning Review, 32*(3–4), 291–310.

Hassan, A. (2017). *Food insecurity and disease: Prevalence, policy, and politics*. Apple Academic Press.

Heltberg, R., Hossain, N., Reva, A., & Turk, C. (2012). *Anatomy of coping: Evidence from people living through the crises of 2008–11* (Policy Research Working Paper No. 5957). World Bank.

Irham,. (2012). Urban Sprawl, Food Security and Sustainability of Yogyakarta City, Indonesia. In S. Rauch & G. Morrison (Eds.), *Urban environment: Alliance for global sustainability book series (Science and technology: Tools for sustainable development)* (Vol. 19, pp. 71–81). Springer.

Jaron, D., & Galal, O. (2009). Food security and population health and well being. *Asia Pacific Journal of Clinical Nutrition, 18*(4), 684–687.

Kazembe, L., Nickanor, N., & Crush, J. (2019). Informalized containment: Food markets and the governance of the informal food sector in Windhoek, Namibia. *Environment and Urbanization, 31*(2), 461–480.

Knorr, D., Khoo, C., & Augustin, M. (2018). Food for an urban planet: Challenges and research opportunities. *Frontiers in Nutrition, 4*, 73.

Mattes, R., Dulani, B., & Gyimah-Boadi, E. (2016). *Africa's growth dividend? Lived poverty drops across much of the continent* (Policy Paper No. 29), Afrobarometer.

Maxwell, D., Ahiadeke, C., Levin, C., Armar-Klemesu, M., Zakariah, S., & Lamptey, G. (1999). Alternative food-security indicators: Revisiting the frequency and severity of "coping strategies. *Food Policy, 24*, 411–429.

Miller, F., Osbhar, H., Boyd, E., Thomalla, F., Bharwani, S., Ziervogel, G., Walker, B., Birkmann, J., van der Leeuw, S., Rockström, J., Hinkel, J., Downing, T., Folke, C., & Nelson, D. (2010). Resilience and vulnerability: Complementary or conflicting concepts? *Ecology and Society, 15*(3), 11.

Moafi, F., Kazemi, F., Samiei Siboni, F., & Alimoradi, Z. (2018). The relationship between food security and quality of life among pregnant women. *BMC Pregnancy & Childbirth, 18*(1), 319.

Morange, M. (2015). Street trade, neoliberalisation and the control of space: Nairobi's Central Business District in the era of entrepreneurial urbanism. *Journal of Eastern African Studies, 9*(2), 247–269.

Nickanor, N., Crush, J., & Kazembe, N. (2019a). The informal food sector and cohabitation with supermarkets in Windhoek, Namibia. *Urban Forum, 30*(4), 425–442.

Nickanor, N., Kazembe, L., & Crush, J. (2019b). *Food security in Africa's secondary cities: No. 2: The Oshakati-Ongwediva-Ondangwa Corridor, Namibia* (Urban Food Security Series No. 28). African Food Security Urban Network (AFSUN).

Nickanor, N., Shinyemba, T., & Kazembe, L. (2021). *Understanding the linkages between urban food systems and non-communicable diseases in Namibia* (Nourishing Spaces Project Working Paper No. 3). University of Cape Town.

Namibia Statistical Agency (NSA). (2014). *Oshana 2011: Census regional profile* (Basic Analysis and Highlights Report). Namibia Statistical Agency.

Nyikahadzoi, K., Zamasiya, B., Muchinako, G., & Dziro, C. (2013). Enhancing social support system for improving food security among the elderly headed household in communal areas of Zimbabwe. *Journal of Food Research, 2*(3), 46–54.

Peyton, S., Moseley, W., & Battersby, J. (2015). Implications of supermarket expansion on urban food security in Cape Town, South Africa. *African Geographical Review, 34*(1), 36–54.

Sassi, M. (2018). *Understanding food insecurity: Key features, indicators, and response design*. Springer.

Shariff, Z., & Khor, G. (2008). Household food insecurity and coping strategies in a poor rural community in Malaysia. *Nutrition Research and Practice, 2*(1), 26–34.

Sharkey, J., Johnson, C., & Dean, W. (2011). Relationship of household food insecurity to health-related quality of life in a large sample of rural and urban women. *Women and Health, 51*(5), 442–460.

Skinner, C. (2008). The struggle for the streets: Processes of exclusion and inclusion of street traders in Durban, South Africa. *Development Southern Africa, 25*(2), 227–242.

Swindale, A., & Bilinsky, P. (2006). *Household Dietary Diversity Score (HDDS) for measurement of household food access: Indicator guide (Version 2).* FHI 360/FANTA.

Tawodzera, G., & Crush, J. (2018). *Inclusive growth and the informal food sector in Cape Town, South Africa* (HCP Report No. 16). Hungry Cities Partnership Project.

United Nations Conference on Trade and Development (UNCTAD). (2008). *The least developed countries report, 2008: Growth, poverty and the terms of development partnership*. United Nations.

UN-Department of Economic and Social Affairs (UN-DESA). (2018). *World urbanization prospects: The 2018 revision*. UN-Department of Economic and Social Affairs.

CHAPTER 10

Analysing Diet Composition and Food Insecurity by Socio-Economic Status in Secondary African Cities

Heather Mackay, Samuel Onyango Omondi, Magnus Jirström, and Beatrix Alsanius

Introduction

In their analysis of the stage of food system transformation in sub-Saharan Africa (SSA), Steven Haggblade et al. (2016) place Ghana and Kenya in the middle tier, with Uganda on the bottom tier at a stage of early

H. Mackay (✉)
Department of Geography, Umeå University, Umeå, Sweden
e-mail: heather.mackay@keg.lu.se

H. Mackay · M. Jirström
Department of Human Geography, Lund University, Lund, Sweden

S. O. Omondi
Department of Agricultural Economics, University of Nairobi, Nairobi, Kenya

B. Alsanius
Department of Biosystems and Technology, Microbial Horticulture Division, Swedish University of Agricultural Sciences, Lomma, Sweden

L. Riley and J. Crush (eds.), *Transforming Urban Food Systems in Secondary Cities in Africa*,
https://doi.org/10.1007/978-3-030-93072-1_10

food system transition. They draw upon the work of Barry Popkin (2001, 2015; Popkin et al., 2012), Carlos Augusto Monteiro et al. (2013), David Tschirley and Thomas Reardon (Reardon et al., 2013; Tschirley et al., 2015), as well as public health research into obesity and other non-communicable diseases (NCDs) such as diabetes, hypertension and heart disease (Abrahams et al., 2011; Mayega et al., 2012; Steyn & Mchiza, 2014) in their analysis. According to Haggblade et al. (2016), a food or dietary transition entails "increased consumption of dairy and meat products, higher demand for fresh fruits and vegetables, and increased demand for processed and prepared convenience foods" (p. 220) and a transformation towards a highly marketized and industrialized agri-food industry. This kind of food system transition draws heavily from the South African experience, which is held out as the most advanced transformation, with much-debated consequences for overweight, obesity, diabetes and hypertension (Haggblade et al., 2016).

Another dimension of this interpretation of a food system transition includes eating outside the home more, consuming more energy-dense snack foods, more fried foods and more carbonated drinks, and greater capture of the food value chain by supermarkets and big food industry (Haggblade et al., 2016; Popkin et al., 2012). As Popkin notes, "Concurrent shifts are seen in two dimensions of the food system: the rapid growth of modern retailing and consumption of consumer packaged foods and beverages and the shift of the food value chain from traditional traders and retailers to one where supermarkets and food manufacturers directly source food from farmers and traders" (Popkin, 2014, p. 91). In addition, while informal small and medium-sized enterprises still dominate the sector of food processing as well as urban retailing of processed food in SSA, the emergence and rapid growth of large-scale enterprises in both sectors have recently been clearly documented (Reardon et al., 2021). Drivers of these food system changes are thought to be urbanization (an increasing share of the population living in urban areas) (Satterthwaite et al., 2010) and per capita income increases (Haggblade et al., 2016), which Julian May terms as important "keystone" socio-economic interactions that work together to "reconfigure the food system in SSA" (May, 2018, p. 6). May describes four keystone influences that are reworking African food systems: economic growth and rapid urbanization; inequality and a growing middle class; obesogenic urban food environments; and a rising prevalence of NCDs (May, 2018). May notes, drawing also from Popkin's work, that urban residents are more prone to purchase convenience foods, highly processed foods, to eat out, to consume more sweetened or salty and fatty foods and to drink more carbonated products (May, 2018). Often these claims of such trends are based on national and international datasets of agricultural production, of supermarket and agribusiness presence, of Gross Domestic Product (GDP) data and of supply within the urban environment, or with data from the capital city (Mackay, 2019b). We explore these claims of food system transition at the finer-grained scale of households in six secondary cities of Ghana, Kenya and Uganda by analysing food sourcing strategies, which food groups were actually consumed, dietary diversity and food insecurity experience, disaggregated by socio-economic status (SES). In addition, we analyse how these three countries and six cities compare, given their experience of diet-related health problems and their purported different stages in food system change.

The chapter continues with a short description of the country and city contexts before describing the methodology. We then present the results of our analysis of household food environments and experience of food insecurity disaggregated by SES. By food environment, we refer to the definition by Christopher Turner et al. (2018, p. 95), with the focus on both access and consumption, whereby the food environment denotes "the interface where people interact with the wider food system to acquire and consume foods". We close the chapter with a consideration of the implications of our findings and cross-country analysis, before making some conclusions.

Positioning of the Research

To further contextualize our case countries of Ghana, Kenya and Uganda, this section provides a brief overview of their urban context, their food security and nutritional status, and their health and NCD experience. Kenya and Uganda border each other in East Africa, whereas Ghana is situated in West Africa. The World Bank considers Ghana and Kenya as lower/middle-income countries and Uganda as a low-income country (World Bank, 2016). The UN's Human Development Index (UN, 2020) puts Ghana at 0.596 (rank 142), Kenya at 0.579 (rank 147) and Uganda at 0.528 (rank 159) in the world, implying they are in fairly similar positions. The study cities included Tamale and Techiman in Ghana, Kisumu and Thika in Kenya, and Mbale and Mbarara in Uganda.

The largest two cities in terms of population were Kisumu and Tamale (Table 10.1). The other cities had about 100,000 inhabitants when surrounding peri-urban and rural areas that are part of official census data are excluded. In terms of character, Thika, only 56 kilometres from the capital of Nairobi, is an industrial town, with a fair share of manufacturing industries (Omondi et al., 2017). Techiman is an agricultural and market town, similar in this way to Mbale. Kisumu, Tamale and Mbarara are regional centres of trade and administration, with significant agricultural trading and agro-processing activity. Food prices are generally cheaper in Techiman than in Tamale, cheaper in Thika than Kisumu, and cheaper in Mbale than Mbarara, due to these areas being closer to agricultural heartlands.

Table 10.1 Population of the study municipalities at time of survey

	Ghana (2010)[a]		*Kenya (2012)*[b]		*Uganda (2016)*[c]	
	Tamale	*Techiman*	*Kisumu*	*Thika*	*Mbale*	*Mbarara*
Population (closest official census data, includes surrounding rural areas)[d]	371,351	206,858	383,444	151,225	92,863	195,160
Estimate of the city population without nearby villages, from local researcher knowledge	250,000	70,000	300,000	100,000	70,000	90,000

[a] *Data source* Ghana Statistical Service (GSS, 2012)
[b] *Data source* Kenya National Bureau of Statistics (KNBS, 2012)
[c] *Data source* Uganda Bureau of Statistics (Uganda Bureau of Statistics, 2016)
[d] These data are from the administrative level at which official census data is collected. The figures therefore include surrounding peri-urban areas and villages in the county or municipality that are not part of the actual city boundaries. Thus, these data over-estimate the size of the actual urban population

As Arabella Fraser et al. (2017, p. 106) note in an analysis of the need for more resilient urban development in SSA, "smaller urban centres are greatly under studied". David Satterthwaite also argues for the need to research patterns and processes in smaller cities, noting that more than half of SSA's urban population lives in cities of fewer than 500,000 and many in urban areas much smaller than this, with these smaller cities often growing most rapidly (Satterthwaite, 2017). About 44% of Ghana's population (data from 2000) lived in urban areas, whereas 23% of Kenya's population (2009 data) and just 16% of Uganda's population (2014 data) were urban residents (Satterthwaite, 2017). Looking at the distribution of this urban population across city size shows the majority living in cities of less than half a million (Table 10.2). In fact, 67% of Uganda's urban population and 64% of Ghana's lived in cities of fewer than 200,000 inhabitants. Kenya has a few more large cities (the capital, Nairobi, in particular) than Ghana and Uganda and thus only 44% of its urban population residing in cities of less than 200,000.

Given this context, and the prediction that a large share of the world's future growth will occur in urban Africa and Asia (Tacoli, 2017), many researchers and organizations are now raising concerns about a growing urbanization of poverty and food insecurity in Africa (Cobbinah et al., 2015; Dodman et al., 2017; Tacoli, 2017). As Cecilia Tacoli (2017) emphasizes, food security in urban Africa is closely tied to income and thus to the employment situation, but non-income dimensions of poverty (such as cooking space, farming space, social networks, rural linkages, infrastructural, environmental or geographic conditions) all play an important role in urban food poverty and in the multiple dimensions of malnutrition. A number of researchers have been raising concerns about an overemphasis on the production side of food security and have called

Table 10.2 Distribution of the urban populations in Ghana, Kenya and Uganda

	Proportion (%) of the population in urban centres of varying size						
	Under 20,000	*20,000–49,999*	*50,000–199,999*	*200,000–499,999*	*0.5–1.99 million*	*2–4.99 million*	*5 million or more*
Ghana	33	15	16	3	34	–	–
Kenya	13	13	18	12	10	35	–
Uganda	11	23	33	6	27	–	–

Source Calculated from census data presented by Satterthwaite (2017, p. 20)

for a disaggregated analysis of actual access to and consumption of food within a household (Crush & Battersby, 2016; Mackay, 2019a; Nichols, 2017; Tacoli, 2017). This chapter forms a contribution to such knowledge gaps on food security, dietary diversity and food sourcing at the secondary city level within SSA. Recent findings have added weight to the view that food can be a useful lens to assess urban food, livelihood and health systems and how they interrelate, providing a possible entry point for improved urban planning (Battersby & Watson, 2019a, 2019b).

Finally, linking to the opening discussion about the links between a diet/nutrition transition and an epidemiological shift from communicable towards non-communicable diseases, what is currently known for cities of Ghana, Kenya and Uganda? In a recent systematic review of 48 studies encompassing more than 48,000 individuals across Ghana, Richard Ofori-Asenso et al. (2016) found overweight prevalence levels in urban Ghana of 27% and obesity prevalence of 21%. The World Health Organization's (WHO's) Global Health Observatory data repository describes adult (over 18 years old) Ghanaian prevalence of raised blood pressure at 19%, diabetes at 5% and obesity at 10% (female obesity 15%). Kenyan prevalence of raised blood pressure was 20%, diabetes 4%, obesity 6% (9% in females). Ugandan prevalence of raised blood pressure was also 20%, diabetes 3% and obesity 4% (7% in females) in 2016 (WHO, 2020). The same database notes that 43% of all deaths in 2016 in Ghana were due to NCDs, with the Kenyan figure at 27% and the Ugandan at 33% (WHO, 2020).

We were not able to find prevalence measures for NCDs disaggregated to the study cities for Ghana and Kenya, but our own measurement of BMI data in the Ugandan cities found a 26% prevalence of overweight in Mbale (number of adults: 1,248), 28% in Mbarara (number of adults: 948). The obesity prevalence was 14% in Mbale and 22% in Mbarara (Mackay et al., 2018). Females were more affected than males, with 18% female obesity in Mbale (of 810 women) and 27% obesity in Mbarara (of 636 women) (Mackay et al., 2018). According to the International Diabetes Foundation's online atlas, both Ghana and Uganda had an age-adjusted diabetes prevalence level of 2.5% in 2019, while the Kenyan figure was 3.1% prevalence (IDF, 2019). These figures give a brief overview of the NCD experiences of these countries. These figures also suggest that, despite Uganda being at a less advanced stage of claimed food system and nutritional transitions, with still high urban food insecurity and low dietary diversity (as discussed in the introduction), the prevalence of NCDs equate with (in some cases even exceed) those of Ghana and Kenya. This suggests a need for caution in making causal assumptions about food system transitions, nutritional transitions and obesogenic food environments causing epidemiologic shifts. Indeed, much research has been investigating the link between early-life and in-utero undernutrition (linked to poverty) and later life predisposition of obesity and other NCDs—known by many as Barker's hypothesis, or the development origins of disease (Barker, 1997).

Methodology

Sampling and Data Collection

The data was gathered as part of a collaborative research project between the Universities of Lund, Umeå and the Swedish University of Agricultural Sciences and the University of Ghana, the University of Nairobi, Kenya and Makerere University, Uganda. Since the Ugandan survey added a health section and included the measurement of adult heights and weights, ethical approval was obtained from the Umeå Regional Ethical Review Board and the Uganda National Council of Science and Technology. The household survey was conducted in October 2013 (Ghana), November and December 2013 (Kenya) and from June to August 2015 (Uganda), with the support of the African Food Security Urban Network (AFSUN). AFSUN granted permission to build from its already tested Household Food Security Baseline Survey instrument, which had been used to survey livelihood circumstances and food access in 11 cities across nine southern African countries in 2008 and 2009 (Crush et al., 2012). The choice of countries and cities in our study was purposive, based on a combination of contacts, practicalities and understanding of local food-related urban conditions. As such, they are not considered representative of all secondary cities in each country, nor of all SSA countries. Thus, we do not generalize our results beyond our study cases, although we believe that our findings may have relevance for other secondary SSA cities.

Further details of the systematic random sampling of every third household across a city neighbourhood, the use of tablet computers and open data kit software as a method of standardizing data entry and minimizing data collection errors and facilitating daily data checking algorithms, as well as the selection of city quadrants in the cases where all neighbourhoods could not be surveyed (specifically Tamale and Kisumu) have been discussed in previous research (Ayerakwa, 2017; Crush et al., 2012;

Mackay, 2018; Omondi et al., 2017). A total of 6,013 households were surveyed, encompassing a total of 20,813 individuals.

Sample and Variables

Table 10.3 provides a summary description of the survey sample from each city as well as the explanatory variables and codes used in our modelling. The "food transfers" variable relates to whether the household reported receiving food from rural- or urban-based friends or relatives during the year preceding the survey. The "engagement in agriculture" variable presents the proportion of households who reported farming or gardening some of their crops or livestock from either an urban area, a rural area, or both, during the year prior to survey. The "income diversity" variable counts the number of different income sources that the household head reported drawing from during the preceding year. The mean number of income sources ranged from 1.3 in Ugandan cities to 1.8 in Kisumu, suggesting that a number of families tried to draw from multiple strategies in these contexts. During the survey, we also asked for the location where the main meal of each household member had been eaten the preceding day and asked respondents about the various places that they sourced food from and what other sourcing strategies they had.

Socio-Economic Status Groupings

The SES indicator in Table 10.3 is based on responses to survey questions about expenditures made monthly (such as rent, food, utilities, transportation costs, etc.) and annually (such as school fees, books, uniforms, medical expenses, insurance costs, funeral- or wedding-related). We calculated a monthly expenditure total (monthly costs, plus one-twelfth of the total annual costs reported for each household) per adult equivalent (deemed as of working age, i.e. 16–60 years of age) in the household. Calculating "per adult equivalent" takes into account the age structure and demographic makeup of the household, including the number of dependents per working-age adult. An adult equivalent score is calculated by assigning adult household members (16–60 years) a value of one, with children (0–15 years of age) given a value of 0.5 and elderly household members (>60 years) given a value of 0.75. Techiman, Kisumu and Mbarara all had a mean adult equivalent score of 3.6, whereas Thika scored 4 and Mbale scored 2.8. Our computed monthly

Table 10.3 Variable descriptions and regression codes

		Ghana		*Kenya*		*Uganda*	
		Tamale (N = *1005*)	*Techiman* (N = *1024*)	*Kisumu* (N = *994*)	*Thika* (N = *1003*)	*Mbale* (N = *1020*)	*Mbarara* (N = *967*)
Variable	*Variable Measurement (also shows dummy variable codes and regression base)*	*Percentage of the households (%) (unless otherwise indicated in variable measurement column)*					
SES quartile* (1 = yes, 0 = no)	Low income (0–25%) (1) [base]	27.0	23.1	22.5	27.6	23.6	26.5
	Lower-mid income (26–50%) (2)	25.8	24.3	23.8	26.2	21.4	28.7
	Upper-mid income (51–75%) (3)	23.9	26.1	25.5	24.6	24.7	25.3
	Upper income (>75%) (4)	23.3	26.5	28.3	21.7	30.3	19.5
		100%	100%	100%	100%	100%	100%
Household structure	Multiple adult household (1)	80.8	63.4	71.6	65.0	63.4	66.4
	Single adult household (0) [base]	19.2	36.6	28.4	35.0	33.6	36.6
		100%	100%	100%	100%	100%	100%
Gender of HH head	Male household head (1)	73.4	72.2	69.7	85.7	79.1	75.5
	Female household head (0) [base]	26.6	27.8	30.3	14.3	24.5	20.9
		100%	100%	100%	100%	100%	100%

(continued)

Table 10.3 (continued)

		Ghana		*Kenya*		*Uganda*	
		Tamale (N = *1005*)	*Techiman* (N = *1024*)	*Kisumu* (N = *994*)	*Thika* (N = *1003*)	*Mbale* (N = *1020*)	*Mbarara* (N = *967*)
Variable	*Variable Measurement (also shows dummy variable codes and regression base)*	*Percentage of the households (%) (unless otherwise indicated in variable measurement column)*					
	Missing data	118	136	5	22	18	34
Work status of HH head (>18 yrs)	Working (self-employed, informally employed, as well as salaried) (1)	89.1	84.4	75.8	73.4	59.6	61.3
	Not working (unemployed, retired, ill, home-maker etc.) (0) [base]	10.9	15.6	24.2	26.6	38.7	40.4
		100%	100%	100%	100%	100%	100%
	Missing data	125	139	40	26	20	41
Food transfers	Yes, received (1)	29.3	32.1	52.6	43.8	51.0	45.5
	No, not received (0) [base]	70.7	67.9	47.4	56.2	49.0	54.5
		100%	100%	100%	100%	100%	100%

		Ghana		Kenya		Uganda	
		Tamale (N = 1005)	*Techiman* (N = 1024)	*Kisumu* (N = 994)	*Thika* (N = 1003)	*Mbale* (N = 1020)	*Mbarara* (N = 967)
Variable	*Variable Measurement (also shows dummy variable codes and regression base)*	*Percentage of the households (%) (unless otherwise indicated in variable measurement column)*					
Engagement in agriculture (1 = yes, 0 = no)	Rural agriculture (1)	19.8	15.3	33.4	25.8	27.4	32.1
	Urban agriculture (2)	14.8	26.2	16.7	17.9	22.5	12.8
	Both rural & urban agriculture (3)	8.0	2.3	7.8	7.7	15.4	6.1
	No agriculture [Base] (4)	57.4	56.2	42.1	48.6	34.8	49.0
		100%	100%	100%	100%	100%	100%
Income diversity	Mean number of income sources (count)	1.6	1.6	1.8	1.7	1.3	1.3
	Missing data					1	2
Age of HH head	Mean age (years)	39.0	44.0	45.9	47.7	37.8	39.5
Mean household size		4.9	4.5	4.7	3.3	3.7	4.6

*The SES quartiles were computed by country, thus why the divisions are not 25% per city

Table 10.4 Cut-off points for SES quartiles

SES quartile	*Monthly equivalent household expenditure in US$, adjusted for adult equivalent per household and to 2015 dollar values*		
	Ghana (US$)	*Kenya (US$)*	*Uganda (US$)*
25th	41	40	17
50th	78	71	34
75th	155	144	73

NB These monthly expenditure figures include one-twelfth of the annual expenditure

expenditure variable was converted to US dollars,[1] then divided by household adult equivalent to allow a fair comparison across households and cities/countries. In order to equate the dollar purchasing power across the countries, the Ghanaian and Kenyan costs were adjusted using 2015 consumer price index values (translated into US dollars) to match to the time of the Ugandan survey to enable fair cross-country comparison.

For each country, households were then grouped into quartiles using this US dollar monthly expenditure per adult equivalent, with the first (0–25%, or the poorest 25% in each city), second (26–50%), third (51–75%) and fourth (>75%, or the best-off 25% in each city) quartiles corresponding to low-income, mid-lower, mid-upper and upper-income households, respectively. The resulting variable is called the SES quartile (Table 10.3) and allows fair cross-comparison across households, cities and countries, and was the variable used as a proxy for the SES of a household, better reflecting the living conditions of a family than an assessment based on self-reported income or on expenditure alone. The cut-off points for the SES quartiles for each country are shown in Table 10.4. Ghana and Kenya had roughly similar percentile cut-off points and the Ugandan levels were approximately half the Ghanaian and Kenyan values. This difference is another indicator of how Uganda is somewhat further behind Ghana and Kenya in its economic development, as noted in our introduction. Using this method, our analysis and modelling thus consider other aspects of well-being, such as household size and number

[1] One US dollar was equivalent to 2.37 Ghanaian cedis (October–November 2013); one US dollar was equivalent to 86.8 Kenyan shillings (November–December 2013); one US dollar was equivalent to 3,421 Ugandan shillings (July–August 2015). The expenditures for Ghana and Kenya were adjusted using the consumer price index to 2015 values.

of dependents, rather than purely monetary indicators, and additionally analyses households' dietary and food security circumstances.

Food Insecurity Data

The questionnaire survey also gathered data on the experience of food insecurity using the internationally validated measure of the Household Food Insecurity Access Score (HFIAS) and the associated grouping in prevalence categories of food secure, mildly food insecure, moderately food insecure and severely food insecure known as the Household Food Insecurity Access Prevalence (HFIAP) recognized by the Food and Agriculture Organization and other international agencies (Coates et al., 2007). We use these categories in the modelling (see statistical modelling section below), but simplified into two groups. Households were grouped as food secure if they fell into the HFIAP categories of food secure or mildly food insecure. The food insecure households consisted of those that were moderately and severely food insecure. Summary description of these data per city is shown in Table 10.4. Notable is that households in Kisumu and Mbale reported greater food insecurity (Table 10.5). This is not surprising: as stated earlier, Uganda is further behind Ghana and Kenya in its urbanization, its dietary transition and general development (Haggblade et al., 2016), and Kisumu is known to struggle with informality, rapid population growth, poor infrastructural and transport connections resulting in higher food prices, as well as being a net importer of food (Wagah et al., 2018).

Dietary Diversity Data

Twelve food groups were used in the Household Dietary Diversity section of the survey. These included: (1) cereals/grains (such as maize, sorghum, millet, rice or wheat); (2) roots and tubers (such as potatoes, sweet potatoes, cassava, yam); (3) vegetables; (4) fruits; (5) meat or meat products, including poultry and game; (6) eggs; (7) fish, shellfish or fish products; (8) legumes (including beans), nuts or seeds; (9) milk or other dairy products; (10) oil, fat and butter (11) sugar, honey or sweeteners; and (12) condiments, spices, tea and coffee. This data allowed calculation of the Household Dietary Diversity Score (HDDS), following the guidelines by the Food and Nutrition Technical Assistance Project (FANTA),

Table 10.5 Descriptive comparison of household food insecurity and dietary circumstances

	Ghana		*Kenya*		*Uganda*	
	Tamale (N = 1005)	*Techiman (N = 1024)*	*Kisumu (N = 994)*	*Thika (N = 1003)*	*Mbale (N = 1020)*	*Mbarara (N = 967)*
% ate main meal in the home*	95	93	95	96	79	80
Mean HH dietary diversity score (HDDS) (0 = no food groups, 12 = maximum)	5.2	6.1	6.7	7.3	4.4	5.2
Mean HFIAS (0 = no lack of food, 27 = often a lack of food in previous month)	3.6	3.9	9.7	5.2	7.1	5.0
HFIAP food insecure (%)	39.7	41.9	81.5	55.1	66.7	55.8
HFIAP food secure (%)	60.3	58.1	18.5	44.9	33.3	44.2
Total (%)	100	100	100	100	100	100

*For Uganda, the majority of those who did not eat their main meal at home were children who ate at school

an international collaborative initiative spearheaded by USAID (Swindale & Bilinsky, 2006). HDDS provides a measure of the diversity of a household's diet, with zero being the minimum score (consumed nothing during the 24 hours prior to the survey) and 12 being the maximum.

Additional only to the Ugandan survey, we did ask specific questions about the consumption of fried snacks, doughnuts, chips and fried meats, and the consumption of sugar and carbonated drinks (sugar-sweetened beverages). We also measured the heights and weights of willing adults in order to gain a body mass index (BMI) measure.

Table 10.5 summarizes the food insecurity and dietary diversity data, allowing a city/country cross-comparison. Notable from this is that the

Ugandan cities had a slightly less diverse diet than the Ghanaian and Kenyan cities, especially Mbale. The low-diversity, carbohydrate-focused diet of Uganda has been found in a number of other studies (Ngaruiya et al., 2017; Raschke & Cheema, 2008).

Analytical Process and Statistical Modelling

Data was analysed using SPSS version 20. The data on the composition of the household food baskets, household experience of food insecurity and household dietary diversity, as well as the main sources of food, was all analysed against our SES proxy. In terms of our analytical process, we first explored the descriptive statistics of household dietary diversity and household food insecurity and compared differences in mean using ANOVA. Furthermore, consumption data, split by food groups and food sources were compared across SES quartiles using the Mann Whitney U test. We then performed a logistic regression model on the household food security, and a Poisson regression model on the household dietary diversity data. Our aim with the modelling was to explore the factors that may be associated with food security and dietary diversity.

In modelling the HFIAP data, we were interested in exploring the relationship between food security status and our SES proxy, as well as other socio-economic and geographic variables. We were also interested in exploring how these factors may differ across the country and city contexts. Our dependent variable is binary, taking the value of one if a household was food secure and zero if a household was food insecure (see Table 10.5 for city-level descriptive). A logistic regression model is appropriate in modelling the relationship between a binary dependent variable and selected explanatory variables (Cramer, 2003; Wooldridge, 2013). A description of the variables used in the models can be viewed in Table 10.3. Inclusion of these variables was motivated by literature on food security.

In modelling household dietary diversity scores, we included all 12 food groups (values 0–12). Thus, the dependent variable is a non-negative count variable that takes relatively few values. An appropriate model to use with such data is a Poisson regression model (Cramer, 2003). The modelling of HDDS is looking to explain factors that may be contributing to an increase in dietary diversity.

Initially, we tested whether consumption of healthier food groups (such as fruits, proteins, vitamin-rich groups) differed from consumption of less healthy food groups (such as oils/fats, sugars/sweeteners, condiments) to explore whether different factors might influence consumption of different kinds of diets. This was in recognition that a straightforward numerical increase in HDDS does not necessarily indicate a *healthier* dietary consumption if many of the additional food groups were oils, sugars or condiments. We also tested whether removing the food group of cereals/grains from the HDDS (where the HDDS dependent variable would thus run from 0 to 11) reduced possible skewing effects since a majority (80–99% of households, see Table 10.6) consumed this food group (with the exception of Techiman, where 78–87% consumed cereals/grains). However, all of these models produced the same significant explanatory variables with the same effect directions, just with a variation in the magnitude of the effect. We concluded that modelling the overall change in all 12 food groups was the best methodological approach, which also did not entail making judgements open to debate in terms of which foods are healthier or less healthy.

Limitations

The data is cross-sectional data at one point in time. As such, it presents a snapshot only and does not show change over time. The use of expenditure data as a proxy for the SES of a household has limitations, but has nevertheless been shown to provide a more accurate assessment of a household's circumstances than self-reporting of incomes. Laura Howe et al. (2008), for example, note that it is "widely asserted that consumption expenditure is a better marker of long-term SEP [socio-economic position] than income", in low-income countries in particular.

The 24-hour recall of diet content has certain limitations in terms of respondent memory, awareness of all household members' food consumption during the preceding day and willingness to divulge information (similar critique may be levied at the household food insecurity questionnaire). Yet, these two internationally tested and validated measures of household food environments are still held to be important diet quality (energy content) and food security tools to assess food access in a review by Jef Leroy et al. (2015).

There were likely some limitations imposed by the means of collecting the food group data: this was not measured but self-reported and therefore vulnerable to variation in respondent recall and by how detailed individual enumerators were when asking about oil, vegetables mixed into sauces, etc. Our data was also limited to the gathering of information under the 12 food group categories and we did not ask how foods were cooked (fried in oil versus boiled, for example). We also recognized that a higher dietary diversity does not necessarily equate linearly to a *healthier* diet, but rather depends on which food groups were being consumed, as well as how they were cooked. Our data could also not account for allocation within a household, nor portion size assessments.

There were also some limitations in the food sourcing data in the way that certain categories in the questionnaire were grouped, in particular putting the "small shop" category together with "restaurant or take-away". It was our feeling from carrying out the surveys, and from our own knowledge of local dietary and food access practices in these cities, that the majority of responses here related to small local shops, not to restaurants, fast food chains or takeaways. Similarly, the merging into one category of "informal markets" with "street foods" as a food source is also slightly misleading. Again, our experience indicates that these responses related to the traditional neighbourhood/city centre marketplaces, where the majority across all countries and cities still bought the bulk of their food.

Findings

In Ghana, Tamale residents spent 41% of their monthly expenditures on food and Techiman 39% (data not shown). The Kenyan cities of Kisumu and Thika had the highest share of monthly expenditure going towards food at 46 and 45%, respectively. Households in Mbale and Mbarara reported the lowest share of food expenditure at 33 and 35%, respectively.

Exploring the descriptive data on which specific food groups were consumed by households of different SES shows that the largest difference between the lowest and the highest SES was in the consumption of milk (+30% points [pp]) and meat (+41 pp) in Tamale, milk (+28 pp in Kisumu) and fruit (+28 pp) and meat (+28 pp) in Thika (calculated from Table 10.6). Whereas in Uganda the greatest difference between the lowest and highest SES was a 28% higher consumption of sugar (calculated from Table 10.6). A majority (>78%) of all SES consumed food made from cereals and grains (Table 10.6 [note that the second and third quartiles are not shown in table]). Table 10.5 also shows low consumption of vegetables in the Ugandan cities and fairly low consumption of fruits across all countries. The proportion of the best-off households (fourth quartile) consuming meat/meat products or fish/fish products

Table 10.6 Food group consumption and difference between the poorest and the best-off

			Proportion (%) of HHs in the quartile having consumed from the food group during the 24 hours prior to survey					
			Ghana		*Kenya*		*Uganda*	
		SES quartile (%)	*Tamale* (N = *1006)*	*Techiman* (N = *1028)*	*Kisumu* (N = *1004)*	*Thika* (N = *1004)*	*Mbale* (N = *1025)*	*Mbarara* (N = *970)*
Staple food groups[a]	Cereals and grains	0–25	96	78	97	98	88	82
		76–100	96	87**	98	99	93*	89*
	Roots and tubers	0–25	42	80	16	47	51	76
		76–100	59***†	82	33***	57*	56	78
	Legumes, nuts, seeds	0–25	15	18	22	50	47	65
		76–100	27**	24	34**	55	54	62
Foods rich in vitamins and minerals	Vegetables	0–25	61	77	83	84	39	29
		76–100	79***	86**	91**	85	50*	40**
	Fruits	0–25	09	29	31	31	10	15
		76–100	25***	51***	55***	59***	29***	34***
Protein-rich foods	Meat and meat products	0–25	29	42	15	15	21	31
		76–100	70***	59***	37***	43***	41***	50***
	Eggs	0–25	05	12	15	11	03	04

			Proportion (%) of HHs in the quartile having consumed from the food group during the 24 hours prior to survey					
			Ghana		*Kenya*		*Uganda*	
		SES quartile (%)	*Tamale* (N = *1006)*	*Techiman* (N = *1028)*	*Kisumu* (N = *1004)*	*Thika* (N = *1004)*	*Mbale* (N = *1025)*	*Mbarara* (N = *970)*
		76–100	22***	26***	22	29***	14***	11**
	Fish	0–25	60	75	28	06	10	11
		76–100	58	86***	34	07	16*	18*
	Milk/milk products	0–25	06	12	41	75	11	49
		76–100	36***	36***	69***	83*	33***	70***
Less healthy (limited nutritional value)	Oil/fat	0–25	20	42	73	75	10	12
		76–100	39***	62***	82*	92***	27***	31***
	Sugar	0–25	59	38	82	93	35	52
		76–100	73**	62***	92**	96	63***	71***
	Condiments, tea, coffee	0–25	22	20	83	89	34	32
		76–100	55***	35***	83	95**	44*	48***

[a]Beans [legumes, nuts, seeds category] are also protein-rich, but since they are quite commonly consumed, in Uganda in particular, we count them as an important staple

†Chi-square statistics were computed using the Mann Whitney U test to check for differences in proportions between the 1st and 4th quartiles

*Significant at 5% confidence level

**Significant at 1% confidence level

***Significant at 0.1% confidence level

NB The 2nd and 3rd quartiles are not shown for purposes of clarity and simplicity

remained less than 50% in Kenya and Uganda. Milk and milk products were increasingly consumed with rising SES across all cities and countries. A higher proportion of households in Kisumu and Thika reported consuming sugar compared to the other four cities, across all wealth groups. The experience of the research team indicates most Kenyans and Ugandans prefer their tea or coffee with sugar, which may explain some of the high proportion consuming sugar. Generally, consumption of oils and fats was reported by a large share of households in Kisumu and Thika compared to other cities. These data already indicate some dietary transformation towards consumption of more fruits, meat and milk products with rising SES, but also towards less healthy food groups as household welfare improves.

In keeping with theories of how dietary diversity and food security might change with improved SES, Table 10.7 shows that the score of food insecurity diminished with increasing SES across all cities, although even the highest SES households in Kisumu and Mbale still experienced some food insecurity, in Kisumu (HFIAS: 7) and Mbale (HFIAS: 5.34) in particular. Table 10.7 also shows that the highest SES households had an approximately two food groups greater mean dietary diversity score than the lowest SES across all cities, which constitutes a reasonable difference within a maximum score of 12.

Table 10.7 Mean household food insecurity and household dietary diversity by SES quartile

		Ghana		*Kenya*		*Uganda*	
	SES quartile (%)	*Tamale*	*Techiman*	*Kisumu*	*Thika*	*Mbale*	*Mbarara*
HFIAS (min. 0, max. 27)	0–25	5.17***	5.04***	12.90***	6.05***	9.79***	6.41***
	26–50	4.31	4.38	10.19	6.35	7.11	4.80
	51–75	2.85	4.07	9.26	4.58	6.54	5.08
	76–100	1.80	2.46	7.00	3.62	5.34	3.25
HDDS (min. 0, max. 12)	0–25	4.23	5.22	5.87	6.74	3.57	4.55
	26–50	4.93	6.00	6.46	7.16	4.13	5.17
	51–75	5.54	6.41	6.81	7.65	4.56	5.39
	76–100	6.37***	7.00***	7.34***	8.01	5.20***	6.02***

*Significant at 5% confidence level between the lowest and highest SES quartile
**Significant at 1% confidence level
***Significant at 0.1% confidence level

We also investigated the ways in which households accessed food (both market and non-market sources). Table 10.8 presents the results, showing that the traditional markets (including neighbourhood markets and city centre markets) were the most commonly used source across most cities, countries and SES. Small local shops were the second most used (note category limitations outlined above).

Table 10.8 Food sourcing and comparison between the lowest and highest SES quartiles

		Proportion of households (%) in each SES quartile using food source					
Food source (sorted by more common to less common)	*SES quartile (%)*	*Ghana*		*Kenya*		*Uganda*	
		Tamale (N = 1006)	*Techiman (N = 1028)*	*Kisumu (N = 1004)*	*Thika (N = 1004)*	*Mbale (N = 1025)*	*Mbarara (N = 970)*
Informal market, wet market, central or neighbourhood market, street food	0–25	71	92	96	97	77	86
	76–100	81*†	97*	94	94	87**	95**
Small shop, restaurant, takeaway	0–25	42	27	80	95	66	77
	76–100	56**	41**	81	90*	67	76
Food transfers from rural- or urban-based relatives or friends	0–25	31	33	53	41	46	39
	76–100	31	32	48	47	55*	53**
Grow it	0–25	48	40	30	34	37	29
	76–100	40	40	41*	39	43	47***
Supermarket	0–25	21	10	30	45	12	16
	76–100	47***	22***	73***	77***	34***	45***

†Chi-square statistics computed from Mann Whitney U test checking for difference between 1st and 4th quartiles

*Significant at 5% confidence level

**Significant at 1% confidence level

***Significant at 0.1% confidence level

NB The 2nd and 3rd quartiles are not shown for purposes of clarity and simplicity

The category of central or neighbourhood market was the main source of food in all the cities, closely followed (in Kenya and Uganda in particular) by buying food from the nearest small shop (Table 10.8). Approximately 30–40% of households across all cities and countries were involved in growing some of their own food and this remained stable across SES (the second and third SES quartiles not shown). The likelihood of purchasing food from supermarkets increased with rising SES, although remaining fairly low (between 22 and 47% of the households in the highest SES quartile in both the Ugandan and Ghanaian cities). Kenyans, across all SES but especially in the highest SES, more commonly sourced food in supermarkets than Ghanaians or Ugandans. However, only a small proportion of the lowest SES households purchased food from supermarkets, especially in Techiman and Mbale. There may be a city-size or outlet availability factor influencing this data. The proportion of the highest SES quartile of households in all cities, except Thika, that purchased food from supermarkets, was more than double the proportion of the lowest SES households, but this better-off SES also continued to source food from the traditional markets (and even increasingly so in Ghana and Uganda).

Food transfers from friends and relatives were a source of food for between approximately 31% and 55% of households (Table 10.8). While in the Ghanaian and Kenyan cities the proportion receiving food transfers remained relatively similar across SES, in the Ugandan cities higher SES households more commonly received food transfers than lower SES households. This trend in Uganda has been discussed by other researchers as relating to the tendency for urban-based Ugandans to invest any spare income into rural land and rural-based farm relationships (Kangave et al., 2016; Reid, 2017). T. S. Jayne et al. (2016) found a similar situation in Kenya, where 36% of urban households in 2014 owned agricultural land, with the corresponding figure in Ghana being just 12%.

Finally, we analysed the BMI data for the Ugandan cities by SES quartiles (Table 10.9). These findings contrast with claims that body mass and problems with overweight and obesity increase with wealth (May, 2018). Instead, they show all SES being affected by overweight and obesity, in Mbarara in particular. This data is also striking when assessed against the household dietary and food insecurity status of the cities shown in previous tables.

Table 10.9 Body Mass Index category[a] prevalence by SES quartile (male and female)

City	*BMI category*	*SES quartile*			
		Low (0–25%)	*Mid-low (26–50%)*	*Upper-mid (51–75%)*	*Upper (76–100%)*
Mbarara (*N* = 948 adults)	Underweight	3.4	2.9	4.1	1.2
	Normal weight	46.6	48.8	46.7	43.5
	Overweight	26.2	34.3	28.2	36.6
	Obese	23.8	14.0	21.0	18.6
Mbale (*N* = 1248 adults)	Underweight	6.9	5.4	4.7	1.9
	Normal weight	57.8	54.1	56.1	54.1
	Overweight	27.0	28.1	30.7	27.4
	Obese	8.3	12.4	8.5	16.5

[a]In accordance with WHO cut-off points of underweight: <18.5 kg/m^2; normal weight: 18.5–24.99 kg/m^2; overweight: 29.99 kg/m^2; obese: 30 kg/m^2 and above (WHO, 2006)

Factors Influencing Household Food Security

The modelling of food security is shown in Table 10.10. It indicates that the SES of a household had the largest magnitude and most significant effect on household food security across the six cities. For example, in the Ghanaian cities, with other factors held constant, a household in the upper SES group had an odds ratio of 3.5, meaning it was 3.5 times more likely to be food secure compared to a low SES household. The upper-mid and mid-lower households were 1.9 and 1.4 times, respectively, more likely to be food secure than the lowest SES households. This pattern and magnitude were broadly similar in both of the Kenyan and Ugandan cities (Table 10.10).

Household involvement in agriculture had varying influence on household food security in the three countries. There was a positive and statistically significant relation between agriculture and food security only in Ghana, and only in households that had been involved in *both* rural and urban agriculture during the preceding year. Such Ghanaian households were 1.8 times more likely to be food secure than those that did not practise any agriculture. The only other statistically significant implication of agriculture on food security was in Uganda where the effect was negative, opposite to that in Ghana. In Uganda the households that engaged *only* in urban agriculture were 0.63 times less likely to be food

Table 10.10 Logistic regression of food security

		Ghana		*Kenya*		*Uganda*	
		Coefficient	*Exp (B) (Odds ratio)*	*Coefficient*	*Exp (B) (Odds ratio)*	*Coefficient*	*Exp (B) (Odds ratio)*
Number of observations		1743		1915		1895	
Constant		−0.631*	0.532	−2.310***	0.099	−1.364***	0.256
Explanatory variable							
SES quartile (base is low income, 1st quartile)	Mid-lower (2nd quartile)	0.333*	1.396	−0.173	0.841	0.391**	1.479
	Upper-mid (3rd quartile)	0.626***	1.871	0.375*	1.455	0.402**	1.495
	Upper (4th quartile)	1.254***	3.506	0.753***	2.123	0.976***	2.653
Income diversity	Number of income sources	0.094	1.098	0.046	1.047	0.067	1.070
Socio-economic characteristics	Household structure: multiple adults (base is single adult)	0.033	1.033	−0.111	0.895	0.212*	1.236
	Gender of head (base is female)	0.072	1.075	−0.086	0.918	0.161	1.175
	Work status of HH head (base is not working)	0.103	1.109	0.108	1.114	0.008	1.008
	Age of HH head	0.001	1.001	0.008**	1.008	0.002	1.002
Food transfers from urban or rural friends or relatives		0.205	1.227	0.040	1.040	−0.169	0.844
Engagement in agriculture (base is no agriculture)	Rural agriculture	−0.200	0.819	0.018	1.019	−0.095	0.910
	Urban agriculture	0.154	1.166	0.101	1.107	−0.456**	0.634
	Both urban and rural agriculture	0.569**	1.767	0.128	1.137	−0.211	0.810
City	Techiman	−0.064	0.938		–	–	
	Thika	–		1.423***	4.150	–	

		Ghana		*Kenya*		*Uganda*	
		Coefficient	*Exp (B) (Odds ratio)*	*Coefficient*	*Exp (B) (Odds ratio)*	*Coefficient*	*Exp (B) (Odds ratio)*
	Mbarara	–			–	0.479***	1.615
Measures of model fit	Nagelkerke R^2	0.079		0.152		0.068	
	−2Log likelihoods	2260		2165		2436	
	χ^2 ($P > 0.001$)	105.749***		219.133***		96.829***	

Dependent variable: is household food secure? Yes/No
*Significant at 5% confidence level
**Significant at 1% confidence level
***Significant at 0.1% confidence level
NB Description of variables used in the regression model is shown in Table 10.3

secure compared to households that did not practise agriculture (Table 10.10).

Other country contrasts appear in the model, but most were not highly significant or had a small magnitude effect, such as the age of the household head in Kenya, or being more food secure in multiple adult households (1.2 times more likely) in Uganda compared to single adult households. City differences were significant in Kenya and Uganda, where those living in Thika had a 4.1 times greater likelihood of being food secure than those who lived in Kisumu (Table 10.10). Similarly, those living in Mbarara were 1.6 times more likely to be food secure than those living in Mbale.

The clearest message from Table 10.9, however, across these three countries and six secondary cities, was that our SES quartile was most significant and had the greatest magnitude of influence on whether a household felt food secure.

Factors Influencing Household Dietary Diversity

The Poisson regression modelling of household dietary diversity is presented in Table 10.11.

Similar to the findings for factors influencing food security, the analysis of factors influencing household dietary diversity in Table 10.11 indicates that, across all three countries and all six secondary cities, being in a higher SES quartile household was significantly and positively associated with better dietary diversity. Such a finding is not unexpected and fits with the literature. To specify, other factors held constant, being in an upper SES household increased the HDDS by 187% in Ghana, by 140% in Kenya and 133% in Uganda compared to the lowest SES quartile. In addition, the diversity of income sources was significantly associated with an increase in dietary diversity, with the effect being 51% and 32% in Uganda and Ghana, respectively, but just 12% in Kenya (Table 10.11).

Yet, a number of other explanatory variables seemed to influence household dietary diversity in comparison with the influences on household food security (previous model, Table 10.10), and there was more variation across the countries on which factors had a significant influence on dietary diversity in comparison with the influences on food security. City variations in modelling dietary diversity were also apparent, significant in all three countries: being a resident of Techiman increased

Table 10.11 Poisson regression model for household dietary diversity score (HDDS)

		Ghana		*Kenya*		*Uganda*	
Variable		*Coefficient*	*Std. error*	*Coefficient*	*Std. error*	*Coefficient*	*Std. error*
Number of observations		1748		1918		1905	
	Constant	3.523***	0.268	5.030***	0.193	2.801***	0.217
SES quartile (base is low-income1st quartile)	Mid-lower (2nd quartile)	0.572***	0.150	0.478***	0.108	0.505***	0.127
	Upper-mid (3rd quartile)	1.226***	0.151	0.914***	0.108	0.773***	0.128
	Upper (4th quartile)	1.865***	0.153	1.396***	0.112	1.329***	0.132
Income diversity	Number of income sources	0.318***	0.068	0.117*	0.048	0.517***	0.059
Socio-economic characteristics	Household structure: multiple adults (base was single adult)	0.232	0.124	0.180*	0.083	0.140	0.095
	Gender of HH head (base is female)	0.085	0.120	0.321**	0.094	0.050	0.109
	Work status of HH head (base is not working)	−0.004	0.158	0.074	0.088	0.019	0.092
	Age of household head	0.006	0.003	0.008**	0.002	0.001	0.003
Food transfers from urban or rural friends or relatives	Received food transfers	0.266*	0.117	0.231**	0.077	−0.010	0.092
Engagement in agriculture (base was no agriculture)	Rural agriculture	−0.493**	0.150	−0.288**	0.094	0.139	0.110
	Urban agriculture	−0.277	0.143	−0.211	0.108	0.009	0.128

(continued)

Table 10.11 (continued)

Variable		*Ghana*		*Kenya*		*Uganda*	
		Coefficient	*Std. error*	*Coefficient*	*Std. error*	*Coefficient*	*Std. error*
	Both rural and urban agriculture	−0.466*	0.205	−0.304*	0.154	0.243	0.160
City	Techiman	0.782***	0.113				
	Thika			0.694***	0.077		
	Mbarara					0.907***	0.092
Measures of model fit	Likelihood ratio χ^2	288***		278***		321***	

Dependent variable: HDDS with 12 food groups (values between 0 and 12)
*Significant at 5% confidence level
**Significant at 1% confidence level
***Significant at 0.1% confidence level
NB Description of variables used in the regression model is in Table 10.3

household dietary diversity by 78% compared to Tamale, while households in Thika were 69% more likely to have a higher dietary diversity score than their Kisumu counterparts. Mbarara households had 90% higher HDDS than Mbale households (Table 10.11), other factors being held constant. In the Kenyan cities, the household structure, gender and age of the household head had significant and positive effects on the HDDS. A Kenyan household with multiple adults increased dietary diversity by 18% compared to households with single adults. Additionally, male-headed households had 32% higher dietary diversity than female-headed households (Table 10.11).

Of interest was the relation of agriculture with dietary diversity. The engagement of households in agricultural activities had a statistically significant and *negative* effect on dietary diversity in both the Ghanaian and Kenyan cities, but no effect in Uganda (Table 10.10). Those households that had engaged in rural agriculture during the preceding year had a lower dietary diversity score by 49 and 29% in Ghana and Kenya, respectively. In the same countries, engagement in both urban and rural agriculture reduced dietary diversity by 47 and 30%, respectively (Table 10.10).

The Complexity of Change in Secondary City Food Systems

In this chapter, we have analysed how diet composition and diversity, food sources and food security varied by the SES of a household (a proxy for well-being calculated from monthly and annual expenditure data and considering the age structure and composition of household members) in six secondary cities across Ghana, Kenya and Uganda. We find that food security and dietary diversity improved with increasing SES. This finding is supported by numerous other studies, for example, in a recent review of urban food environment change in Africa by Michelle Holdsworth and Edwige Landais that also highlights the importance of "wealth" (Holdsworth & Landais, 2019) and Tacoli's study noting the importance of "income" (Tacoli, 2017). We suggest, more precisely, that what is important here is actually cash availability, that is disposable income. Such a growth in disposable income is why we think that having multiple income sources was positively correlated with dietary diversity.

From the modelling, we find an association of better food security and higher dietary diversity with greater SES, across all three SSA countries and all six secondary cities. The SES variable alone accounted for a significant magnitude of the difference in the experience of food security in all contexts: the more money a household spent, the more food secure they felt. The significantly lower level of food security found in Kisumu in comparison to Thika (Table 10.5) was expected, since Kisumu is characterized by a high number of informal settlements and is a net food importer and thus food is more expensive, as noted in the introduction and found by Wagah et al. (2018).

The effect of engagement in agriculture (either in rural areas or within an urban area, or both) by urban-based households did not have a statistically significant positive impact on food security in our model (Table 10.10), except in Ghana and then only for households who farmed in both a rural and an urban area (thus households more likely to have more resources, assets and/or contacts). In Uganda, households that practised urban agriculture had a negative association with food security (Table 10.10). Although these findings do not necessarily provide strong support for the thesis that own crop or livestock production improves food security, the practice should not be regarded as unimportant. As shown already in Table 10.8, approximately 30–40% of all households across all SES groups noted growing their own food as an important food source, and authors' previously published studies in these cities note the common and persistent engagement in agriculture across SES, with indications that this even increases with SES whereby the better-off urban residents invest in agriculture as a diversification and a livelihood security strategy (Mackay, 2019a; Omondi et al., 2017; Turner & Jirström, 2014). Had farming households not engaged in agriculture, their food security status would likely have been much worse.

Considering the modelling of the factors influencing household dietary diversity, we found additional important influences beyond our SES proxy. Consistent across all countries was the role of the number of income sources. For every numeric increase in income sources, a household experienced a 12% increase in HDDS in Kenya, 32% in Ghana and a full 52% in Uganda (Table 10.11). This was after controlling for the number of adults and dependents in the household, the number of working-age members and other socio-economic and geographic factors. Households with more diversified income sources are likely more cushioned from shocks, job loss or other such change in circumstance, thus they might be expected to have been able to purchase a greater variety of food groups (resulting in higher dietary diversity scores) than those with fewer income sources. Why the role of having multiple income sources seems to have lower magnitude in the Kenyan context in comparison with the Ghanaian and Ugandan situations is intriguing and would benefit from further investigation.

The other, perhaps somewhat counter-intuitive finding relating to the modelling of household dietary diversity was the negative relationship with engagement in agriculture in Ghana and Kenya, and the non-significant effect of agriculture on dietary diversity in Uganda. Our contextual experience and other studies (Mackay, 2018, 2019a, 2019b), together with findings from other literature describing a varied and uneven link between farm production diversity and household dietary diversity (Sibhatu & Qaim, 2018), support the case that the lack of positive relation between engaging in agriculture and higher dietary diversity is likely because much of this agricultural activity is concentrated around producing the most commonly consumed food groups, in particular maize, cooking bananas, sweet potatoes and green leafy vegetables. Consumption of this agricultural produce would not broaden the dietary diversity score, since these are the most commonly eaten food groups across all contexts and all SES, and even for households not engaging in agriculture.

In summary, our findings do indicate a small decline in food insecurity and a more considerable increase in dietary diversity with a rise in SES, as theory and many other studies, such as the work of Haggblade et al. (2016), and Popkin et al. (2012; Popkin, 2001, 2015) predict. As Table 10.7 shows, the difference in mean food insecurity scores from the lowest to the highest SES amounted to 3–4 HFIAS points, and the difference in mean household dietary diversity between the lowest and the highest SES amounted to two (out of a total of 12) food groups across all cities and countries. Greater cash availability and a general increase in household circumstances (both indicated in our SES quartile variable) do seem to translate into more food secure circumstances and a greater diversity of food groups being consumed (not necessarily directly equivalent to healthy foods, as previously noted).

Yet our findings on the sourcing of food, the continuing use of the traditional marketplaces even for better-off households, the similar level of receipt of food transfers regardless of SES and the consistent engagement (even slight increase) in agriculture with greater SES may nuance how ideas of nutrition transition, or of the supermarketization of Africa, of fundamental shifts in urban food systems, of the declining relevance of land to farm, or of increasing separation of urban and rural livelihoods play out in the SSA urban context (Haggblade et al., 2016; May, 2018; Popkin, 2001, 2015; Reardon et al., 2003). Our Kenyan cities, however, do show a greater patronage of more Western-style supermarket shopping than our Ghanaian and Ugandan cities. Our focus on secondary cities may explain some of these differences in food systems. Our secondary study cities were of a rather small size compared to many capital cities or megacities, which are often the focus of research. They also lacked, except in Kisumu, the large concentrations of abject poverty (slum areas) often to be found in capitals and megacities. Our secondary cities also, due to their smaller areal spread, their infrastructure and linkages, may also allow easier access to the peri-urban and nearby rural areas than capital and megacities. These factors may signify some reasons for a possible modification of trajectories of food system change compared to those identified from large cities or from national aggregate data. However, as Reardon et al. (2021) recently note, the traditional small-scale retail and food service sectors, which our study concurs are dominant food sources in these SSA cities, are beginning to supply more ultra-processed (packaged and unpackaged) food. Our study shows that small neighbourhood stores are an important food source, and previous work by Mackay (2019a) also notes that local shop owners, with whom an individual has built a personal relationship, play a crucial role in providing food access on credit in times of stress. Thus, as Reardon et al. (2021) also note, research investigations and policy interventions should not focus only on supermarkets and large-scale processors/retailers but also consider the role of small shops and small and medium enterprises in SSA in food system change.

In addition, our descriptive data on consumption of the different food groups by SES quartile generally show an increase in consumption of all food groups with rising SES. In contrast to some of the postulations reported in May's (2018) analysis of keystones that are reconfiguring African food systems, we do not see evidence that consumption of staple foods declines with improved circumstances in our city/country contexts (these rather increased, Table 10.5). May's paper does recognize that consumption patterns will vary by country context and tradition. However, our data is limited by not being able to distinguish between different types of staple cereals and grains (more nutritious millets and sorghums to less nutritious or polished wheats or rice), nor the level of polishing and processing. Reardon et al. (2021), in their overview of African food system change, also point out the importance of distinguishing between types of staples, their nutrient content and their degree of processing.

Cross-city, cross-country comparison of the food consumption data by SES revealed similar broad patterns and trends of rising consumption with rising SES, despite the countries being positioned differently regarding their stage of economic development and food system transformation in the literature (Haggblade et al., 2016). However, some country differences were apparent, with the Ugandan cities reporting significantly less consumption of fruits and vegetables (remaining low even at the highest wealth quartile) than their Ghanaian (specifically Techiman) and Kenyan counterparts, the Ghanaian cities reporting higher consumption of fish than the Kenyan and Ugandan households, and the Kenyan households reporting significantly higher consumption of lower nutritional value foods (Table 10.5). While some of these differences may be a feature of data collection variations, some are indicative of different socio-cultural context and traditional dietary behaviours. The higher use of oils/fats, sugar/sweeteners and condiments, tea and coffee in Thika and Kisumu deserves further investigation.

Finally, although there was insufficient space within this chapter to analyse links to obesity or other NCDs in detail, and we did not have such data for our Ghanaian and Kenyan samples, our findings should be viewed together with the obesity/NCD data that were presented in the introduction and within Table 10.9. Our findings from the BMI data in Uganda do not strongly support claims that body mass rises with wealth and related overconsumption of particularly less healthy foods (Table 10.9). This data is striking when viewed against the household dietary and food security status of these cities presented earlier, showing more than half of households experiencing food insecurity (Table 10.4) and dietary diversity scores averaging 3–6 across all SES (Table 10.6) in Mbale and Mbarara. Viewing these BMI data also against the responses gathered in relation to our additional questioning regarding the consumption of sugar-sweetened beverages, fried foods, fast foods, regular snacking and eating out in the Ugandan cities revealed these to be uncommon habits (Mackay et al., 2018). These findings do not match well with postulations from some research on nutrition transition and food system transformation of keystone factors contributing to rising NCD experience, such as those framed by May (2018). Our findings support others who have raised the importance of understanding the social environments, local perceptions and contextual factors and of the need for deep qualitative investigation (Holdsworth & Landais, 2019; Mackay, 2020).

Conclusion

Food insecurity in secondary cities in Ghana, Kenya and Uganda is a serious challenge. In the six cities studied, a large share—between 40 and 80% (Table 10.5)—of the households felt themselves to be either moderately or severely food insecure. The Ghanaian cities were more food secure than the Kenyan (Kisumu in particular) and Ugandan cities. While variation among households was found to be clearly associated with their SES, the overall picture was some degree of food insecurity being an experiential reality, even for residents in the better-off strata.

Food security and having a diverse diet are multidimensional and require a multipronged approach, yet we find cash availability is one of the most important drivers of food security, in agreement with findings from other researchers (Holdsworth & Landais, 2019; May, 2018; Tacoli, 2017; Tschirley et al., 2015). Thus, efforts must be made to ensure that populations are able to have a reliable and liveable source of income or—even more importantly in the case of dietary diversity—multiple sources of income. This links to labour market conditions and employment and entrepreneurial opportunities, and the conditions for informal livelihoods (Mackay, 2019a; Tacoli, 2017). Important policy implications here thus relate to removing restrictions on, or punishments for, informality; working creatively with informal livelihoods; as well as trying to influence food consumption patterns by nudging consumer behaviour through subsidizing healthier food products; or by taxation and hierarchical marketing charges based on nutritional content, as Holdsworth and Landais (2019) and Reardon et al. (2021) also note.

While the effects of own food production (crop or livestock) on food security and dietary diversity was weak in our models, this does not mean it is not important. The food security situation of those engaged in agriculture would likely have been worse had they not farmed. Policies supportive of small-scale farming, as well as those increasing the possibility for multiple livelihood opportunities, offer a strategy to combat food insecurity and may encourage more diverse diets.

Our study provides a detailed analysis of how food system and nutritional change are manifesting at the household level in secondary cities of three SSA countries at slightly different stages of development. While we find processes in line with key tenets of the concept of food system and nutritional transition, we also find difference in terms of food sourcing strategies and eating behaviours. We see reason to be cautious (from the

Ugandan data) about making direct causal claims regarding consumption change and obesogenic urban environments being the major contributors to a rising obesity and NCD burden in these developing countries. There needs to be an awareness among researchers, planners and decision-makers of the wider social and macro-environmental determinants of food environments and of health and disease than simply individual behaviours (food consumption) (Holdsworth & Landais, 2019). There is a need to recognize how poverty, insecure livelihoods and "unequal and unjust socioeconomic and health systems" (Mackay, 2020, p. 13) operate, together with features of the built environment, and in interaction with an individual's (historic and present) experience of hunger and undernutrition (Barker's development origins theory), to influence NCD expression *over and above* current food intake. A holistic perspective is essential.

Acknowledgements The authors would like to thank AFSUN for permission to use the AFSUN Household Food Security Baseline Survey instrument as part of this research. We would especially extend our gratitude to Cameron McCordic, Balsillie School of International Affairs, Canada, for coordinating the survey in Ghana and Kenya and his training to Heather Mackay to run the survey in Uganda. Our thanks go also to Frank Mugagga in Uganda, Daniel Bruce Sarpong and Fred Dzanku in Ghana, and Willis Oluoch-Kosura and Stephen Wambugu in Kenya for support in running the field surveys. We also thank Hayford Ayerakwa for feedback on an early draft. Last but not least, we are grateful to Agneta Hörnell, from the Department of Food, Nutrition and Culinary Science, Umeå University, Sweden, for comments and advice.

Funding Sources This work was supported by the Swedish International Development Cooperation Agency [grant number SWE-2011-028] and the Swedish Research Council for Environment, Agricultural Sciences and Spatial Planning [grant numbers 250-2014-1227 and 225-2012-609].

References

Abrahams, Z., Mchiza, Z., & Steyn, N. (2011). Diet and mortality rates in Sub-Saharan Africa: Stages in the nutrition transition. *BMC Public Health, 11*(1), 801.

Ayerakwa, H. (2017). Urban households' engagement in agriculture: Implications for household food security in Ghana's medium sized cities. *Geographical Research Special Issue, 55*(2), 217–230.

Barker, D. J. (1997). Maternal nutrition, fetal nutrition, and disease in later life. *Nutrition, 13*(9), 807–813.

Battersby, J., & Watson, V. (2019a). The planned 'city-region' in the New Urban Agenda: An appropriate framing for urban food security? *Town Planning Review, 90*(5), 497–518.

Battersby, J., & Watson, V. (Eds.). (2019b). *Urban food systems governance and poverty in African cities*. Routledge.

Coates, J., Swindale, A., & Bilinsky, P. (2007). *Household food insecurity access scale (HFIAS) for measurement of household food access: Indicator guide (version 3)*. FHI 360/FANTA.

Cobbinah, P., Erdiaw-Kwasie, M., & Amoateng, P. (2015). Africa's urbanisation: Implications for sustainable development. *Cities, 47*, 62–72.

Cramer, J. (2003). *Logit models from economics and other fields*. Cambridge University Press.

Crush, J., & Battersby, J. (Eds.). (2016). *Rapid urbanisation, urban food deserts and food security in Africa*. Springer.

Crush, J., Frayne, B., & Pendleton, W. (2012). The crisis of food insecurity in African cities. *Journal of Hunger and Environmental Nutrition, 7*(2–3), 271–292.

Dodman, D., Leck, H., Rusca, M., & Colenbrander, S. (2017). African urbanisation and urbanism: Implications for risk accumulation and reduction. *International Journal of Disaster Risk Reduction, 26*, 7–15.

Fraser, A., Leck, H., Parnell, S., Pelling, M., Brown, D., & Lwasa, S. (2017). Meeting the challenge of risk-sensitive and resilient urban development in sub-Saharan Africa: Directions for future research and practice. *International Journal of Disaster Risk Reduction, 26*, 106–109.

Ghana Statistical Service (GSS). (2012). *2010 population and housing census final results*. Government of Ghana.

Haggblade, S., Duodu, K., Kabasa, J., Minnaar, A., Ojijo, N., & Taylor, J. (2016). Emerging early actions to bend the curve in sub-Saharan Africa's nutrition transition. *Food and Nutrition Bulletin, 37*(2), 219–241.

Holdsworth, M., & Landais, E. (2019). Urban food environments in Africa: Implications for policy and research. *Proceedings of the Nutrition Society, 78*(4), 513–525.

Howe, L., Hargreaves, J., & Huttly, S. (2008). Issues in the construction of wealth indices for the measurement of socio-economic position in low-income countries. *Emerging Themes in Epidemiology, 5*(1), 3.

International Diabetes Foundation (IDF). (2019). *IDF diabetes Atlas Ninth Edition 2019*. International Diabetes Foundation.

Jayne, T., Chamberlin, J., Traub, L., Sitko, N., Muyanga, M., Yeboah, F., Anseeuw, W., Chapoto, A., Wineman, A., Nkonde, C., & Kachule, R. (2016).

Africa's changing farm size distribution patterns: The rise of medium-scale farms. *Agricultural Economics, 47*(S1), 197–214.

Kangave, J., Nakato, S., Waiswa, R., & Zzimbe, P. (2016). *Boosting revenue collection through taxing high net worth individuals: The case of Uganda* (International Centre for Tax and Development [ICTD] Working Paper No. 45). Institute of Development Studies.

Kenya National Bureau of Statistics (KNBS). (2012). *2009 Kenya population and housing census-analytical report on urbanization* (Vol. VIII). Republic of Kenya.

Leroy, J., Ruel, M., Frongillo, E., Harris, J., & Ballard, T. (2015). Measuring the food access dimension of food security: A critical review and mapping of indicators. *Food and Nutrition Bulletin, 36*(2), 167–195.

Mackay, H. (2018). Mapping and characterising the urban agricultural landscape of two intermediate-sized Ghanaian cities. *Land Use Policy, 70*, 182–197.

Mackay, H. (2019a). Food sources and access strategies in Ugandan secondary cities: An intersectional analysis. *Environment and Urbanization, 31*(2), 375–396.

Mackay, H. (2019b). *Food, farming and health in Ugandan secondary cities* (PhD dissertation). Department of Geography, Umeå University.

Mackay, H. (2020). Of fatness, fitness and finesse: Experiences and interpretations of non-communicable diseases in urban Uganda. *Cities and Health* (published online).

Mackay, H., Mugagga, F., Kakooza, L., & Chiwona-Karltun, L. (2018). Doing things their way? Food, farming and health in two Ugandan cities. *Cities and Health, 1*(2), 147–170.

May, J. (2018). Keystones affecting sub-Saharan Africa's prospects for achieving food security through balanced diets. *Food Research International, 104*, 4–13.

Mayega, R., Makumbi, F., Rutebemberwa, E., Peterson, S., Östenson, C., Tomson, G., & Guwatudde, D. (2012). Modifiable socio-behavioural factors associated with overweight and hypertension among persons aged 35 to 60 years in eastern Uganda. *PLoS One, 7*(10), e47632.

Monteiro, C., Moubarac, J., Cannon, G., Ng, S., & Popkin, B. (2013). Ultra-processed products are becoming dominant in the global food system. *Obesity Review, 14*, 21–28.

Ngaruiya, C., Hayward, A., Post, L., & Mowafi, H. (2017). Obesity as a form of malnutrition: Over-nutrition on the Uganda "malnutrition" agenda. *Pan African Medical Journal, 28*(1), Article 49.

Nichols, C. (2017). Millets, milk and maggi: Contested processes of the nutrition transition in rural India. *Journal of Agriculture and Human Values, 34*(4), 871–885.

Ofori-Asenso, R., Agyeman, A., Laar, A., & Boateng, D. (2016). Overweight and obesity epidemic in Ghana—A systematic review and meta-analysis. *BMC Public Health, 16*(1), 1239.

Omondi, S., Oluoch-Kosura, W., & Jirström, M. (2017). The role of urban-based agriculture on food security: Kenyan case studies. *Geographical Research, 55*(2), 231–241.

Popkin, B. (2001). The nutrition transition and obesity in the developing world. *Journal of Nutrition, 131*(3), 871S-873S.

Popkin, B., Adair, L., & Ng, S. (2012). Global nutrition transition and the pandemic of obesity in developing countries. *Nutrition Reviews, 70*(1), 3–21.

Popkin, B. (2014). Nutrition, agriculture and the global food system in low- and middle-income countries. *Food Policy, 47*, 91–96.

Popkin, B. (2015). Nutrition transition and the global diabetes epidemic. *Current Diabetes Reports, 15*(9), 64.

Raschke, V., & Cheema, B. (2008). Colonisation, the New World Order, and the eradication of traditional food habits in East Africa: Historical perspective on the nutrition transition. *Public Health Nutrition, 11*(7), 662–674.

Reardon, T., Timmer, C., Barrett, C., & Berdegue, J. (2003). The rise of supermarkets in Africa, Asia, and Latin America. *American Journal of Agricultural Economics, 85*(5), 1140–1146.

Reardon, T., Tschirley, D., Minten, B., Haggblade, S., Timmer, C., & Liverpool-Tasie, S. (2013). *The emerging 'quiet revolution' in African agrifood systems*. Brief prepared for Harnessing Innovation for African Agriculture and Food Systems: Meeting Challenges and Designing for the 21st Century; African Union Conference Centre; 25–26 November, Addis Ababa, Ethiopia.

Reardon, T., Tschirley, D., Liverpool-Tasie, S., Awokuse, T., Fanzo, J., Minten, B., Vos, R., Dolislager, M., Sauer, C., Dhar, R., Vargas, C., Lartey, A., Raza, A., & Popkin, B. (2021). The processed food revolution in African food systems and the double burden of malnutrition. *Global Food Security, 28*, 100466.

Reid, R. (2017). *A history of modern Uganda*. Cambridge University Press.

Satterthwaite, D. (2017). The impact of urban development on risk in sub-Saharan Africa's cities with a focus on small and intermediate urban centres. *International Journal of Disaster Risk Reduction, 26*, 16–23.

Satterthwaite, D., McGranahan, G., & Tacoli, C. (2010). Urbanization and its implications for food and farming. *Philosophical Transactions of the Royal Society B: Biological Sciences, 365*(1554), 2809–2820.

Sibhatu, K. T., & Qaim, M. (2018). Meta-analysis of the association between production diversity, diets, and nutrition in smallholder farm households. *Food Policy, 77*, 1–18.

Steyn, N., & Mchiza, Z. (2014). Obesity and the nutrition transition in sub-Saharan Africa. *Annals of the New York Academy of Sciences, 1311*(1), 88–101.

Swindale, A., & Bilinsky, P. (2006). *Household dietary diversity score (HDDS) for measurement of household food access: Indicator guide (version 2).* FHI 360/FANTA.

Tacoli, C. (2017). Food (in) security in rapidly urbanising, low-income contexts. *International Journal of Environmental Research and Public Health, 14*(12), 1554.

Tschirley, D., Reardon, T., Dolislager, M., & Snyder, J. (2015). The rise of a middle class in Eastern and Southern Africa. *Journal of International Development, 27*(5), 628–646.

Turner, C., Aggarwal, A., Walls, H., Herforth, A., Drewnowski, A., Coates, J., Kalamatianou, S., & Kadiyala, S. (2018). Concepts and critical perspectives for food environment research: A global framework with implications for action in low- and middle-income countries. *Global Food Security, 18*, 93–101.

Turner, C., & Jirström, M. (2014). Staple foods in urban and peri-urban agriculture. In U. Magnussson & K. Follis Bergman (Eds.), *Urban and peri-urban farming in low-income countries—Challenges and knowledge gaps* (pp. 37–41). SLU Global.

Uganda Bureau of Statistics. (2016). *National population and housing census 2014.* Uganda Bureau of Statistics. Main Report.

United Nations (UN). (2020). *Global human development indicators.* United Nations Development Programme.

Wagah, G., Obange, N., & Okello Ogindo, H. (2018). Food poverty in Kisumu, Kenya. In J. Battersby & V. Watson (Eds.), *Urban food systems governance and poverty in African cities* (pp. 223–235). Routledge.

World Health Organization (WHO). (2006). *Global database on body mass index: BMI classification.* http://www.assessmentpsychology.com/icbmi.htm.

World Health Organization (WHO). (2020). *Global health observatory (GHO) data.* https://www.who.int/data/gho/info/gho-odata-api.

Wooldridge, J. (2013). *Introductory econometrics: A modern approach* (5th ed.). Nelson Education Canada.

World Bank. (2016). *New country classifications by income level: 2016–2017.* World Bank Data Team.

CHAPTER 11

Household Dietary Patterns and Food Security Challenges in Peri-Urban South Africa: A Reflection of High Unemployment in the Wake of Rising Food Prices

Gamuchirai Chakona

Introduction

Urbanization is one of the major social changes sweeping the globe, with rapid growth of the urban population and stagnating growth of the rural population, especially in developing regions (UN-Habitat, 2020). In sub-Saharan Africa, rapid urbanization and poverty are the major fundamental development challenges that are perpetuating and deepening the crisis of food and nutrition insecurity in urban areas (Battersby, 2012). Many people living in urban areas face under-nutrition, mainly due to their lack

G. Chakona (✉)
Environmental Learning Research Centre (ELRC), Department of Education, Rhodes University, Makhanda/Grahamstown, South Africa
e-mail: G.Chakona@ru.ac.za

L. Riley and J. Crush (eds.), *Transforming Urban Food Systems in Secondary Cities in Africa*,
https://doi.org/10.1007/978-3-030-93072-1_11

of income rather than to a lack of capacity to produce food (Satterthwaite et al., 2010). The health and nutritional status of urban populations with very low incomes are at risk from rising prices in staple foods. This became evident with the rising hunger among urban populations after the food price rises in 2007 and the first half of 2008 (Cohen & Garrett, 2009). In South Africa, poverty, unemployment and high food prices are dominant and influence dietary change, which subsequently increases urban food insecurity and malnutrition (Battersby, 2012).

Although South Africa is food secure at a national level, large numbers of households within the country are food insecure. Statistics South Africa (2019) has reported that about 6.8 million South Africans experienced hunger (not having enough food intake) in 2017, with 1.7 million households across the country being affected. Due to low incomes, about 30–50% of the population have insufficient food or consume imbalanced diets (Hendriks, 2014). More than 43% of households are vulnerable to poverty, with almost 28% living in extreme poverty, below the food poverty line (SSA, 2017). With more than two-thirds of the South African population living in urban areas (SSA, 2014), and food prices and unemployment rates continually increasing (SSA, 2019), chronic poverty and household food insecurity become worse as the informal safety nets are weakened, especially in urban areas (Drimie et al., 2013).

According to the 2018 Global Nutrition Report (Development Initiatives, 2018), South Africa is among those countries with a triple burden of malnutrition, where high rates of childhood stunting, anemia and overweight in adult women are dominant. This is mainly because poor people tend to cope with poverty by adopting unvaried diets mainly of starch staples, such as maize-meal, with limited vegetable and fruit intake (Chakona & Shackleton, 2017, 2018; Schönfeldt et al., 2010). The triple burden of malnutrition is especially acute in urban areas, where the greater proportion of the population are more reliant on food purchases and tend to consume energy-rich and processed foods, including refined grains and foods higher in saturated fat, sugar and salt. These foods are cheap to purchase, but can lead to individuals being overweight (Drimie et al., 2013).

Poor households who depend on less formal means of securing food, such as from the informal markets, are the most vulnerable to nutrition insecurity. They mostly rely on informal social safety nets (Battersby, 2011). For example, in many cases, purchased food constitutes up to 90%

of household diets, accounting for 60–80% of total household expenditure (Chakona & Shackleton, 2017). Due to reduced purchasing power, "the poor are facing higher food prices but no greater income, and they begin to starve" (Sen, 2008). In South Africa, the estimated average income of the poor is less than R524 (US$35.70) per month per person (NPC, 2012) and a quarter of the population live below the food poverty line of R561 (US$38.22) per month (SSA, 2019). This has a negative impact on a large proportion of households already vulnerable to food insecurity, thereby increasing hunger and malnutrition (Altman et al., 2009).

This situation undermines the efforts to have a country with zero hunger and good health and well-being for all by 2030 and thus has dire consequences for most of the Sustainable Development Goals (SDG) targets. For example, food insecurity has a negative influence on good health and well-being (SDG 3), as well as human development, through its effects on nutrition and as a component of overall family stress. Furthermore, with good nutrition and living healthy lives, adults may not lose employment due to illness, which may form a vicious cycle of poverty and food insecurity. With better nutrition, children may experience less sickness and therefore have more days in school, increasing their chances of attaining higher academic achievement (SDG 4). A good quality education increases their intellectual growth and knowledge; with excellent academic grades, they can get decent employment, thus increasing the nation's economic growth (SDG 8). This in turn may reduce poverty (SDG 1), leading to reduced inequalities in the country (SDG 10), which are the major challenges South Africa is currently facing.

Food insecurity has often misleadingly been seen as an issue that affects only rural populations, especially in South Africa, where studies have commonly noted the highest levels of food insecurity in these settings (De Cock et al., 2013). There has been a strong bias in promoting food security programs among rural populations (Crush & Frayne, 2010, 2011), including support of smallholder agriculture, and neglecting urban populations. Yet, due to increasing urbanization and rising food prices, food insecurity is increasingly prominent in urban settings, especially within informal peri-urban areas (Chakona & Shackleton, 2017, 2018; Crush & Frayne, 2010). With two-thirds of the South African population living in urban areas and the majority being dependent on cash income to purchase food, increases in food prices directly affect their food security status. In South Africa, food prices have been on the rise since 2015, forcing poor urban dwellers to consume lower quantities of food and adopt diets that are cheap, affordable and of low nutritional quality (Jonah & May, 2020). Despite high levels of poverty and rising food prices, urban agricultural practices are limited in poor urban households. Many households rely on social grants from the government (Chakona & Shackleton, 2019), while limited access to land, labour and inputs constrain the contribution of agriculture to household food security (Crush et al., 2011). As a result, malnutrition remains one of the major nutritional disorders in the country (Chakona & Shackleton, 2018), which prevents the country from achieving the 2030 Agenda for Sustainable Development Goals.

The prime causes of household food insecurity in South Africa are widespread chronic poverty, rising food prices and unemployment, which also weaken informal safety nets, especially in urban areas (Drimie et al., 2013). In South Africa, most households are female-headed, and urban food insecurity is a gendered problem (Caesar & Riley, 2018) related to higher poverty levels (57.2%) among women than their male counterparts (53.7%) (SSA, 2017). Women are generally more vulnerable to food insecurity, in part because the responsibility to make sure the entire household is fed is typically assigned to women. When faced with inadequate income, women who are household food managers trade off food quality for quantity in their own diets to prevent household members, especially children, from feeling persistently hungry (Kuku et al., 2011). Monotonous diets, based mainly on grains and lacking vegetables, fruits and animal-source foods, dominate their diets. These diets are closely associated with food insecurity and malnutrition (Kennedy, 2009). Some women may rely on less expensive foods that are energy-rich but nutrient-poor (Chakona & Shackleton, 2017), thus jeopardizing their health.

While there has been some research in larger urban centres in South Africa (Caesar & Riley, 2018; Drimie et al., 2013; Oldewage-Theron et al., 2006), information is scarce on dietary patterns and food security status among female-headed households in peri-urban settlements surrounding South Africa's smaller urban settlements. This chapter analyzes household food insecurity and dietary diversity of households residing in the peri-urban areas of small urban centers in South Africa, where higher poverty levels limit households' ability to access food. The study examined and compared the determinants of food security among male- and female-headed households in these settlements, focusing on household employment and income status, food expenditure, geographical location, household size, educational level and access to food. The novelty of this study is in using a large sample size to carry out comparative analyses on food security of households in peri-urban settlements in four urban centers across three South African provinces.

Study Area

The study was carried out in Richards Bay, Dundee, Harrismith and Balfour. Richards Bay in Kwa-Zulu Natal Province (population 57,000) is characterized by highly developed urban areas surrounded by poor and undeveloped peri-urban and rural areas. The municipality is faced with high levels of unemployment (40%), mainly due to lack of skills and widespread poverty, especially in peri-urban and rural areas (uMhlathuze Municipality, 2012). Average annual household income for the municipality is R121,177 (US$8,243.67) and 15.2% of people do not have any source of cash income (SSA, 2012). The majority of the population are poorly educated: 52.4% of adults do not have secondary qualifications and only 8.5% have higher education. Most households are male-headed (57.7%) and 42.3% are female-headed (SSA, 2012).

Dundee in Kwa-Zulu Natal Province (population 35,000) has a well-developed urban area supporting the agricultural, tourism and manufacturing economic base of the district (uMzinyathi District Municipality, 2012). The municipality is predominantly urban with only 16.8% of the population living in non-urban areas. Dundee has significantly higher educational levels, with about 31.9% of people having completed secondary education and only 7% with no education. However, there are high levels of unemployment in the municipality (26.4%), with the unemployment rate among economically active youth (15–34 years) being 36.2%. Average annual household income for the Endumeni municipality is R87,430 (US$5,848.30) (SSA, 2012) and about 12.4% of households do not have any cash income (SSA, 2012). More than 60% of the population are living below the poverty line. Most households in the municipality (59.3%) are male-headed and only 40.7% are female-headed (SSA, 2012).

Harrismith in the Free State Province (population 28,000) has a central location in relation to the large national cities of Johannesburg, Durban and Bloemfontein. Its economy is mainly influenced by various forms of movement flows, transport and related industries (Maluti-a-Phofung Municipality, 2014). Harrismith is in the most poverty-stricken area in the Free State, with approximately 60% of households earning below

R1,650 (US$112.33) monthly and 40% earning an average monthly income of R600 (US$40.85) (Maluti-a-Phofung Municipality, 2014). Approximately 13.5% of households have no cash income (SSA, 2012). Education levels are low: Only 26.8% of people have secondary qualifications; 7.9% have higher education; and 9% have no education. The unemployment rate is high (41.8%), and among the economically active youth (15–34 years) the rate is 53% (SSA, 2012). The majority of the population depends on subsistence farming and backyard gardens for their food supply and livelihood. Half of the households in the municipality are female-headed (50.9%).

Balfour is a town situated along the Kat River Valley in Raymond Mhlaba municipality in the Eastern Cape Province. The area is home to about 50,000 people, mainly IsiXhosa (84%) (SSA, 2012). Education levels are poor, with approximately 9.7% of the population having completed secondary education (SSA, 2012). The area is characterized by extremely low formal employment rates, with 90% of households receiving one or more monthly old age pension or disability grants of R1,570 (US$105) or monthly child support grants of R480 (US$32.50) (SSA, 2012). There are high unemployment rates and government social grants are the most common sources of cash income. The majority of the households are female-headed and many households rely on food purchasing.

Research Methods

Data were obtained through administering questionnaires to randomly selected households within different neighborhoods in each town. A total of 78 households in Balfour, 76 in Richards Bay; 75 in Dundee and 85 in Harrismith participated in the surveys. All interviews were conducted with the person who prepares most of the meals in their preferred language of IsiXhosa (Balfour), IsiZulu (Richards Bay and Dundee) and Sesotho (Harrismith) or English. Women were targeted because in most African households, women are more knowledgeable about household food dynamics and the individual dietary intake occurring within their households.

The questionnaire measured food insecurity in a South African context using household dietary diversity and food insecurity measures reflecting household food availability and food access (Coates et al., 2007; FAO, 2011). Household Dietary Diversity Scores (HDDS) were determined from the respondents' recall of all food and drink items consumed by the household members during the previous 24 hours. The HDDS tool reflects household food availability and food access, on the premise that households consume a variety of foods when they have the means to acquire them. The tool uses 12 different food groups for analysis and the sum of the food groups consumed by household members from the total of 12 makes the HDDS. Depending on the number of food groups consumed, households were classified into classes of low dietary diversity ($\leq$5 food groups), medium dietary diversity (6–7 food groups) and high dietary diversity ($\geq$8 food groups) (FAO, 2008).

The Household Food Insecurity Access Scale (HFIAS) forms a continuous measure of the degree of household food insecurity, mostly related to access in the past 30 days (Coates et al., 2007). A standardized HFIAS questionnaire composed of nine specific questions with three themes of food insecurity was used (Coates et al., 2007). The nine questions represent a generally increasing level of severity of food insecurity and nine "frequency-of-occurrence" questions were asked as a follow-up to each occurrence question to determine how often the condition would have occurred in the past 30 days. For each frequency-of-occurrence question, a score was assigned to each household: (0) if the answer was "never"; (1) if the response was rarely (condition having happened once or twice in the past 30 days); (2) if it occurred sometimes (three to ten times in the past 30 days) or (3) if the answer was often (occurred more than 10 times in the past 30 days) (Coates et al., 2007). A score was assigned at the end of the nine questions for each household and this ranged from 0 (if answer was "never" for all households) to a maximum of 27 (if answer was "often" for all nine questions). Households were then grouped into different levels of food insecurity, with a high HFIAS score indicating a household's poor access to food and significant household food insecurity (Chakona & Shackleton, 2018). The food insecurity classes were: food secure (HFIAS 0–1), mildly food insecure (HFIAS 2–7), moderately food insecure (HFIAS 8–11) and severely food insecure (HFIAS > 11).

Some explanatory variables, including household characteristics such as income, ages of interviewee and household head, gender of household head, education status, household size, number of children and adults within the household, employment status of household head and other members, sources of food and household's monthly food expenditure, were captured in the interviews.

Data were entered and cleaned using Microsoft Excel, and all statistical analyses were performed using Statistica version 13 (StatSoft Inc.). All descriptive data are presented as means and SDs (mean ± sd) and percentages, and this was presented per study site and overall. The differences in HDDS and HFIAS between study sites and gender of household head were tested using two-way ANOVA with Bonferroni's correction for multiple comparisons. The relationship between HDDS and HFIAS with food expenditure, gender of household head and study site were examined through Spearman correlation tests, and HDDS and HFIAS were used as response variables analyzed as a function of food expenditure, gender of household head and study site. The association between HDDS and HFIAS was also tested using Spearman correlation. Statistical significance was set at $p < 0.05$.

Description of Households

The sample consisted of 314 households with the overall mean age of the interviewee and household head being 32.3 ± 10.8 and 50.8 ± 15.6 years, respectively (Table 11.1). Overall, 63% of the households were female-headed. The mean household size for the study sites ranged from 4.9 ± 2.4 to 7.7 ± 4.7 persons, with the overall mean of 6.4 ± 4.1 persons. Most interviewees had finished secondary education, with an insignificant percentage having a post-secondary qualification of a certificate, diploma or degree. Balfour had the lowest education level, with only about a quarter of interviewees having matriculated. Overall, food expenditure was low, with households in Richards Bay spending less cash per month on food than those in other locations, and households in Balfour spending more money on food. Food expenditure was generally lower in female-headed households than in male-headed households. All households were dependent on food purchasing and about a third farmed some of their food. Households in Richards Bay were the most likely to use means other than purchasing to acquire food. Most households were relying on income from different social grants and pensions, while a few had members with part-time or full-time employment. Only Harrismith (close to 50%) and Richards Bay (almost 40%) had a higher percentage of employed household heads.

Table 11.1 Sample distribution by household characteristics

Indicator	*Category*	*Balfour* (N = *78)*	*Dundee* (N = *75)*	*Harrismith* (N = *85)*	*Richards Bay* (N = *76)*	*Overall* (N = *314)*
Interviewee	Mean age in years	36.6 ± 12.8	32.6 ± 10.5	31.3 ± 9.3	28.6 ± 8.5	32.3 ± 10.8
Population	Sample size (%)	25	24	24	27	–
Household characteristics						
Household size (mean number of persons per household)	Overall	7.1 ± 4.3	7.7 ± 4.7	4.9 ± 2.4	6.1 ± 4.0	6.4 ± 4.1
	Male adults per household	1.4 ± 1.8	1.4 ± 1.1	1.0 ± 0.9	1.3 ± 1.3	1.3 ± 1.3
	Female adults per household	2.5 ± 2.1	2.3 ± 1.7	1.8 ± 1.0	2.1 ± 1.3	2.2 ± 1.6
	Children per household	3.1 ± 1.9	4.0 ± 3.0	2.0 ± 1.7	2.7 ± 2.9	2.9 ± 2.5
Age and gender of household head	Mean age in years	59.7 ± 15.7	49.9 ± 13.6	48.6 ± 15.1	44.8 ± 13.5	50.8 ± 15.6
	Female-headed (%)	62	47	75	68	63
	Male-headed (%)	38	53	25	32	37
Employment status (mean number of employed household members)	Males with full-time jobs	0.1 ± 0.3	0.3 ± 0.6	0.4 ± 0.5	0.4 ± 0.6	0.3 ± 0.5
	Females with full-time jobs	0.1 ± 0.3	0.2 ± 0.5	0.3 ± 0.05	0.4 ± 0.6	0.3 ± 0.5
	Males with part-time jobs	0.1 ± 0.3	0.2 ± 0.5	0.1 ± 0.3	0.2 ± 0.6	0.2 ± 0.5
	Females with part-time jobs	0.1 ± 0.4	0.3 ± 0.5	0.1 ± 0.3	0.1 ± 0.4	0.2 ± 0.4
Food security indicators						

(continued)

Table 11.1 (continued)

Indicator	*Category*	*Balfour* (N = *78*)	*Dundee* (N = *75*)	*Harrismith* (N = *85*)	*Richards Bay* (N = *76*)	*Overall* (N = *314*)
Mean household food expenditure per month (R)	Overall mean	955 ± 402	293 ± 576	231 ± 160	173 ± 171	412 ± 483
	Female-headed households	917 ± 431	224 ± 161	242 ± 161	150 ± 140	377 ± 28
	Male-headed households	1017 ± 341	353 ± 770	200 ± 153	222 ± 214	471 ± 56
HDDS	Overall mean HDDS	6.2 ± 1.3	7.6 ± 1.6	7.4 ± 1.6	8.3 ± 1.6	7.4 ± 1.7
	Female-headed households	6.0 ± 1.4	8.0 ± 1.4	7.5 ± 1.6	8.3 ± 1.7	7.5 ± 0.1
	Male-headed households	6.4 ± 1.1	7.2 ± 1.7	6.7 ± 1.5	8.2 ± 1.4	7.2 ± 0.1
HFIAS	Overall mean HFIAS	10.7 ± 6.2	11.4 ± 7.1	8.3 ± 6.4	5.5 ± 6.8	8.9 ± 7.3
	Female-headed households	10.8 ± 6.2	11.8 ± 7.3	7.9 ± 6.6	5.9 ± 6.7	8.8 + 0.5
	Male-headed households	10.4 ± 6.1	11.0 ± 7.0	9.6 ± 5.8	4.6 ± 7.0	9.2 ± 0.7

Household Dietary Diversity

Household diets were similar across all four sites, with more than 50% of households consuming food groups that included: cereals; spices, beverages, condiments (mostly salt and curry powder); oils and fats; sweets (mostly sugar and sweetened juice); and vegetables (mostly cabbage and onion). Meat (mostly chicken) and milk and milk products were consumed by more than 50% only in Dundee, Harrismith and Richards Bay. Furthermore, white roots and tubers (mostly potatoes) and legumes, nuts and seeds (mostly beans) were consumed by more than 50% of households only in Balfour and Richards Bay, respectively (Fig. 11.1). Other food groups such as fruits, eggs, and fish and seafoods were hardly consumed by households in any of the towns, with the lowest percentage observed in Balfour.

The HDDS were generally low in all towns, with an overall mean of 7.4 ± 1.7 (Table 11.1). Richards Bay had the highest mean HDDS (8.3 ± 1.61) and Balfour had the lowest (6.2 ± 1.3). There was a significant difference in HDDS between the four locations ($F_{6,309} = 2.45$, $p < 0.005$) which was higher in Richards Bay than in the other towns. Significant differences were also observed in HDDS between Balfour and the other two towns, which were not significantly different from one

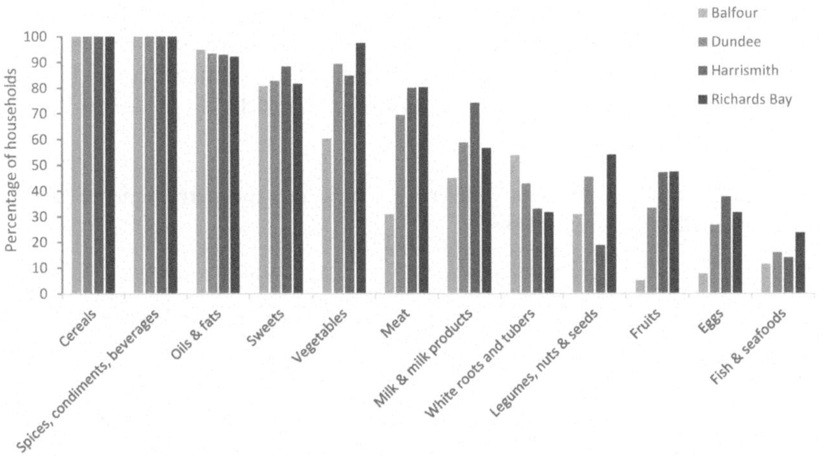

Fig. 11.1 Percentage of households consuming different food groups

another (Table 11.1). Most households in these locations were classified in the high-HDDS group, except for Balfour, which had half the households classified as medium and 29% as low-HDDS groups. No significant differences were observed in HDDS between gender of household head ($F_{2,312}$ = 2.96, $p > 0.05$), although HDDS were generally higher in female-headed than male-headed households.

Household Food Access and Food Security

The HFIAS was generally low in Richards Bay and Harrismith and moderately high in Balfour and Dundee, with a mean of 9.0 ± 7.0 for the sample (Table 11.2). Overall, 23% of the households were food secure, 22% were mildly food insecure, 31% were moderately food insecure and 24% were severely food insecure. Households in Dundee and Balfour had the highest HFIAS scores, of 11.4 ± 7.1 and 10.7 ± 6.2, respectively, showing low food access, while Richards Bay had the highest level of food access (HFIAS score of 5.5 ± 6.8). There was a significant difference in HFIAS between these peri-urban locations ($F_{6,309}$ = 44.9, $p <$ 0.001). Significant differences were observed between Richards Bay and the other three towns and between Dundee and Harrismith; no significant differences were observed between Balfour and these two towns. The prevalence of household food insecurity was much higher in Balfour, with almost 50% of the households being severely food insecure (Table 11.2). About 50% of households in Richards Bay were food secure, with only 8% being severely food insecure. No significant differences in HFIAS were observed between gender of household head ($F_{2,312}$ = 49.7, $p >$

Table 11.2 Household dietary diversity scores and the percentage of households grouped in different classes

Town	*HDDS (mean ± sd)*	*Percentage of households*		
		Low	*Medium*	*High*
Balfour	6.2 ± 1.3[a]	29	50	21
Dundee	7.6 ± 1.64[b]	9	35	56
Harrismith	7.4 ± 1.64[b]	9	42	48
Richards Bay	8.3 ± 1.61[c]	5	46	49
All	7.4 ± 1.7	13	43	44

NB Unlike superscripts indicate significant differences

Table 11.3 Percentage of food insecure and food secure households as classified using HFIAS

Town	*HFIAS (mean ± sd)*	*Percentage of households*			
		Severely food insecure	*Moderately food insecure*	*Mildly food insecure*	*Food secure*
Balfour	10.7 ± 6.2^{a}	49	23	19	9
Dundee	$11.4 \pm 7.1^{a/b}$	27	41	19	13
Harrismith	$8.3 \pm 6.3^{a/c}$	13	36	29	21
Richards Bay	5.5 ± 6.8^{d}	8	22	20	50
Overall mean	9.0 ± 7.0	24	31	22	23

NB Unlike superscripts indicate significant differences

0.05), although food insecurity was generally higher in male-headed than in female-headed households.

There was a significant negative correlation between HDDS and HFIAS, HDDS and town as well as HDDS and household food expenditure. Positive significant correlations were also observed between HFIAS and food expenditure and HFIAS and town (Table 11.3). No significant correlations were observed between the food security indicators and the gender of household head.

Discussion

South Africa has the second largest economy in Africa and has adequate food supply at the national level, but this is not translated into the food security status of many households (Shisana et al., 2014). This is also reflected in the study, as households in all four peri-urban locations had poor dietary diversity, especially in Balfour, where only one in five households had good quality diets. In the other three study sites, more than 50% of households had diets lacking good quality and diverse foods rich in all essential nutrients. Similar findings have been reported in other studies (Chakona & Shackleton, 2017, 2018; Labadarios et al., 2011). Furthermore, about 77% of the households were food insecure, with one in four households being severely food insecure (Table 11.4). It may be that most households have food on the table but are consuming cheap diets based on grains and lacking essential nutrients, as was also noted

Table 11.4 Spearman correlations between HDDS and HFIAS and selected socio-economic indicators

Variables	*HDDS*	*HFIAS*
HFIAS	−0.45	–
Town	−0.43	0.21
Food expenditure	−0.22	0.63

NB All correlations are significant at $p < 0.001$

by Gina Kennedy (2009). Balfour had the highest percentage of severely food insecure households (49%), with households consuming fewer food groups, which could be attributed to high unemployment rates.

Surprisingly, given reports that vulnerability to food insecurity is more pronounced in female-headed households (DoA, 2002); in this study, food insecurity was higher and dietary diversity lower in male-headed households, even though male-headed households spent more money on food. This could be because more female-headed households had diversified means of obtaining food other than food purchasing. Of the households who obtained some of their food from farming, collection from open spaces, and donations, 74%, 74% and 71%, respectively, were female-headed. The addition of food from these other sources acted as "women's safety net" and could have increased their dietary diversity, thus lowering their food insecurity scores. Previous research (Chakona & Shackleton, 2017, 2019) reported that households who supplemented their food with farm produce and wild foods had a lower risk of food insecurity than those who did not, as these alternative sources would lower dependence on food purchasing. The advantage for female-headed households could also be partly due to social grants, a factor that Belinda Dodson et al. (2012) speculated accounted for better relative outcomes for low-income, female-headed households when compared to low-income nuclear households.

The study also showed strong correlations between household dietary diversity, food access and monthly food expenditure. As household dietary diversity increased, monthly food expenditure decreased, while the ability of a household to access food decreased. This was mostly evident in Balfour, where food insecurity was high, with less food diversity, although households in this study site spent more on food. This could be because households in Balfour relied mostly on food purchasing, mostly of cereals (maize-meal and rice), spices, condiments, beverages, oils and fats and sweets, which could not increase their dietary diversity and food security status. However, it has been observed that within the study sites, more female-headed households would only purchase food that is important when necessary. A greater percentage of households in the study sites consumed a number of similar food groups, irrespective of geographical location, income status or level of education. More than 50% of the households across all towns consumed similar food groups such as cereals, spices, beverages, condiments oils and fats, sweets and vegetables. Other food groups consumed by households were limited to the location of the town. For example, food groups such as meat, and milk and milk products, were consumed by more than 50% of households only in Dundee, Harrismith and Richards Bay, while roots and tubers, and legumes, nuts and seeds were consumed by more than 50% of households only in Balfour and Richards Bay. Other food groups such as fruits, eggs, and fish and seafoods were seldomly consumed, especially in Balfour, which explains the significant differences in HDDS evident across the four peri-urbans locations. Furthermore, dietary diversity was strongly negatively correlated with living in a town, also reflecting that household characteristics (e.g., food expenditure) and the food environment within a town have significant effects on dietary diversity and food security.

The levels of food insecurity were higher in the peri-urban areas characterized by low education levels and high unemployment rates, with most households relying heavily on social grants for daily cash needs, as was evident in Balfour. On average, households in Balfour consumed six food groups that lacked the essential nutrients found in fruits, vegetables and animal protein. Within the peri-urban settings, the poorest households may have limited access to food due to high levels of poverty and unemployment, possibly due to household members' limited education, which affects their ability to get formal employment. This study concurs with Oldewage-Theron et al. (2006) and Labadarios et al. (2011), who also found that poor socio-economic status has an impact on household food insecurity in peri-urban locations in South Africa.

Food insecurity in the study sites is closely linked to poverty. Poor households coped with poverty and rising food prices by reducing the quality, quantity and number of meals consumed per day, as reported in earlier research (Chakona & Shackleton, 2018). John Hoddinott and Yisehac Yohannes (2002) also noted that higher dietary diversity is positively associated with higher socio-economic status and household food security. However, different households have different capacities to adapt and mitigate these effects, therefore umbrella intervention approaches are not enough in addressing food insecurity challenges, as some households can be left behind. Through a gender-based analysis, it is more apparent in the study that non-commercial food sources and the "informal safety net" are important factors determining household food security status. Affordability has been the major constraint to the consumption of good quality diets, in adequate amounts, in these peri-urban areas. With the global climate change impacts affecting food systems and impacting food prices, coordinated research programs that emphasize nutrition-sensitive approaches may help address these challenges. Adopting more localized and diverse food systems that are context-specific, sustainable, healthy, and available at local markets at affordable prices, is an important component in addressing poverty, food and nutrition insecurity in peri-urban settings of South Africa. This is essential for creating opportunities for increased access to locally grown foods and improved local economies through income generation and employment creation in these settings.

References

Altman, M., Hart, T., & Jacobs, P. (2009). Household food security status in South Africa. *Agrekon, 48*(4), 345–361.

Battersby, J. (2011). Urban food insecurity in Cape Town, South Africa: An alternative approach to food access. *Development Southern Africa, 28*(4), 545–561.

Battersby, J. (2012). Beyond the food desert: Finding ways to speak about urban food security in South Africa. *Geografiska Annaler: Series B, Human Geography, 94*(2), 141–159.

Caesar, M., & Riley, L. (2018). *Gender inequality, poverty and urban household food security in Cape Town* (HCP Discussion Paper No. 18). Hungry Cities Partnership, Waterloo, ON.

Chakona, G., & Shackleton, C. (2017). Minimum dietary diversity scores for women indicate micronutrient adequacy and food insecurity status in South African towns. *Nutrients, 9*, 812.

Chakona, G., & Shackleton, C. (2018). Household food insecurity along an Agro-Ecological gradient influences children's nutritional status in South Africa. *Frontiers in Nutrition, 4*, 72.

Chakona, G., & Shackleton, C. (2019). Food insecurity in South Africa: To what extent can social grants and consumption of wild foods eradicate hunger? *World Development Perspectives, 13*, 87–94.

Coates, J., Swindale, A., & Bilinsky, P. (2007). *Household food insecurity access scale (HFIAS) for measurement of household food access: Indicator guide (version 3)*. FHI 360/FANTA.

Cohen, M., & Garrett, J. (2009). The food price crisis and urban food (in)security. *Environment and Urbanisation, 22*(2), 467–482.

Crush, J., & Frayne, B. (2010). *The invisible crisis: Urban food security in Southern Africa* (Urban Food Security Series No. 1). African Food Security Urban Network (AFSUN), Kingston, CA and Cape Town, SA.

Crush, J., & Frayne, B. (2011). Urban food insecurity and the new international food security agenda. *Development Southern Africa, 28*(4), 527–544.

Crush, J., Hovorka, A., & Tevera, D. (2011). Food security in southern African cities: The place of urban agriculture. *Progress in Development Studies, 11*(4), 285–305.

De Cock, N., D'Haese, M., Vink, N., van Rooyen, C., Staelens, L., Schönfeldt, H., & D'Haese, L. (2013). Food security in rural areas of Limpopo province South Africa. *Food Security, 5*, 269–282.

Development Initiatives. (2018). *Global nutrition report: Shining a light to spur action on nutrition* (Technical Report). Development Initiatives, Bristol, UK.

Department of Agriculture (DoA). (2002). *Integrated food security strategy*. Department of Agriculture.

Dodson, B., Chiweza, A., & Riley, L. (2012). *Gender and food insecurity in Southern African cities* (Urban Food Security Series No. 10). African Food Security Urban Network (AFSUN).

Drimie, S., Faber, M., Vearey, J., & Nunez, L. (2013). Dietary diversity of formal and informal residence in Johannesburg, South Africa. *BMC Public Health, 13*(1), 911.

Food and Agriculture Organization (FAO). (2008). *Report on use of the household food insecurity access scale and household dietary diversity score in two survey rounds in Manica and Sofala provinces, Mozambique, 2006–2007* (FAO Food Security Project GCP/MOZ/079/BEL Report). Food and Agriculture Organization.

Food and Agriculture Organization (FAO). (2011). *Guidelines for measuring household and individual dietary diversity*. Food and Agriculture Organization.

Hendriks, S. (2014). Food security in South Africa: Status quo and policy imperatives. *Agrekon, 53*(2), 1–24.

Hoddinott, J., & Yohannes, Y. (2002). *Dietary diversity as a household food security indicator*. FANTA Project, Academy for Educational Development.

Jonah, C., & May, J. (2020). The nexus between urbanization and food insecurity in South Africa: Does the type of dwelling matter? *International Journal of Urban Sustainable Development, 12*(1), 1–13.

Kennedy, G. (2009). *Evaluation of dietary diversity scores for assessment of micronutrient intake and food security in developing countries* (PhD thesis). Department of Human Nutrition, University of Wageningen.

Kuku, O., Gundersen, C., & Garasky, S. (2011). Differences in food insecurity between adults and children in Zimbabwe. *Food Policy, 36*(2), 311–317.

Labadarios, D., Mchiza, Z., Steyn, N., Gerike, G., Mounder, E., Davids, Y., & Parker, W. (2011). Food security in South Africa: A review of national surveys. *Bulletin of the World Health Organization, 89*(12), 891–899.

Maluti-a-Phofung Municipality. (2014). *Maluti-a-Phofung municipality integrated development plan* (2013/14 Annual Report). Maluti-a-Phofung Municipality, Phuthaditjhaba, SA.

National Planning Commission (NPC). (2012). *National development plan.* National Planning Commission, Office of the Presidency, Pretoria, SA.

Oldewage-Theron, W., Dicks, E., & Napier, C. (2006). Poverty, household food insecurity and nutrition: Coping strategies in an informal settlement in the Vaal Triangle, South Africa. *Public Health, 120*(9), 795–804.

Satterthwaite, D., McGranahan, G., & Tacoli, C. (2010). Urbanisation and its implications for food and farming. *Philosophical Transactions of the Royal Society B: Biological Sciences, 365*(1554), 2809–2820.

Schönfeldt, H., Gibson, N., & Vermeulen, H. (2010). The possible impact of inflation on nutritionally vulnerable households in a developing country using South Africa as a case study. *Nutrition Bulletin, 35*(3), 254–267.

Sen, A. (2008, May 28). The rich get hungrier. *The New York Times*. https://www.nytimes.com/2008/05/28/opinion/28sen.html.

Shisana, O., Labadarios, D., Rehle, T., Simbayi, L., Zuma, K., Dhansay, A., Reddy, P., Parker, W., Hoosain, E., Naidoo, P., Hongoro, C., Mchiza, Z., Steyn, N., Dwane, N., Makoae, M., Maluleke, T., Ramlagan, S., Zungu, N., Evans, M., Jacobs, L., Faber, M., and SANHANES-1 Team. (2014). *South African national health and nutrition examination survey (SANHANES-1).* HSRC Press.

Statistics South Africa (SSA). (2012). *Census 2011.* Statistics South Africa, Pretoria, SA.

Statistics South Africa (SSA). (2014). *Poverty trends in South Africa: An examination of absolute poverty between 2006 and 2011* (Report No. 03-10-06). Statistics South Africa, Pretoria, SA.

Statistics South Africa (SSA). (2017). *Poverty trends in South Africa: An examination of absolute poverty between 2006 and 2015* (Report No. 03-10-06). Statistics South Africa, Pretoria, SA.

Statistics South Africa (SSA). (2019). *Towards measuring the extent of food security in South Africa: An examination of hunger and food adequacy* (Report No. 03-00-14). Statistics South Africa, Pretoria, SA.

uMhlathuze Municipality. (2012). *uMhlathuze integrated development plan* (2012/2017 Annual Report). City of uMhlathuze, Richards Bay, SA.
uMzinyathi District Municipality. (2012). *uMzinyathi district municipality integrated development plan* (2012/13 Annual Report). uMzinyathi District Municipality, Dundee, SA.
UN-Habitat. (2020). *Mainstreaming urban-rural linkages in national urban policies: National urban policy guide*. United Nations Human Settlements Programme.

CHAPTER 12

"We Eat Every Day, but I'm Perpetually Hungry": Interrogating Food System Transformation and (Forced) Dietary Changes in Tamale, Ghana

Issahaka Fuseini

Introduction

Food system transformation and dietary changes are occurring ubiquitously across much of Africa's urban areas (Battersby, 2017; Reardon, 2015). While many factors interact to influence these changes, urbanization remains one of the active influencers of the continent's food systems transformation. High concentrations of people in urban areas increase demand for food, a situation that induces commercial inflows of food in

I. Fuseini (✉)
Department of Adult Education and Human Resource Studies, University of Ghana, Accra, Ghana
e-mail: isfuseini@ug.edu.gh

L. Riley and J. Crush (eds.), *Transforming Urban Food Systems in Secondary Cities in Africa*,
https://doi.org/10.1007/978-3-030-93072-1_12

urban settings (Worku et al., 2017). There is evidence that rapid urbanization in countries with low human development, which describes many African countries, has a negative impact on urban food security in these countries (Szabo, 2016). Given high demand for food in urban areas, and the corresponding systems of supply, food prices tend to be higher in urban areas, making food less accessible to segments of the urban population as access to food in urban settings is largely a function of income (Bush, 2010; Worku et al., 2017). The result is that poor urban households spend a disproportionate share of their income on food, which does not even guarantee them adequate consumption levels (Worku et al., 2017). It thus remains a concern in rapidly urbanizing cities in the Global South that the food security situation of poor and vulnerable groups could be impacted negatively by the affordability, stability and safety of food (Wenban-Smith et al., 2016).

The importance of food preference cannot be underestimated in the food security equation. Attention to food preference and the social elements of food and nutrition extend the definition of food security beyond the material and biological realms to include a suite of culturally embedded, context-relevant subjective well-being under the rubric of the human well-being approach to food security (CFS, 2017; Noack & Pouw, 2015). Pursuant to this line of argument, Anna-Lisa Noack and Nicky Pouw (2015, p. 172) contend that "food and nutrition security can only be achieved when 'sufficient culturally adapted food' is available and accessible to a household or community to meet physiological and social needs." Food is "linked to identity and social relationships and the subjective and emotional components of food and eating should not be ignored" (CFS, 2017, p. 78). The statement quoted in the title, "We eat every day, but I'm perpetually hungry," was taken from an interview with an elderly man in Tamale, a secondary city in Ghana. It captures the paradox of hunger amidst plenty that characterizes urban food insecurity across Africa, while highlighting the subjective well-being of an urban resident. The respondent's explanation of what he meant—that he does not get to eat the foods he loves stuck with me and almost always springs to mind when someone suggests dissatisfaction with what they have to eat.

In seeking to understand the effect of food system changes on urban Ghanaians, this chapter draws together two areas of focus: the issue of food preference, and to what extent the food available today is subjectively satisfying; and food utilization, in terms of food procurement decisions and preparation. In doing so, the chapter takes a qualitative approach to describe the experiences of people in the fast-changing food environment, especially the elderly. The intention is to characterize subjective well-being from the viewpoint of the city's population as it relates to food preference and utilization. Thus, the chapter is distinguished from a public health-centred technical analysis, which employs a suite of measures to analyze food utilization and well-being, measures that are often external to the people involved. The chapter argues that the social function of food is as important as its biological function as far as food preference and utilization are concerned (De Groote & Kimenju, 2012; Quisumbing, 2013; Webb et al., 2006).

Drivers of Food System Changes in Urban Africa

The literature on urban food system change in Africa has outlined many factors interacting variously to influence the changes occurring in African cities and towns. These factors range from urbanization related to spatial development, globalization and supermarket expansion. Globalization has intensified the diffusion and spread of food products and cuisines, and the latter has arguably transcended cultures as never before. To varying degrees across Africa, globalization has reshaped urban food systems through the integration of African urban markets with transnational food distribution channels (Battersby, 2017). The increasing availability of foreign foods in local markets alters household food choices, and thus food consumption patterns.

One sign of globalization's effect on urban food system transformation in Africa is the spread of international fast-food outlets (Tsrah et al., 2020). Backed by international capital investment and product promotion, fast-food chains convey an image of urban life as more "comfortable" than rural life in terms of greater diversity and convenience in food choices. These images bolster the long-held rural view that urban life is luxurious. Public messaging, coupled with rising urban incomes, means that in many African cities these fast foods are redefining taste, consumption habits and group identity formation. Writing for the *New York Times* on the evolution of the fast-food industry in Ghana, Dionne Searcey and Matt Richtel (2017) noted how the "finger lickin' good" slogan of KFC attracted a young man with a well-paying job in Ghana's booming oil sector:

> He has an apartment, a car, a smartphone and a long-distance girlfriend he met on a dating website. So he had reasons and the means to celebrate his 27th birthday in late July. His boss and co-workers joined him for an evening of laughter and selfies, lingering over dinner at his favorite restaurant: KFC. [He] first learned about the fried chicken chain on Facebook. The "finger lickin' good" slogan caught his attention and it has lived up to expectations. 'The food is just …' he said, raising his fingertips to his mouth and smacking his lips. 'When you taste it you feel good.' (Searcey & Richtel, 2017, paras. 1 and 2)

Another sign of globalization's impact on Africa's urban food system transformation is the expansion of food retail through supermarket chains. Supermarkets are found in most major African cities, after decades of rapid expansion (Crush & Frayne, 2011; Reardon et al., 2007; Smit, 2016; Weatherspoon & Reardon, 2003). South African supermarket giants Shoprite and Game and Indian-owned Melcom, MaxMart and Citydia are some of the popular supermarket chains operating in Ghana. Supermarkets can create formal sector employment, offer new food choices, and offer lower prices on some items due to competition and economies of scale, which are some of the advantages raised in arguments for policies promoting supermarket expansion (Andam et al., 2018; Demmler et al., 2018; Rischke et al., 2015). On the other hand, supermarkets promote the sale and consumption of processed foods, which is associated with high risk of food-related non-communicable diseases (Demmler et al., 2017; Rischke et al., 2015). It is also argued that supermarkets do not necessarily align with the food consumption strategies of the poor segments of Africa's urban populations (Peyton et al., 2015). Some of these food sourcing strategies include accessing food on credit in times of financial crisis, making food items available in smaller quantities commensurate with the incomes of poor households, and favourable working hours (Skinner, 2016; Skinner & Haysom, 2016). For example, while many formal supermarkets' business hours end around 7 or 8 p.m., the neighbourhood and home-based informal food outlets may operate until later in the evening, allowing more hours of access to food in poor neighbourhoods.

Perhaps it is these food sourcing strategies, based on social networks, that make traditional local markets the preferred option for Ghanaian urbanites (Oltmans, 2013). Regardless of the inconclusive verdict on whether supermarkets offer more positives than negatives to urban food security, the phenomenon of supermarketization is having a significant effect on food system transformation and dietary transition in African urban settings. In the long run, people remain food insecure in spite of food being widely available in the market because they simply cannot afford the types of food they prefer to eat. These drivers of urban food system transformation are discussed later in terms of how they interact to affect people's subjective well-being in Tamale.

Methods of Investigation and Data Sources

The chapter is based on a qualitative approach and the original empirical data comes largely from in-depth interviews with six people in Tamale, combined with secondary literature and personal observations. The interviews were conducted in February 2020. Four respondents were purposively selected because they were knowledgeable about food system changes and dietary changes in Tamale. Some were participants in past studies and others were selected based on snowball sampling. Two female respondents were selected through convenience sampling to share their experiences of household food purchases, food preparation and consumption practices. This sample was deemed adequate for the purpose when saturation was reached after the fourth respondent. The personal observations included unstructured observing of what people eat, and informal interaction and noting of what people say about what they eat. This activity spanned a longer period than the interviews (both before and after), to allow ample time to observe how people interacted with their food environment.

The interviews were recorded and transcribed to facilitate analysis of the responses. The data was coded manually and thematically on the historical dimension of food system changes, the nature of food system changes, and changing gender roles within the transforming food system. It is important to note that the historical accounts of food system changes in Tamale depended solely on the recollection of the elderly respondents, as there is a lack of documented evidence to corroborate these. The results of the interviews and personal observations are presented in the sections below.

The Decline in the Traditional Food System

The historical perspective of food system change in Tamale is told through recollections of elderly interviewees and the documented history of Dagbon, a traditional area controlled by the Dagbamba people, of which Tamale is the largest settlement. According to accounts by the elderly respondents, a noticeable, recallable event that ushered in changes in the food system of Dagbon was a drought-induced famine in the late 1940s. Until that famine, the food system of Dagbon and Tamale was anchored on native foods, including sorghum, millet, yam, beans, bambara beans (*Vigna subterranean*) and other foods harvested from the wild. These were made into diverse dishes using native spices and condiments, including ***dawadawa***, ***nili*** (melon seeds), ***bungu*** (sesame), pepper and ***kantong (condiment made from baobab and kapok seeds)***. However, the roots of the centuries-old indigenous food system were shaken in the late 1940s when the long drought and locust invasion triggered serious famine in the Dagbon area. This incident is referred to in the history of Dagbon as ***kanchaglanchu yuuni***, meaning "the year of white maize," referring to the new crop that was introduced in the area as part of food aid from international sources.

The people found it difficult to process the maize into food, as their rudimentary food-processing technology at the time—stone mills that were used to grind cereals—did not allow for turning the maize into their staple porridge meal. The traditional porridge meal was known in the Dagbanli language as *saɣim*, which is popularly known in the broader Ghanaian context as TZ, based on the Hausa word *tuo zaafi*. As a result, people employed different methods to turn the maize into food, including roasting the grains and eating them like roasted bambara beans. The introduction of hammer mills made grinding easier, but it did not make consumption of maize-meal popular. People resented preparing *saɣim* with maize because they felt it did not taste like the sorghum or millet *saɣim* they were used to. But by the late 1980s and 1990s, maize had become the mainstay for *saɣim* in Dagbon and Tamale. This outcome owed much to the severe drought and consequent famine that ravaged Ghana and West Africa, with food aid, again, coming to the rescue. This local perspective positions famine as the trigger of food system and consumption changes in the area.

These natural events were coupled with a policy focus on agricultural intensification. Present policy focus and investment promote maize production under the aegis of a new green revolution in northern Ghana (Nyantakyi-Frimpong & Bezner Kerr, 2015; Vercillo et al., 2020). For example, a collaborative project known as the Savannah Zone Agricultural Productivity Improvement Project, between the Government of Ghana and the African Development Bank, is being implemented in northern Ghana. Maize and soya are the priority crops during the first phase of the project, and rice may be the next crop added (personal communication with an official at regional directorate of Ministry of Food and Agriculture, May 18, 2021, Tamale). This indicates that the policy orientation regarding modernization or commercialization of agriculture in northern Ghana favours these "non-native" crops.

Maize production in Ghana was vigorously pursued under the Sasakawa Agricultural Global 2000 Project to boost farmer yields and achieve food security. Small-scale farmers recorded significant improvement in their per acre yields (Dowswell, 1989). This project was also associated with the introduction of early maturing sorghum known in Dagbanli as *kajia*, literally translated as "short sorghum," which refers to both its short gestation and the height of the stalk at maturity. Given where Ghana was coming from in terms of periodic famine, increased yield of maize per capita meant that maize became a very popular food crop in Dagbon and Tamale. Thus, maize production improved stability of the food system leading to increased consumption by households.

Although cassava had been introduced in Dagbon earlier on, it did not immediately gain wide acceptance in the area. Its cultivation was limited to plantings at the edges of yam farms. However, the crop gained acceptance and relevance as one of the strategies to diversify the area's food system and thus contribute to long-term solutions to famine and food shortages. Unlike maize and other cereal crops, cassava requires low inputs to grow, so it was easy to establish large cassava farms or plantations, and became an important food item beginning in the 1970s and 1980s. A 70-year-old male interviewee summarized his recollection of the introduction of the "new" crops in Tamale as follows:

> I will say that crops like maize and cassava have been around for about 40 to 50 years. Growing up, we did not know about these. I recall that cassava was introduced in my lifetime and it had very low acceptance. I remember that some elderly people used to say that cassava was the same as the roots of trees, so they would not eat it. Let me say that persistent food shortage or poverty is a recent development in Dagbon, it was not

> common. Therefore, the new crops were introduced to help increase crop yields in order to help address food shortage and food poverty. (Interview, February 19, 2020, Tamale)

However, deteriorated soil quality and environmental pressures, including urbanization, have affected the cultivation of cassava too, reducing the crop's importance. A 64-year-old male respondent captured these dynamics as follows:

> Many factors affect our crop yields in Dagbon, especially in the outlying areas of Tamale. First, we have lost our farmlands to urbanization. Droughts have been common because the rainfall pattern has changed significantly; we can no longer predict it with accuracy. Adoption of new farming methods has also affected crop yields in this area. The use of tractors, for example, has meant that some people farm more than they can manage, leading to poor harvests. But the worst of all is the introduction and acceptance of agrochemicals [weedicides, pesticides, herbicides]. These chemicals have caused the impoverishment of the soils and this has affected the cultivation of native crops. (Interview, February 22, 2020, Tamale)

Table 12.1 summarizes the key indigenous and exotic food crops in Tamale and the meals made from them.

New and early maturing beans were introduced around the late 1980s and early 1990s, and this was associated with high pesticide use. But because agrochemicals had not yet been made popular at the time, suitable pesticides for food crops were either not available to farmers or not popular. As a result, many farmers diverted the DDT chemicals supplied them to control pests on their cotton farms to the control of pests on the early maturing beans. However, people soon noticed that they became sick when they consumed the beans. Respondents now interpret this as the effect of high doses of DDT on the beans, but they said that at the time people could not make this connection. They simply acknowledged the effect that consumption of the beans had on their physical performance and gave it a derogatory name: *ɡa mpabɡi*, which means "eat and become physically weak," but it also has a derogatory connotation with eating filth. Accordingly, this breed of newly introduced beans did not gain ground in the way that maize did. However, it is resurfacing now as there are specific pesticide chemicals available to be used on such crops.

The respondents explained that the increased use of agrochemicals, including fertilizers, has led to alteration of the physical properties of

Table 12.1 Key food crops in Tamale and meals made from them

Food crop/item	*Meals made*	*Still readily accessible?*
Indigenous foods and meals		
Sorghum	*Sayim*, *koko*, brewing of local malt drink, snacks	No
Millet	*Sayim*, koko, snacks with fresh milk, *fula*	No
Yam	Eaten boiled, roasted or fried; *fufu*	Yes, but very limited
Beans	Eaten boiled, *gablee*, *tubani*, snacks	No
Bambara beans	Eaten boiled and roasted, *gablee*, *tubani*, snacks	Yes, but very limited
Bungu (sesame)	Eaten as snack with millet, used as condiment to make soup for *sayim*	No
Nili (melon seeds)	Used primarily to make soup for *sayim*	No
Adua (pigeon pea)	Eaten boiled	Yes, but becoming limited
Newly introduced foods and meals		
Maize	Has replaced sorghum and millet as principal material to make *sayim*, *koko*	Yes
Sorghum (*kajia*)	Mainly *koko*, snacks, brewing of local malt drink	Yes
Rice	Many dishes: plain, jollof and fried rice, *waakye* and rice balls	Yes
Groundnuts/peanuts	Replaces *bungu* and *nili* as the main material for soup making	Yes, but with declining harvest
Beans (early maturing)	Eaten boiled, *gablee*, *tubani*, snacks	Yes, but with declining popularity
Soybeans	Grown largely as cash crop but used occasionally for *gablee*, snacks	Yes
Cassava	Mixed with maize to make *sayim*, mixed with bean flour to make *gablee*, processed into *gari*	Yes, but with declining harvest

the soil and the environment generally, and that has affected the yield of native crops. Inappropriate use of fertilizers, weedicides and pesticides was said to have affected soil quality so much that, together with pressures from urbanization, native crops such as sorghum, millet, beans and yam can no longer be grown within the peri-urban and outlying rural communities of Tamale. The reduced supply within the peri-urban communities of Tamale and other adjoining districts means that these local or native products have to be brought in from far-away areas at high cost, which in turn affects the price of these commodities. This means that poor households are unable to access these products and this impacts negatively on the welfare of the elderly, whose preference is still very high for the local foods. With no other options, households turn to the globalized food system for cheap imports from elsewhere, but these do not necessarily promote the subjective well-being of the people. The elderly in particular complain about the lack of "proper" food available to them, saying that they eat these foods for survival because ***ninsal noli yaɣi mori ŋubbu***: "a human cannot live on grass like animals." The following section discusses people's feelings of being food insecure in terms of their food preferences and utilization.

Exclusion of the Elderly in the New Food System

The above discussion of historical events and processes that have interacted to change the food system of Tamale provides context for discussion linking individual food preferences with subjective well-being. Households in both the urban core and peri-urban areas of Tamale now depend on the market for their food needs. However, the respondents mentioned that some households draw on rural–urban linkages to access some of these foods, either through migrant farming at distant farming communities or through family members who engage in farming at such distant farming areas. They said that it was still possible to produce some of the local food crops at farming communities beyond a 60-km radius of Tamale. However, this type of migrant farming required a lot of capital outlay, and many disadvantaged households rendered landless by urban expansion in the peri-urban areas cannot garner such capital. Therefore, the market becomes the only source of food for these households. Given their low incomes, dependence on the market exposes these households and their members to food insecurity, in terms of access to the type and quantity of foods they would love to eat.

The older members of poor urban households are affected more by limited access to the foods that households would prefer to eat. The respondents bemoaned the situation by using appropriate Dagbanli proverbs. For example, the 70-year-old respondent remarked:

> Complete dependence on the market for food is a very big inconvenience. There is no way one can eat well while depending on the market. As you know, our elders say that *ŋunda ndi ku tooi ŋmani kamboŋlana*, meaning "whoever depends on the market for their food needs cannot attain the same food security status as one who eats from their barn." (Interview, February 19, 2020, Tamale)

This expression has deep and varied meanings in the cultural context of the elderly respondents. First, it emphasizes that the people's cultural pride in food sufficiency is such that buying food from the market is seen as a sign of failure as a farming household. Second, it implies that it is very difficult to depend on the market and still achieve food security in terms of eating the right quantities and the type of food one would want to eat at any given point in time. While this reasoning is culturally embedded, it also reflects the low-income levels of the people. While the proverbial reference implies that it is not possible to depend on the market for one's food needs, many people can depend on the market and even overconsume because they have enough income to buy whatever food they need from the market. It therefore has a more direct reference to the worldview of elderly residents and low-income households.

Expanding on the idea that "we eat every day, but I am perpetually hungry," a 63-year-old male respondent had this to say:

> It is true, I have nostalgic feeling all the time about those good old days when we ate real food. We did not know Maggi [popular industrial spice in the market] or all of these funny products. We ate good soup made from nili [melon seeds] and *kpalgu* [dawadawa, from the African locust plant], *tubani* [dumplings made from beans or bambara], *koko-nyina* [porridge breakfast made from millet, sorghum or in later times maize], you name them. Now we do not even know what we are eating, we just swallow in stuff like grazing cattle. Can you compare soup made of groundnuts and Maggi to one made from nili, kpalgu or kantong? No way! So, why won't one be hungry all the time when he or she reminisces about these things? (Interview, February 16, 2020, Tamale)

The soup made of nili is one of the foods the respondents said they missed the most. They explained that it was not only nutritious but also medicine for many minor diseases. They believed that regular eating of soup made from nili with kpalgu gave them good health. They explained further that due to its nutritional value, women who had just given birth were fed exclusively with nili soup made with pepper to help heal their wounds as well as to help restore their general vitality after childbirth. While the local people had no scientific know-how to tell the exact nutritional value or properties of nili and other indigenous food products, there is scientific evidence that nili and other melon seeds contain vital nutritional properties that are good for human health (Mensah, 1986; National Research Council, 2006). Soup made from nili thus assumed not only nutritional significance, but also cultural importance.

The respondents independently expressed the same view that the changes that have occurred within their food system have generally lowered the status and living standards of the elderly. They explained that in Dagbon culture, the elderly and postpartum women enjoyed preferential treatment with respect to food. For example, historically meat has not been a regular part of the diet of an average Dagbamba household. However, the elderly and postpartum women were served meat regularly. Cured meat was added to the special soup prepared for new mothers, while the elderly were treated to the same cured meat in their regular mid-morning meals. An animal or guinea fowl was killed for the purpose, as culturally the Dagbamba do not kill a fowl for meat except on special occasions such as when receiving a visitor, when celebrating a marriage, when doing sacrifice or during funerals and festivals. While the practice of serving meat was meant to help restore the vitality of women after they had given birth, enjoying these treats was a mark of status for the elderly. The mid-morning meal for the elderly was also made from millet porridge mixed with fresh milk. However, the prevailing urban food system transformation does not make it possible for the elderly to eat the foods that once characterized their "pension lives" and made them enjoy their old age. Some of these foods can be found in the market, but the elderly poor cannot afford them as they would like. As such, they are eating like anyone else. As a 72-year-old respondent put it:

> *Gumachuɣu porisirila o nini ni nye sheli* [a chameleon changes to the colour of what it sees], so we are eating what we get rather than what we would like to eat. I think we are not enjoying our old age as much as our fathers did, and it has also meant that we do not command respect from society as much as we should. I feel this is partly due to how and what we eat. (Interview, February 10, 2020, Tamale)

Another respondent was of the view that the issue of urban food system transformation is very complex. In addition to the physical conditions and urbanization pressures that make it difficult to produce one's own food, globalization of cultures, lifestyles and consumption also has an impact:

> For example, the youth of today have taken on lifestyles that alienate them from their parents such that the latter are less catered for. The youth now eat out a lot of the time and eating a lot of the foods they should not be eating at their age [meat]. These days, one finds kebab joints everywhere in Tamale and it is the youth who patronize these with their girlfriends, to the neglect of their supporting role to their elderly parents. Gone are the days when a young person could not eat meat at the expense of their elderly parents. Now things are different. (Interview, February 19, 2020, Tamale)

A respondent narrated an incidence of a father rejecting a son's gift in public because he felt the son had neglected him for too long. The story has it that a father stopped by a "chopbar" (local restaurant) to eat. While there, the son came in with his friends, apparently to drink. On seeing the father, the son quickly walked to him and handed him GHC10 (about US$2) to buy food. But the father rejected the money, announcing to anyone listening that the son had abandoned him and was only giving him money as a public display of care.

These are some food-specific examples that interviewees gave to demonstrate that the elderly no longer enjoy the respect they deserve in society. Food cultures and traditions are very important elements in the food security equation in terms of food preference, identity and subjective well-being (CFS, 2017; Noack & Pouw, 2015). As shown below, decision making around household food purchases and preparation may also contribute to the worsening situation of people not eating what they want to eat.

Intersections of Gender and Generations in the Changing Food System

Food utilization is one of the key pillars in the Food and Agriculture Organization (1996) definition of food security, along with availability, access and stability. According to Polly Ericksen (2008), food utilization has three elements: nutritional value (how much of the daily nutrient requirements we obtain); social value (the cultural aspects of our consumption practices); and food safety (how safe or otherwise is the food we eat, as a result of its preparation or processing). It is the second of the three elements that is of interest here, one that has been given little attention in the food and nutrition security equation (Noack & Pouw, 2015). The cultural value of food utilization is important because observing culturally accepted norms and practices in our eating habits promotes subjective well-being in some ways: people feel really satisfied if food is served and eaten the way it ought to be, and in the right environment (CFS, 2017). This, then, borders on decision making regarding what is bought, prepared and served, and by whom. How and when the food is eaten are shaped by culture and they also express cultural values embedded in food utilization. Decision making and the observance of some of these cultural values are particularly complex in urban settings where cultural values, the nature of work, household gender and age dynamics, and social norms are rapidly changing (Riley & Dodson, 2019).

In traditional Dagbon culture, men and women have distinct roles in food preparation and utilization. Culturally, it used to be men's role to provide and ration out groceries for household use, while women processed the groceries into meals and served household members according to cultural norms. The traditional way of serving was that toddlers and the household head (whether male or female) ate alone, while any others ate in groups. Women, men and children would eat in groups and the size of the group would be determined by the total number of household members in each gender and age category. While serving food was generally considered part of food preparation, meat would be served by the household head or the eldest son. After the other food was served, the household head or eldest son was called upon to share the meat to every bowl or plate. The household head was supposed to start eating before anyone else. The elderly respondents gave two reasons for this. First, it was a sign of respect for the household head to do so. Second, the household head had to "verify" that the food was safe for the consumption of his or her household members. It was believed that the household head had the experience and "powers" to detect bad food and protect other members from eating it. In essence, he or she functioned as a laboratory and a "guinea pig" for the household food.

However, many of these cultural practices no longer apply in urban settings, significantly altering decision making related to food utilization. For example, it is no longer mandatory that men ration out groceries for household use. This was easier when a greater proportion of food came from own production and was stored in traditional granaries and barns knowns as *chenchaŋkuma* and *kambona,* and yam barns (*nyukori*). With the current dependence on the market, most urban households do not store grains in quantities greater than 100-kg bags, and most of this is purchased by women. The responsibility to ration food for the household now belongs to women. There are two reasons for this. First, it is believed that women interact with the market more often than men, and so have better knowledge of the food geography therein. The second reason relates to status. As reported earlier, the respondents believed that whoever depends on the market for food cannot enjoy the same food security status as the one who eats from his or her granaries. It is therefore seen as embarrassing for the man to be seen buying groceries from the market. What it means is that women now make the most important decisions that drive food utilization, from purchase and preparation to serving. This new role sounds empowering for women in Tamale, given that in most societies women decide what is eaten in the household (CFS, 2017). Women also now serve meat for household members. But behind all this is the increased role of women in the household food provisioning. A 64-year-old male respondent elaborated on this as follows:

> How do you dictate to her when she is the one that often buys the food [with her own money]? I told you that some of us have no work to do, we earn no income at all. And if you do not have older sons working to feed the family, then your wife takes the mantle by default because she cannot watch her children starve. Women are more resilient because they always have something to do to earn some income, petty trading. So I think it is useless to try and control her decision making because if she was not a good and responsible wife and mother, she would not work to feed the family. (Interview, February 22, 2020, Tamale)

Generally, men appreciate and respect their wives' efforts. They also feel, however, that they can only eat whatever the women decide to cook. They cannot really make a choice unless the woman seeks their opinion. Sometimes the men know that the women are struggling to buy basic food items, so there is no need to bother her to buy specific foods that might be very expensive. Some respondents expressed the opinion that women having control over household food provisioning is one of the factors that makes husbands and elderly men lose their status and self-esteem. They believe that women become pompous and disrespectful when they know they have more economic power than the men in their households.

Women's control over household food provisioning and utilization decisions has implications for household food security outcomes in Tamale. First, women's control over household food sourcing and utilization helps speed up the food system transformation unfolding in the city. This is seen in women using convenience as a factor in deciding which foods to buy and cook for the household; they tend to avoid more traditional foods, which many contemporary women struggle to cook well. For example, the elderly male respondents were of the view that many of the indigenous foods listed in Table 12.1 are no longer part of their regular meals because many of the young women who now take charge of cooking for their families do not know how to cook such foods. Instead, these young women prefer convenient foods, principally rice, which was never part of the regular dishes of the people. Rice was formerly a cash crop and only eaten occasionally during festivals and other social gatherings. Women also prefer modern, industrial spices and condiments instead of the traditional, more healthy ones such as dawadawa and nili.

The women who participated in this study, aged 30 and 42 years, explained that because of schooling and urban life, they did not get enough training on how to cook some of these foods. They also did not have adequate time to prepare some of the foods, and with respect to the popular nili, they claim that soup made from it easily turns watery, so they prefer using groundnut paste. The responses from one of these women were very interesting in that her husband boasted that he does not eat food cooked with Maggi or groundnut paste because the latter causes piles. He goes to the market himself to buy the nili and fish, and he knew his wife knew how to cook with these. The wife admitted receiving these ingredients from the husband but said she does not use them (but the husband did not know this); when she used them, she said, the children did not enjoy the food because they found it less tasty. She also saw on TV advertisements that these foods are good, otherwise why would they be advertised to people?

Another implication of women's control of food utilization decisions is that it exposes households to diverse foods, including meat. This is where women's empowerment has a positive effect on household nutritional outcomes, especially among children (CFS, 2017). Even though Table 12.1 shows diverse traditional dishes in Dagbon, the reality was different; the meals of households revolved largely around koko (breakfast) and saɣim (lunch and dinner). It is generally believed in the area that women tend to want to eat diverse, "nice" foods more so than men. A respondent remarked that a man could quarrel with his wife on suspicion that the wife spent her working capital on eating sweets and other foods considered luxuries, such as meat, cake and snacks. This happened when a wife sought the husband's financial support to boost her dwindling trade. It means that with women in control, they will likely diversify the household meals away from the koko and saɣim that appeared so monotonous. This has had significant effects on children, who now eat meat and eggs to boost their growth. In typical Dagbon cultural practices, it was not encouraged for children to eat meat and eggs, as it was believed that it would make them become thieves. So, when a fowl or guinea fowl was killed in the household, the meaty parts were consumed by adults, leaving only the feet, head and wings for children. However, with women controlling what is eaten in the household, households are now served diverse meals and children eat meat and eggs on a regular basis.

Conclusion

The food system in Tamale is transforming very fast. Looking back over several decades, this change has been brought on by several factors, including famines, the promotion of maize, urbanization pressures on land use, global food market integration and cultural globalization. On balance, these factors create vulnerability for some elderly and low-income households in Tamale who can no longer access the traditional foods they were used to, and the problem is compounded by the inability to prepare these traditional foods even where households are able to access some of them. Using the subjective well-being of four elderly men in Tamale as an analytical frame to understand the impact of food system transformations provided rich insights into a population who feel left behind by rapid change. Their food needs in terms of preference are seldom met, and they feel they are "living in the shadow of themselves." They feel their dignity and social status have been undermined by the inability to eat what they should be eating at their age. They didn't lack food to eat, but they also never felt satisfied with the foods they were eating, leaving them in perpetual hunger (subjectively). On the other hand, the food system changes are benefiting children, who used to be at the periphery of household food consumption. Young household members now have a variety of foods to choose from and the agency to make food choices. However, the long-term effects of children's exposure to the changing food system are beyond the scope of this chapter. For example, a growing preference for convenience foods may predispose children to poor nutritional outcomes such as overweight and obesity.

The study has policy and programmatic implications if the tenets of inclusive urban development are to be achieved. Working to achieve two of the Sustainable Development Goals (SDG 2: Zero hunger; and SDG 11: Make cities inclusive, safe, resilient and sustainable) requires attention to subtle differences in experience of different segments of the urban population in terms of vulnerability and opportunities, and attempting targeted interventions accordingly. In this regard, a practical starting point to address exclusionary urban development in the Ghanaian context will be to broaden the scope of Ghana's Livelihoods Empowerment Against Poverty (LEAP) programme, which was designed to alleviate poverty but which appears overly focussed on rural poverty. Among other goals, the LEAP programme seeks to provide social protection to vulnerable and poor Ghanaians (including the elderly poor) who lack productive capacity, by granting these unconditional cash transfers to help improve their living conditions (Government of Ghana, 2007). Given increasing urban poverty, together with the evidence presented in this chapter, it is imperative to broaden the focus of the programme to target the urban elderly, who have huge unmet needs regarding their food preferences.

From the perspective of practitioners, the evidence presented calls for a rethinking of how utilization is measured in the food security equation. For example, biological and anthropometric measures are often employed to gauge utilization (Swindale & Bilinsky, 2006); however, the elderly's subjective feeling of being food insecure suggests that the biological and anthropometric measures might not make for a comprehensive assessment of a population's or an individual's food security status. While the standard tests for food insecurity (Household Food Insecurity Access Scale, Months of Adequate Household Food Provisioning, and Household Dietary Diversity Score) include questions about household experiences of food inadequacy, the experiences of the elderly in this study show a highly complex set of experiences within households that may not be readily measurable. This is why it is recommended that practitioners and researchers take a comprehensive approach to assessing food security by broadening the indicators to include food preference, social and cultural dimensions and notions of subjective well-being (CFS, 2017; De Groote & Kimenju, 2012; Quisumbing, 2013; Webb et al., 2006). For example, assume that a nutritionist independently applies the biological and anthropometric measures to an individual and the results show the person is food secure in terms of utilization.

Yet the same person says he is perpetually food insecure because his desire to eat certain foods has not been met. Whose argument should the analyst take and why? Perhaps the non-alignment or low correlation between the Household Food Insecurity Access Prevalence results that indicated 70% of sampled households were food insecure, compared to only 2% of the sampled households falling below the acceptable threshold with respect to the Food Consumption Score in a study by Joanna Van Asselt et al. (2018) in Accra, may be explained by the subjective element of saying that one is food insecure. In short, the component of utilization of food that concerns cultural values and subjective well-being makes it difficult to quantitatively measure household or individual food security. Perhaps a food system assessment might prove a useful alternative in the context of the evidence in this chapter, as it not only captures the as-it-is situation of food security problems "but also provides an understanding of the context and dynamics that have led or are leading to a crisis" (Haysom & Tawodzera, 2018, p. 122).

References

Andam, K., Tschirley, D., Asante, S., Al-Hassan, R., & Diao, X. (2018). The transformation of urban food systems in Ghana: Findings from inventories of processed products. *Outlook on Agriculture, 47*(3), 233–243.

Battersby, J. (2017). Food system transformation in the absence of food system planning: The case of supermarket and shopping mall retail expansion in Cape Town, South Africa. *Built Environment, 43*(3), 417–430.

Bush, R. (2010). Food riots: Poverty, power and protest. *Journal of Agrarian Change, 10*(1), 119–129.

Committee on World Food Security. (2017). *Nutrition and food systems* (HLPE Report 12). High Level Panel of Experts on Food Security and Nutrition (HLPE), Committee on World Food Security (CFS).

Crush, J., & Frayne, B. (2011). Supermarket expansion and the informal food economy in Southern African cities: Implications for urban food security. *Journal of Southern African Studies, 37*(4), 781–807.

De Groote, H., & Kimenju, S. (2012). Consumer preferences for maize products in urban Kenya. *Food and Nutrition Bulletin, 33*(2), 99–110.

Demmler, K., Ecker, O., & Qaim, M. (2018). Supermarket shopping and nutritional outcomes: A panel data analysis for urban Kenya. *World Development, 102*, 292–303.

Demmler, K., Klasen, S., Nzuma, J., & Qaim, M. (2017). Supermarket purchase contributes to nutrition-related non-communicable diseases in urban Kenya. *PLoS ONE, 12*(9), 1–18.

Dowswell, C. (1989) *Feeding the future: Agricultural development strategies for Africa*. Centre for Applied Studies in International Negotiations (CASIN).

Ericksen, P. (2008). Conceptualizing food systems for global environmental change research. *Global Environmental Change, 18*(1), 234–245.

Food and Agriculture Organization (FAO). (1996). *Rome declaration on world food security and world food summit plan of action* (World Food Summit Report). FAO.

Government of Ghana. (2007). *The national social protection strategy (NSPS): Investing in people*. Government of Ghana, Ministry of Manpower, Youth and Employment (MMYE).

Haysom, G., & Tawodzera, G. (2018). "Measurement drives diagnosis and response": Gaps in transferring food security assessment to the urban scale. *Food Policy, 74*, 117–125.

Mensah, E. (1986). *Chemical, functional and processing characteristics of varieties of melonseeds: Agushie (Cucumeropsis Edulis Hook.F) and Neri (Citrullus Lanatus Var. Neri. Thunb)* (Master's thesis). Department of Nutrition and Food Service, University of Ghana.

National Research Council. (2006). *Lost crops of Africa: Volume II: Vegetables*. The National Academies Press.

Noack, A., & Pouw, N. (2015). A blind spot in food and nutrition security: Where culture and social change shape the local food plate. *Agriculture and Human Values, 32*(2), 169–182.

Nyantakyi-Frimpong, H., & Bezner Kerr, R. (2015). A political ecology of high-input agriculture in northern Ghana. *African Geographical Review, 34*(1), 13–35.

Oltmans, S. (2013). *A case study on the food retail environment of Accra, Ghana* (Master's thesis). Department of Community and Regional Planning, Iowa State University.

Peyton, S., Moseley, W., & Battersby, J. (2015). Implications of supermarket expansion on urban food security in Cape Town, South Africa. *African Geographical Review, 34*(1), 36–54.

Quisumbing, A. R. (2013). Generating evidence on individuals' experience of food insecurity and vulnerability. *Global Food Security, 2*, 50–55.

Reardon, T. (2015). The hidden middle: The quiet revolution in the midstream of agri-food value chains in developing countries. *Oxford Review of Economic Policy, 31*(1), 45–63.

Reardon, T., Henson, S., & Berdegue, J. (2007). 'Proactive fast-tracking' diffusion of supermarkets in developing countries: Implications for market institutions and trade. *Journal of Economic Geography, 7*, 399–431.

Riley, L., & Dodson, B. (2019). *The interface between urbanization, gender and food in the Global South* (HCP Discussion Paper No. 36). Hungry Cities Partnership.

Rischke, R., Kimenju, S., Klasen, S., & Qaim, M. (2015). Supermarkets and food consumption patterns: The case of small towns in Kenya. *Food Policy, 52*(C), 9–21.

Searcey, D., & Richtel, M. (2017, October 2). Obesity was rising as Ghana embraced fast food. Then came KFC. *The New York Times*.

Skinner, C. (2016). *Informal food retail in Africa: A review of evidence* (Consuming Urban Poverty Working Paper No. 2). University of Cape Town, Cape Town, SA.

Skinner, C., & Haysom, G. (2016). *The informal sector's role in food security: A missing link in policy debates?* (Working Paper No. 44). PLAAS, UWC and Centre of Excellence on Food Security.

Smit, W. (2016). Urban governance and urban food systems in Africa: Examining the linkages. *Cities, 58*, 80–86.

Swindale, A., & Bilinsky, P. (2006). *Household dietary diversity score (HDDS) for measurement of household food access: Indicator guide (version 2)*. FHI 360/FANTA.

Szabo, S. (2016). Urbanisation and food insecurity risks: Assessing the role of human development. *Oxford Development Studies, 44*(1), 28–48.

Tsrah, P., Quarpong, W., & Laar, A. (2020). Healthiness of foods on promotional flyers of fast-food outlets located within Accra-based shopping malls. *World Nutrition, 11*(3), 51–61.

Van Asselt, J., Masias, I., & Kolavalli, S. (2018, November). *Measures and determinants of urban food security: Evidence from Accra, Ghana* (GSSP Working Papers). IFPRI.

Vercillo, S., Weis, T., & Luginaah, I. (2020). A bitter pill: Smallholder responses to the new green revolution prescriptions in northern Ghana. *International Journal of Sustainable Development & World Ecology, 27*(6), 565–575.

Weatherspoon, D., & Reardon, T. (2003). The rise of supermarkets in Africa: Implications for agrifood systems and the rural poor. *Development Policy Review, 21*(3), 333–355.

Webb, P., Coates, J., Frongillo, E. A., Rogers, B. L., Swindale, A., & Bilinsky, P. (2006). Measuring household food insecurity: Why it's so important and yet so difficult to do. *Journal of Nutrition*, 136, 1404S–1408S.

Wenban-Smith, H., Faße, A., & Grote, U. (2016). Food security in Tanzania: The challenge of rapid urbanisation. *Food Security, 8*(5), 973–984.
Worku, I., Dereje, M., Minten, B., & Hirvonen, K. (2017). Diet transformation in Africa: The case of Ethiopia. *Agricultural Economics, 48*(S1), 73–86.

CHAPTER 13

Understanding Food Security and Hunger in Xai-Xai, Mozambique

Inês Macamo Raimundo and Mary Caesar

> What we Xai-Xai dwellers need now is a dish of *xima* meal on our households' tables. 'Food security' is your matter. (Respondent in Xai-Xai during 2017 field work)

Introduction: The Practical and Symbolic Dimensions of Food Security

Despite the availability of food in the local markets of Xai-Xai, Mozambique, people who participated in the qualitative study that inspired this chapter agreed unanimously that what they need is a dish of *xima*

I. M. Raimundo (✉)
Faculty of Arts and Social Sciences, Department of Geography and Center for Policy Analysis, Eduardo Mondlane University, Maputo, Mozambique

M. Caesar
Balsillie School of International Affairs, Wilfrid Laurier University, Waterloo, ON, Canada

L. Riley and J. Crush (eds.), *Transforming Urban Food Systems in Secondary Cities in Africa*,
https://doi.org/10.1007/978-3-030-93072-1_13

(pronounced sheema). They encapsulated the experience of food insecurity and hunger by stating their desire for this ubiquitous staple dish. What does this seemingly ordinary statement indicates about the importance of *xima* in the Xai-Xai household diet and, more specifically, the ways in which people articulate their experiences of hunger and food insecurity?

At the most basic level, expressing a desire for *xima* may indicate that low-income households often find themselves without enough cash to buy adequate food, including the ingredients needed for a dietary staple such as *xima*. It is well documented that low-income households may go for days or months without adequate food, even while formal and informal markets and bazaars are fully stocked with foods that are not sold and eventually rot (Chikanda & Raimundo, 2017; McCordic, 2016; McCordic & Abrahamo, 2019).

Another possible explanation is that the desire for *xima* is a matter of food preference, meaning that *xima* has a cultural importance that trumps the availability and accessibility of other food items. This would not be altogether unexpected. Variations of *xima* are found across southern and eastern Africa, and prepared and consumed according to local customs. In Xai-Xai, maize, cassava, sorghum and millet are used to prepare the stiff porridge normally consumed with green leafy vegetables, fish or other seafood, or with meat. Maize has become the preferred ingredient for *xima* in Mozambique, as elsewhere in the region, because it is widely available, relatively easy to cook and its flavour is preferred (Ekpa et al., 2019; Lusambili et al., 2020).

A third way to understand the respondent's statement is the possibility that Xai-Xai residents understand food security as the ability to cultivate the foods they need, including maize for *xima*, in a backyard garden. The disappearance of the backyard garden may be causally connected to a variety of factors, including increasing political and social pressures for urban-dwellers (whether established or newly arrived from rural areas) to conform to expectations of modern, "big city" norms, such as keeping gardens of flowers rather than fruits and vegetables, obtaining food exclusively from retail sources and eating a diet of processed foods purchased at chain grocery stores. Decades of increasing climate shocks, such as cyclones, floods, coastal erosion and droughts, have also devastated food production in Xai-Xai, including at the scale of the backyard garden.

Food security studies on secondary cities in Mozambique are scarce, which makes it difficult to accurately understand food access, availability and especially food stability in these geographical areas. In addition, the quantitative methodologies applied to understand food security in primary or capital cities such as Maputo do not allow us to understand the cultural experiences of food security in secondary cities. A review of case study evidence from Mozambique suggests an overreliance on quantitative methods and assessment tools. These methods and tools allow for a high-level view of food insecurity issues, resulting in comparative studies that can be delivered with time- and cost-effectiveness, but they may come at the expense of discovering the real-life experiences of hunger and food security in local communities and households.

For the Government of Mozambique, the goal of understanding food security is reducing poverty. To achieve this goal, various ministries compile data on food status. One of the main information tools is the Integrated Agricultural Survey (IAS), which collects and processes data on the value chain of agricultural activity, and food and nutrition security, including: the number and size of farms; the number and size of farming households; agricultural activity (livestock or agricultural production); the types of crops grown and yield per farm. When President Filipe Nyusi launched the IAS in 2020, he stated that the primary goal of the survey was to collect "reliable, good quality national statistics" to inform planning, implementation and assessment of development programmes (allAfrica, 2020). The Ministry of Agriculture and Rural Development, together with the National Statistics Institute, are the main implementers of the IAS. Another source of government information is the Food and Nutrition Security survey conducted by the Technical Secretariat for Food and Nutrition Security (Secretariado Técnico de Segurança Alimentar e Nutricional), under the direction of the Ministry of Agriculture and Food Security (Ministério da Agricultura e Segurança Alimentar [MASA]). The latter publishes national- and provincial-level information, but excludes secondary cities and small administrative units. The MASA website provides information on nutrition, hunger and household income with a view to measuring chronic food and nutritional insecurity.

As two of the principal sources of government information on nutrition, food production and the value chain, these instruments and reports contain significant gaps regarding hunger and food insecurity experiences. The IAS does not gather data on vegetables and indigenous and/or locally cultivated food, which are the very foods that households typically consume and are most concerned about. Yanick Borel Kamga (2021, this volume) demonstrates the importance of indigenous foods in the diets of households in a secondary city in Cameroon, and his findings resonate with the experience of co-author Inês Raimundo in Xai-Xai. The official data gap in Mozambique is precipitated by the employment of and overreliance on quantitative methods.

Similarly, academic research on food security in Mozambique continues to focus mainly on metropolitan areas and/or large cities and tends to employ quantitative as opposed to qualitative research methods. Two international urban food security research projects, the African Food Security Urban Network (AFSUN) and the Hungry Cities Partnership (HCP), conducted multiple representative surveys on urban household food security and the urban food system, beginning in 2008. The survey instruments employed by both projects were the food security measurements designed by the Food and Nutrition Technical Assistance Program, including the Household Food Insecurity Access Scale (HFIAS), the Household Dietary Diversity Score (HDDS), the Months of Adequate Household Food Provisioning (MAHFP), the formula determining food poverty lines, and an income and expenditure survey (Battersby & Crush, 2016; McCordic, 2016; Raimundo et al., 2016; Riley & Caesar, 2017). These research projects continue to make valuable contributions to understanding, on a large scale, urban food security. They make it possible to measure numerous aspects of the urban food system in Mozambique, and in particular its major cities, such as Maputo and its suburbs. However, like the government information sources, the projects focus primarily on the major urban centres and use quantitative surveys as the dominant research method.

Methodologically, a mixed methods approach to complex or multifaceted problems such as hunger and food security will yield a richer dataset than a single method. In light of the preponderance of quantitative research on the subject, this chapter aims to highlight the invaluable role of qualitative research methods, with a special emphasis on autoethnography. The goal is to adopt an approach, or approaches, that can deepen our understanding of hunger and urban food security at the scale of human experience. For example, qualitative research methods can answer the following questions:

- What really matters to people with regard to food and food security?
- What exactly does it mean for them to be food secure or insecure?

- How do households remain food secure in the context of climate shocks?
- How do cultural norms and historical relationships with food inform people's understanding of hunger and food security?

This chapter is an attempt at critical engagement with the widely used concept of food security and how food security experiences are informed by climate shocks and by people's cultural experiences of food. Additionally, and perhaps because of this context, we suggest qualitative as opposed to quantitative research methods would be more effective to achieve this goal. The dichotomy between qualitative and quantitative methods is not novel in feminist scholarship (Oakley, 1998; Westmarland, 2001); however, the dominance of the latter method has led us to conclude that the time has come to employ qualitative research methods to round out and expand our understanding of food security experiences in Mozambique and in its secondary cities in particular.

The chapter is informed by empirical research conducted by Inês Raimundo in Xai-Xai in 2017 and by her recollections of growing up in Mozambique. Dr. Raimundo was born and raised in Xai-Xai and spent many hours as a child listening to the stories told by her grandmother and mother about the history of Xai-Xai and about people's daily lives. In addition, as an adult, she makes regular trips back to Xai-Xai to visit relatives and also as a scholar. She has witnessed significant transformations in food access, food availability and dietary changes, more often than not imposed by external forces.

The next section of the chapter begins with a discussion of the tools available to measure food security, focusing on the fact that the use of certain methods will yield a certain understanding of food security. This is followed by contextual information on Xai-Xai. The chapter then discusses the research evidence around hunger and food security experiences, as expressed by household members, and historical ways in which community members have navigated food insecurity. The final two sections consider the broader context that continues to impact food security and hunger: first, "planting cement," instead of vegetable gardens, as a marker of urbanization and modernization; and second, contending with climate shocks.

Measuring Food Security: The Results Follow the Tools

One way of measuring food security is by employing quantitative research methods. These can be used to understand, for example, the extent of food security in a specified research location or multiple locations for comparative purposes, and to enumerate some of the reasons why households experience food insecurity. With quantitative methods, the notion of food security is predetermined and coded, with no room for explanation and subjectivity. When using qualitative research methods, the researcher allows the participant to articulate their food security experience in their own words and what emerges is a more nuanced understanding of food security.

Starting in 2008, researchers and graduate students associated with AFSUN and HCP conducted urban food security research. As the research project leader for Mozambique, Raimundo collaborated with a number of these national and international researchers to design research projects, facilitated fieldwork on urban food security studies and co-published a number of publications (Raimundo et al., 2014, 2018). As a result of this level of involvement in research projects, she became familiar with the technical language of food security employed by quantitative surveys such as the HFIAS, HDDS and the MAHFP. These measurements are not universal, but have become widely used indicators to measure the extent of household food (in)security. By employing these tools, the experiences of research participants are reduced to pre-coded answers of "yes," "no," "often," "sometimes" and "I don't know," or other scales such as "often," "randomly," "sometimes" and "never." In addition, translating descriptions such as "food secure," "mildly food insecure," "moderately food insecure" and "severely food insecure" into Mozambican national languages complicates the data collection due to the insignificant difference that exists between these categories.

For these reasons, this chapter argues in favour of interviews and life histories to complete the map of food insecurity. These methods are the only way for the researcher to gain insight into the food production and food security situation over the last three decades, a time period used by meteorologists to calculate significant changes in patterns. Statistical analyses and mathematical predictive models do not bring to light what is going on in reality, nor do they deliver a full representation of food insecurity. To what extent can the data produced by these methods respond to the statement: "We don't have *xima* on our plates, therefore we are not food secure"? How can one be food secure if accessing food depends on having hard currency? What happened to the backyard fruit and vegetable production that once formed the foundation of food security for city dwellers? The interviews conducted in Xai-Xai in 2017 and the auto-ethnographic methods employed by Raimundo come much closer to understanding the subjective experience of hunger and food insecurity.

Understanding Xai-Xai Through Interviews and Auto-Ethnography

This chapter relies on qualitative research conducted by Raimundo during a 2017 African Urban Research Initiative (AURI) research project entitled "Understanding Informal Systems in a Medium Sized City: A case study of Xai-Xai." The project was part of a larger research agenda entitled "Fostering a Comparative Research Agenda in African Cities: Urban Spatial Inequalities and the New Urban Agenda," which was sponsored by the African Centre for Cities at the University of Cape Town and the Ford Foundation.

Raimundo's project investigated the impact of rapid urbanization in small and medium-sized cities, which has been a concern in Mozambique. These cities have shown faster growth since the late 1980s (INE, 1980, 1999, 2009, 2019). This rapid growth reflects the effects of political instability (a civil war that lasted 16 years), the post-war reconstruction period following the signing of the General Peace Agreement in 1992, and the many floods and drought that have hit the country over the past four decades (Muanamoha & Raimundo, 2018a, 2018b; Raimundo, 2020). Having the right to live where one chooses (such as in a city), along with political reforms, natural disasters and the impact of Structural Adjustment Programs, have had profound impacts on migration patterns and fertility rates, which have, in turn, made small and medium-sized cities unsustainable to live in. Urbanization brings challenges to all who must permanently deal with the phenomenon. However, social cohesion cannot exist while large sections of a country's population are deprived of basic needs and live under permanent stress due to the vulnerabilities inherent in an informal economy, a shortage of land for new building, and environmental problems (such as major storms leading to massive coastal erosion) that pose threats to human security and food security.

Xai-Xai represents challenges that city dwellers face on a medium scale. Xai-Xai has been an official autarchy within Mozambique since 1997, but the management system is still centralized and participatory processes are not visible. Clearly, the various priorities of city dwellers, city managers and civil society are not aligned. Medium-sized cities are neglected in various studies on Mozambique, since they are not seen to produce significant wealth through either economic activity or taxation. They are seen as less essential than larger centres, thus their invisibility to policymakers. In this context, the study looked at how city dwellers cope with living in a coastal city that experiences yearly cyclones and floods, dangerous levels of coastal erosion and a high level of informality in terms of settlement and the economy. It also looked at how residents of Xai-Xai cooperate with city managers in the process of city management.

The research was developed using qualitative tools such as interviews and focus group discussions. Raimundo interviewed those who expressed interest in the study, and sought out key informants who could provide information about the city, including municipal authorities (an urban councillor and the councillor for education and housing), and in the neighbourhoods she interviewed the heads and secretaries of the wards. Interviewees were organized into the following groups: councillors and bureaucrats from the municipal council; local administrative leaders, including the head of the administrative post, the secretary of the wards and the chiefs of the *quateiroes*; city dwellers in their respective wards; and traders. The interviews took place from October 5 to 10, 2017.

In addition to the interviews, the tools of auto-ethnography were used to shape the recollections of Raimundo, of food and hunger in Xai-Xai when she was a child. Auto-ethnography, first employed in 1979, is "an approach to research that seeks to describe and systematically analyse personal experience in order to understand cultural experience" (Ellis et al., 2011). It is particularly appropriate in this case because it is an approach to research that "challenges canonical ways of doing research and representing others and treats research as a political, socially-just and socially-conscious act" (Ellis et al., 2011). Given the climate disasters that plague Mozambique, the south in particular, reducing respondents' experiences of food security to predetermined quantitative measurements, a method that has come to dominate food security research, fails to uncover the cultural and personal meaning of food and thus the entirety of the food security experience.

Raimundo was born and raised in Xai-Xai and spent hours listening to her grandmother talk about the past and her mother talk about the present. This is how she came to learn the history of Xai-Xai and heard descriptions of people's daily lives. Even after leaving Xai-Xai to study, get married and work in Maputo, she remained connected to this city, as the family still maintains the house of their late parents and regularly visits. In addition to these family ties, Raimundo has come to view Xai-Xai through the eyes of a scholar and has witnessed significant transformations in food access (including changes in backyard production), food availability and diets, and the impact of external factors such as climate shocks.

Xai-Xai: A Place Made Food Insecure by Climate Shocks

Xai-Xai, the capital city of Gaza Province, is located in the south of Mozambique and borders on Chibuto and Mandlakaze Districts to the north, Chokwe District to the west, Limpopo and Bilene Districts to the south and the Indian Ocean to the east. The city has four administrative areas, covering about 132 square kilometres at an average altitude of 9 m above sea level. The main economic activity in Xai-Xai is agriculture (Mazuze, 2019; Xavier, 2019).

The city is divided into a lower zone and an upper zone by the Nguluzangue River, which flows towards the Indian Ocean where it meets the Limpopo River. The city is situated in a zone that is prone to cyclones and floods, below the Limpopo River and about 50 kms from the mouth of the Indian Ocean. Table 13.1 shows that the population of this secondary city doubled between 1980 and 1997; from 1980 to 2017, it grew faster than both the provincial and the national population. However, the growth rate slowed down between 2007 and 2017

Table 13.1 Population Growth of Xai-Xai, Gaza Province and Mozambique between 1980 and 2017

Location	*1980*	*1997*	*2007*	*2017*
Xai-Xai	44,000	99,442	115,752	141,963
Gaza Province	982,603	1,062,380	1,236,284	1,446,654
Mozambique	12,500,000	15,278,334	20,252,223	27,909,798

Sources DNE (1980, 1999, 2009, 2019)

as a result of a decline in fertility rates, according to local demographers (Arnaldo, 2013; Arnaldo & Muanamoha, 2013).

Climate Shocks and Food Security: The Floods of 2000 and 2013

Any understanding of food security and any programmes aimed at improving household food security should take into account the impact of climate shocks, especially the devastating effect on urban agriculture and, more generally, food production and supply. The impact of climate shocks on Xai-Xai is evident in massive destruction of infrastructure and death. In 2000, heavy rainfall lasted five weeks and caused the Limpopo River to overflow, flooding villages and cities. Cyclone Leon-Eline caused further flash floods. A total of 700 people died and more than 500,000 lost their homes, becoming internally displaced (UNICEF, 2000). In addition to the loss of life and destruction of infrastructure, large stretches of fertile farmland and crops were destroyed: about 12% of cultivated land and 90% of irrigated land in five provinces, with the largest impact on Gaza Province (USAID, 2002). The affected crops included maize, beans, rice, sweet potatoes and peanuts, and more than 50,000 head of cattle died (UNICEF, 2000). The economic impact was marked by a loss of US$600 million, which caused the gross domestic product to drop from 7.5% growth in 1999 to 1.6% in 2000 (República de Moçambique, Conselho de Ministros, 2017).

Xai-Xai households lost the ability to achieve and/or maintain food security at least two decades ago, and households throughout the country have struggled to recover from the 2000 floods and cyclone. In a city that was transformed by a massive climate shock, Xai-Xai residents now consider themselves food secure only when they have an *escudo* (the Portuguese colonial currency in use until 1980) to buy a bag of maize or they have a dish of *xima* to consume. In the past, Limpopo Valley was considered the granary of the country, with communities producing surpluses of crops; now the same communities are dependent on food aid (Mazuze, 2019; Raimundo, 2019). It has become a struggle to cultivate crops for household, local and national consumption between waves of water on the one hand, and droughts and cyclones on the other. The Mozambique Master Plan for Disaster Risk Reduction 2017–2030 indicates that from 1980 to 2016, the country was ravaged by floods, droughts, tropical cyclones and epidemics (Table 13.2). These events disrupted the economy, eliminated any economic growth that had been

Table 13.2 Natural Disasters and Epidemics between 1980 and 2016

Number	*Event*	*Number of occurrences*
1	Floods	27
2	Drought	12
3	Tropical Cyclones	16
4	Epidemics	27

Source República de Moçambique, Conselho de Ministros (2017, p. 10)

forecast, and caused infrastructure damage that extends from household assets to state assets.

Gaining a Cultural Understanding of Hunger and Food Security

> Food, food and food is part of our life, but which food? Are you telling me that mayonnaise, *pallone* (bologna), *Boerewors* (sausage), hamburgers, fried chicken, is food? (Interview, October 7, 2017, Patrice Lumumba Neighbourhood, Xai-Xai)

Raimundo's siblings and herself grew up in a suburb of Xai-Xai where their parents' backyard garden had papaw trees, lemon trees, *chimunhamunwane* (a small, black, chewy wild fruit), blackberries, mangoes, cocoa and *matsawo* (pumpkin leaves). As a matter of fact, blackberries, *chimunhamunwane* and papaws constituted their fence. At that time, her mom sometimes used to go to the bazaar to buy fresh fish or dried prawns to prepare *mathapa* (smashed cassava/manioc leaves, cooked with coconuts, groundnuts and fresh or dried prawns) or to buy butter beans for soup. The bean soup was considered a luxury as it contained olive oil, some beef bones and cabbage, and it was only served for Sunday lunch. During this period of her childhood, her mom sourced most of their fresh seafood from a lady who lived by the sea, while fruit and vegetables were grown in the backyard. They could seldom eat apples, grapes, peaches and kiwis, as her parents could not afford to buy these Portuguese fruits and because they had blackberries, *chimunhamunwane* and papaws in their yard. The only fruits that her parents could afford to buy were tangerines, oranges and bananas.

This is the city where she was raised about 50 years ago. She recalls the 1978 floods, but never heard about cyclones or drought in this region. Nowadays, the landscape of the city has changed dramatically. The population has grown rapidly and together with certain notions of modernization and urban development, the *chimunhamunwane*, mango trees, papaws and blackberries have been replaced by bricks. Xai-Xai has become a place where people "plant cement" instead of fruits and vegetables. In addition, heavy rains, floods and cyclones have become commonplace events in the twenty-first century:

> Floods, cyclones and hailstorms never seem to let up. We are busy trying to fix the roof tiles broken by hailstorms or that were thrown away by cyclones or washed away by floods. Look over there at the lower areas. They are permanently flooded as the Nguluzangue River is no longer taking its water to the mouth. People have blocked the river with new construction. There are some people who settled here during the Matsanga War[1] as displaced persons. They occupied the entire valley and obstructed the way out of the river. (Interview, October 5, 2017, Patrice Lumumba Neighbourhood, Xai-Xai)

The first time she saw her mom going to a bazaar to buy some vegetables was during the 1980s because, firstly, the FRELIMO-led, post-independence government's discourse regarding urbanization and modernization dictated that people living in the city were not meant to grow any vegetables. This was also the time when she witnessed her parents cutting down the *chimunhamunwane* and blackberries and replacing them with a backyard fence made of reed. Secondly, the country was in the midst of a devastating civil war that severely interrupted food production and supply. Their Portuguese pastor said that the country was so cursed that even a chicken could not lay an egg.

Through the stories told by her grandmother, usually around the fire, she learnt about hunger in remote areas of Mozambique. Some of these

[1] After more than 450 years of Portuguese colonial rule, two factions, the National Resistance of Mozambique (RENAMO) and the Liberation Front of Mozambique (FRELIMO), emerged in Mozambique. First they fought for liberation and second, between 1977 and 1992, for dominance. André Matsangaíssa was the first leader of RENAMO. Locally, people do not refer to this conflict as a civil war, but rather as the Guerra de Matsanga, which means the War of Matsangaíssa. See, for example, Emerson (2014) and Morier-Genoud et al. (2018).

tales remained with her throughout her childhood and shaped her views on how food matters and how different phases of food insecurity can drive people to do silly things.

The Man Who Stole Food from His Wife and Children's Pot

One story that her grandmother told was about a married man who, at the time of hunger in the remote districts of Chibuto-Manjacaze and Panda in Gaza and Inhambane provinces, respectively, was summoned by the chief of the village to be part of the gazelle hunt. After the hunt, the chief distributed the meat to each family and also slaughtered some cows to distribute among his vassals. That man was concerned about high rates of hunger, so he did not inform his wife and children about the meat distributed by the chief. Rather, he told them that the chief advised them to store food because they would face hunger sometime in the future. The man further advised his wife and children to take care of *cacana*, a local name for the plant *Momordica balsamica*, of which the fruit and leaves are edible; he told them to dry *matsawo* or *nhemba* beans or whatever food items they managed to collect. He told them they were not to hunt animals or kill any cattle for consumption. The wife survived because she fed her children *cacana*. This man, however, hid in the bushes and stole food that the wife prepared for herself and the children. Later on, his family discovered that he had deceived them. As she was taught to be an obedient and good wife, she pardoned her husband and even prepared *cacana* for him as part of the family.

The Son-In-Law Who Stole the Chicken Leg

Another story the grandmother told her was of a son-in-law who visited his wife's family for the first time and learned that one should not show hunger in that family. If he was served chicken, the only piece that he should eat was the gizzard (the muscular, walled part of the stomach) However, this man liked the leg so he made a plan to steal the chicken leg during the night. When he was caught, he hid the drumstick in his hut, but the groundnut curry sauce that the chicken had been cooked in was flowing down his face. He argued that it was a kind of sweat characteristic in his homeland.

The Man Who Hid the Bread Under His Armpit

At one point her mom hired a man to paint their house. This gentleman, who was their neighbour, was very greedy. Her mom went to his house early one morning while he was eating a sandwich made of fresh bread and sardines. When he heard her mom asking to enter, he hid the bread under his armpit. Her mom saw it and she was embarrassed because she had learned from her parents that food was meant to be shared. However, she also understood that this man had very little food and feared he would starve if he shared the bread with her mom.

These stories demonstrate the significance of food, not merely to people who are hungry but to the community. For example, it highlights some of the strategies that people have used to avoid hunger in order to guarantee food security. It is also these accounts of hunger and food security that question the value of a purely technical understanding of food security.

According to MASA (2016), the country recorded higher rates of acute food insecurity in 2005 and 2016 due to drought in the central and southern regions, which means that for thousands of households, food was and remains a luxury. The commercialization of the entire food system has only made the situation worse for residents in Xai-Xai.

> Nowadays you are no longer authorized to enter into someone else's *machamba* (land) and take some manioc or any fruit without being accused of theft. You have to buy. These are some changes that we are facing here. In the past it was accepted that someone could enter any *machamba* and take a piece of whatever he or she found to kill hunger. (Interview, October 5, 2017, Neighbourhood of Coca-Missava, Xai-Xai)

Nowadays, in this city where Raimundo was born, the fruits and vegetables that she had access to have been replaced by imported apples, grapes, peaches and kiwis—some of the very same fruit that her mother found unaffordable. One is barely able to find fish, mussels, clams or prawns. These have been replaced by pallone (bologna), Boerewors (sausage) and "plastic" chicken brought by *mukheristas* (cross-border traders) and sold at supermarkets and informal markets.

Yet the question remains whether food security and hunger as experienced by communities and households are the same concepts measured by researchers and governments, or are they two faces of the same coin?

More important is the cumulative effect of a changing food system—the dominance of food retail; the disappearance of the backyard garden and, with it, affordable, locally produced maize; increased access to previously unaffordable food items, which are still unaffordable; and climate shocks—submitting local households and Xai-Xai residents to the mercy of market forces and disrupting cultural attachments to specific food items? Since consuming food goes beyond filling an empty stomach, for Xai-Xai residents and households, food security and hunger consist of more than accessing any food or adapting diets to eat food items available at local markets.

In recent decades, multiple stresses, including climate shocks, have severely impacted food security in Xai-Xai. Members of the Xai-Xai community are forced to bear the cost of these changes on food access and diet. Whereas Raimundo's family and nearby neighbours managed to eat maize and rice daily, and chicken or beef on Sundays or special holidays such as Christmas, Easter or New Year's Eve, on ordinary days they would eat *matsawo* and *cacana*. These vegetables were not purchased, as they grew in their backyards.

Conclusion

It is clear that food security in Mozambique poses a considerable challenge, made worse by climate shocks. It is imperative that we in the academy develop a clear understanding of hunger and food security within households and communities. Fieldwork done through the AFSUN and HCP research networks has generated information regarding the status of food security in Maputo and Matola, the largest cities of Mozambique. The Food, Urbanization, Environment and Livelihoods (FUEL) project is the first AFSUN project and HCP-associated project that focuses specifically on food security and hunger in secondary cities. Using their 2017 research done under the auspices of AURI, the authors took the opportunity provided by FUEL to revisit their findings from Xai-Xai, which sought to understand food security in secondary cities by employing qualitative methodologies.

The current analysis goes beyond the dominant traditional indicators used to analyze food security, such as the HFIAS, HDDS, MAHFP and scales of measurement that include "often," "randomly," "sometimes" and "never" in relation to eating food. The inspiration to adopt a set of qualitative research tools arose when some of the participants in the 2017 focus group discussions in Patrice Lumumba and Coca-Missava neighbourhoods opined that food security is *your* matter, that is, a researcher's matter or the government's matter, because Xai-Xai dwellers only need a dish of *xima* meal for their households. This statement was rooted in the respondents' experience of hunger and starvation, and their acknowledgement that they no longer know how to access food outside the markets as they would have done in the past. Many Xai-Xai residents experience hunger and consider themselves food insecure despite the fact that the markets are stocked with diverse types of food ranging from vegetables, meat, cereals, tubers and all types of fruits. While it is true that hunger and poverty are part of the food security experience, it should also be true that city dwellers of Xai-Xai (and elsewhere) have the right to define their food security status for themselves. While the use of quantitative tools and assessments can be very useful for large-scale studies of food security, this chapter is an attempt to challenge scholars to broaden the concept of food security and expand the methodological tools employed.

To conclude, in Xai-Xai, measuring food security has to do with understanding the importance of *xima*, *cacana* and *matsawo* in the household diet and not necessarily asking about food items such as beef, bread, cooking oil or household income and expenditure. The changes in the Xai-Xai diet were also influenced by the city's vulnerability to what have become commonplace climate shocks and by misguided notions of modernization and urban development where households "plant cement" instead of backyard gardens.

References

allAfrica. (2020, September 30). *Mozambique: Nyusi launches integrated agricultural survey*. allAfrica report. http://www.allafrica.com/stories/202009300880.html

Arnaldo, C. (2013). Fecundidade me Moçambique nos Últimos 50 Anos: Alguma Mudança? In C. Arnaldo & B. Cau (Eds.), *Dinâmicas da População e Saúde em Moçambique* (pp. 37–60). CEPSA.

Arnaldo, C., & Muanamoha, R. (2013). Tendências e Desafios do Crescimento da População em Moçambique. In C. Arnaldo & B. Cau (Eds.), *Dinâmicas da População e Saúde em Moçambique* (pp. 7–36). CEPSA.

Battersby, J., & Crush, J. (2016). The making of urban food deserts. In J. Crush & J. Battersby (Eds.), *Rapid Urbanisation* (pp. 33–46). Springer.

Chikanda, A., & Raimundo, I. (2017). Informal entrepreneurship and cross-border trade between Mozambique and South Africa. *African Human Mobility Review, 3*(2), 943–972.

Ekpa, O., Palacios-Rojas, N., Kruseman, G., Fofliano, V., & Linnermann, A. (2019). Sub-Saharan African maize-based foods: Processing practices, challenges and opportunities. *Food Reviews International, 35*(7), 609.

Ellis, C., Adams, T., & Bochner, A. (2011). Autoethnography: An overview. *Forum: Qualitative Social Research (FQS), 12*(1), Article 10.

Emerson, S. (2014). *The Battle for Mozambique: The Frelimo-Renamo Struggle, 1977–1992*. Helion and Company Ltd.

DNE. (1980, 1999, 2009, 2019). Instituto Nacional de Estatística (INE). www.ine.gov.mz

Lusambili, A., Naanyu, V., Manda, G., Mossman, L., Wisofschi, S., Pell, R., Jadavji, S., Obure, J., & Temmerman, M. (2020). Nutritional influences on the health of women and children in Cabo Delgado, Mozambique: A qualitative study. *International Journal of Environmental Research and Public Health, 17*(17), 6205.

Mazuze, A. (2019). *Transição agrária e transformações sociais locais nos países em vias de desenvolvimento: O caso do projecto WAMBAO no Distrito de Xai-Xai, 2012–2014* (Master's thesis). Faculdade de Ciências Sociais e Filosofia, Universidade Pedagógica de Maputo.

McCordic, C. (2016). *Urban infrastructure and household vulnerability to food insecurity in Maputo, Mozambique* (PhD dissertation). School of Environment, Resources and Sustainability, University of Waterloo.

McCordic, C., & Abrahamo, E. (2019). Family structure and severe food insecurity in Maputo and Matola, Mozambique. *Sustainability, 11*(1), 267.

Ministério da Agricultura e Segurança Alimentar (MASA). www.agricultura.gov.mz

Morier-Genoud, E., Cahen, M., & Do Rosário, D. (2018). *The War Within: New Perspectives on the Civil War in Mozambique, 1976–1992*. Boydell and Brewer.

Muanamoha, R., & Raimundo, I. M. (2018a). Cartografia das migrações internas de Moçambique. *Revista Interdisciplinar da Mobilidade Humana, 26*(54), 31.

Muanamoha, R., & Raimundo, I. M. (2018b). Migrações Campo-Cidade e Integração no Meio Urbano. In J. Feijó e I. Raimundo (Eds.), *Migrações internas para áreas de concentração de Grandes Projectos*. Publifix.

Oakley, A. (1998). Gender, methodology and people's ways of knowing: Some problems with feminism and the paradigm debate in social science. *Sociology, 32*(4), 707.

Raimundo, I. (2019, September 18–20). *The Lower Limpopo Valley communities: Paying the price for climate change*. Paper presented at "Source, Infrastructures and the Anthropocene: Dialogues between the Global-North and the Global-South conference", Lisbon, Portugal.

Raimundo, I. M. (2020). O Ciclo Vicioso de Deslocamentos Forçados e a Formação de Espaços Incompletos em Moçambique. In *Olhares Geográficos sobre o Moçambique* (organizado por Edvaldo Cesar Moretti). Geo UERJ (e53912).

Raimundo, I., Crush, J., & Pendleton, W. (2014). *The State of Food Insecurity in Maputo, Mozambique* (Urban Food Security Series No. 20). African Food Security Network (AFSUN).

Raimundo, I., Crush, J., & Pendleton, W. (2016). Food insecurity, poverty and informality. In J. Crush & J. Battersby (Eds.), *Rapid Urbanisation* (pp. 71–84). Springer.

Raimundo, I., McCordic, C., & Chikanda, A. (2018). *The State of Household Food Security in Maputo, Mozambique* (HCP Report No. 10). Hungry Cities Partnership Project.

República de Moçambique, Conselho de Ministros. (2017, October 17). *Plano Director para a redução do risco de desastres 2017–2030*. Aprovado pela 36a Sessão Ordinária do Conselho de Ministros, Maputo, MZ.

Riley, L., & Caesar, M. (2017). *Comparing Household Food Security in Cities of the Global South through a Gender Lens* (HCP Report No. 9). Hungry Cities Partnership Project.

UNICEF. (2000, May 26). *Mozambique Situation Report*. http://www.reliefweb.int/report/mozambique/mozambique-situation-report-26-may-2000

US Agency for International Development (USAID). (2002). *Mozambique, 1999–2000 Floods. Impact Evaluation: Resettlement Grant Activity.*

Westmarland, N. (2001). The quantitative/qualitative debate and feminist research: A subjective view of objectivity. *Forum Qualitative Social Research, 2*(1).

Xavier, E. (2019). *Orçamento participativo como ferramenta da Gestão Municipal. O caso do Conselho Municipal da Cidade de Xai-Xai (2013–2016)* (Master's thesis). Faculdade de Ciências Sociais e Filosofia, Universidade Pedagógica de Maputo.

PART III

Environments, Linkages and Mobilities

CHAPTER 14

Hunger in an Agricultural City: Exploring Vulnerability in Dschang, Cameroon

Alexander Legwegoh and Liam Riley

Introduction

This chapter provides a case study of household food security in Dschang, Cameroon. Cameroon faces a dangerous combination of rapid urbanization, an inadequate system of urban governance, and a lack of information about urban socio-economic trends that threaten the sustainable development of its cities. The urban research available shows that many urban residents struggle to make ends meet: unemployment is common, food prices are high relative to incomes, and, with population growth, it is increasingly difficult to survive by producing one's own food (Fon, 2011; Krishna Bahadur et al., 2018; Legwegoh & Fraser, 2017; Sneyd, 2013).

A. Legwegoh
London, ON, Canada

L. Riley (✉)
Balsillie School of International Affairs, Wilfrid Laurier University, Waterloo, ON, Canada
e-mail: lriley@balsillieschool.ca

L. Riley and J. Crush (eds.), *Transforming Urban Food Systems in Secondary Cities in Africa*,
https://doi.org/10.1007/978-3-030-93072-1_14

Cameroon's population is currently 56% urban and projected to be 73% urban by 2050 (UN-DESA, 2018). The rapid expansion of the urban population presents an urgent need for investments and policies that will provide a framework for sustainable and inclusive growth, with food security a central pillar of this vision. With over half the urban population (55%) living in secondary cities with fewer than 500,000 people, the challenges facing secondary cities such as Dschang will have to be addressed.

Dschang is a sub-regional hub with a national university and a population of around 200,000. A household food security survey conducted in 2017 found high rates of food insecurity, which is ironic because Dschang has been known as an "agricultural city" that boasts a strong tradition of blending urban amenities with agricultural productivity and bountiful forests that provide wild foods (ADEID, 2011; UCCC, 2021). The case of Dschang, in keeping with many African secondary cities (Battersby & Watson, 2019; Mackay, 2019), informs theoretical debates about informal urban food systems because its entire food system epitomizes what is usually classified as "informal." Periodic open-air markets are the main source of food purchases by far, and they are important sources for households across the social strata. Most households produce some of their own food and few buy food at supermarkets.

Comparing the anecdotal evidence of Dschang in the past, including from the personal recollections of the first author and memories of residents, with survey findings, suggests a decline in the effectiveness of traditional informal food sources and local agricultural production to meet the food security needs of residents. Following a description of Dschang and a review of the survey findings, we turn to three political economic factors to explain the high rates of household food insecurity: the political organization of local governance in Cameroon; the anglophone crisis and the integration of internally displaced persons (IDPs) in Dschang households; and the "formal" food system that extracts food from Dschang to feed petrol-cash-rich cities in the region. These factors serve to illustrate that vulnerability to food insecurity in Dschang is the outcome of complex causal factors. Drawing these connections through a case study approach reveals part of a broader set of structural issues that need to be addressed across sub-Saharan Africa.

Dschang Profile

Dschang is situated in Cameroon's Western Highlands, which are characterized by their moderate climate, uneven topography, and rich vegetation. The city centre, which occupies the small street grid built a century ago, is along a regional highway that leads to Douala in one direction and Bafoussam in the other. Within a kilometre of the central roundabout are the bus depot, the two central markets, several shops and hotels, and a large cathedral. Further along the road toward Bafoussam are the city hall, hospital, the Université de Dschang campus, and the dam and reservoir built by the German administration more than a century ago. Surrounding this core area, in all directions, are burgeoning informal developments such as Fouréké and Foto, where there is an assortment of modern houses, multi-story apartment buildings, open fields, grass-thatched buildings, and traditional chieftaincy courts. Dschang includes a diverse built environment made up of an urban area with 20 neighbourhoods and a rural area with five groups and 96 villages (ADEID, 2011).

Dschang's written historical record begins with a visit in 1895 by a German government representative called Eugen Zintgraff (UCCC, 2021). It became an administrative and civil city in 1903 and, after World War I, when Cameroon was split between the British and the French, came under French control. Dschang became the capital of the Bamiléké region in 1920, but at independence in 1960 was replaced as regional capital by Bafoussam. The administrative census of 1956 listed Dschang as having a population of 3,000. While under colonial rule, Dschang played an important role in the provision of commercial, educational, and administrative facilities to the adjacent rural areas of Fongo-Ndeng, Fotetsa, Fossong-Wentcheng, and Fontem. Dschang is now the administrative centre of the Menoua Division in the West Region. The third General Census of Population and Housing in 2005 (the most up-to-date source of local population figures) lists the Dschang municipality's total population at 120,207 (Government of Cameroon, 2005). Extrapolating based on the rate of increase of Cameroon's urban population used in the World Bank (2020) calculations, the 2020 population is about 200,000 and by 2040 the population is projected to be close to 400,000.

Dschang has a particularly dynamic demography, as its growth is mainly due to the presence of the Université de Dschang, which was established in 1994 and enrolls thousands of new undergraduate and graduate students annually. The Université de Dschang is part of the national system of tertiary education and, as such, many civil servants are posted to Dschang from other parts of the country. This system of public sector employment contributes to a highly mobile and nationally connected population (Legwegoh et al., 2020). While the university has helped to diversify Dschang's economy, agriculture continues to be the mainstay for most households and businesses (UCCC, 2021). The geography is "dominated by low plateaus, highly dissected by small valleys which are sometimes marshy" (Temgoua et al., 2012). The topography of alternating hills and valleys means that there are many ecological zones of different altitudes within the municipality, and this provides an environment that can produce an abundance of diverse foods. The climate is also favourable for agriculture, with one dry season (mid-November to mid-March) and a much longer rainy season (mid-March to mid-November). Most of the soil is brown earth derived from volcanic basalt rock, which is beneficial for agriculture (ADEID, 2011). This soft soil and the rugged terrain create physical constraints and an elevated risk of flooding and landslides. These risks pose limits to scaled-up agricultural development and building urban and rural infrastructure, but they are ideally suited to small-scale production in valleys, terraced slopes, and diversified farms.

Findings from the Household Food Security Survey

Methodology

Dschang was selected as one of three secondary city case studies for the Food, Urbanization, Environment and Livelihoods project (FUEL) (in addition to Mzuzu, Malawi, and Oshakati-Ongwediva-Ondangwa, Namibia). The exploration of Dschang was an extension of the authors' previous work in Southern Africa (Legwegoh & Riley, 2014; Riley & Legwegoh, 2014). It marked the Central African introduction of the AFSUN approach to measuring urban food security in Africa, and it therefore provides household food security metrics comparable to surveys in other FUEL and African Food Security Urban Network (AFSUN) studies (Crush & Frayne, 2010; Legwegoh et al., 2020; Nickanor et al., 2019; Riley et al., 2018). Part of the interest in Dschang was due to its agricultural character, its rapid growth since the university opened in the 1990s, and its remote location relative to other secondary cities in the area such as Bafoussam and Bamenda. The anglophone crisis, which escalated after the FUEL project was initiated, also made Dschang an interesting case study because its immediate service area straddles the anglophone/francophone divide.

The data presented in this chapter was gathered through a door-to-door survey of 964 households conducted in December 2017 across the urban, peri-urban, and rural areas within the municipal boundaries of Dschang (Legwegoh et al., 2020). The survey instrument covered experiences of food insecurity using measurement tools developed by the Food and Nutrition Technical Assistance project (FANTA) (Coates et al., 2007; Swindale & Bilinsky, 2006); access to basic goods and services; food sources; and livelihood-sustaining activities. The survey included questions on household members (defined as people eating from the same pot and sleeping in the same dwelling for at least six months of the year on average). Researchers administered the survey in French, the predominant language in Dschang, and in Yemba, the most common mother tongue in the area.

The average household size was 4.5 people. About one in four (27.6%) households had one or two members. The population is very young: one-quarter (25.1%) of all household members are under age 10 and nearly three in four (70.3%) are under age 30. Nearly three-quarters (72.7%) of household members were born in Dschang, and most of the population born outside of Dschang was born in another urban area in Cameroon (23.2%). Only four percent were born in a rural area of Cameroon, and 0.2% were born outside of Cameroon.

Household Food Security

Household food insecurity was widespread in Dschang. The Household Food Insecurity Access Scale (HFIAS) average score in Dschang was 8.0 and the median was 7. The distribution of scores reflected a highly unequal city where 28% of households had scores above 12, meaning they were highly food insecure, and one-third of households had very low scores of 3 or less and thus were rarely faced with food insecurity. The overall HFIAS score does not differentiate the severity of different kinds of experiences indicative of food insecurity, in effect giving equal weight to experiences such as going a whole day and night without eating anything, and worrying about not having food. Taking into account the different levels of severity of the components of the HFIAS series of questions, the Household Food Insecurity Access Prevalence (HFIAP) indicator provides a complementary interpretation of the data (Coates et al., 2007). According to the HFIAP, the largest share of Dschang households belong to the category of severely food insecure (66.9%), followed by food secure (18.4%), moderately food insecure (10.8%), and mildly food insecure (3.9%). The HFIAP results highlight the fact that many households that do not frequently experience food security nonetheless have members who sometimes experience extreme forms of deprivation.

The second series of questions used to assess household food security in Dschang was the Household Dietary Diversity Score (HDDS) (Swindale & Bilinsky, 2006). Households indicated whether any of their members consumed foods in the previous 24 hours from 12 food categories: (1) grains; (2) roots/tubers; (3) fruit; (4) vegetables; (5) meat and meat products; (6) eggs; (7) fish and shellfish; (8) nuts and legumes; (9) milk and dairy products; (10) foods made from oil and fat; (11) sugar and sweets; and (12) other foods (including spices, condiments and drinks). The HDDS is the sum of the food categories consumed by households, with higher scores representing more diverse diets and suggesting higher food security. The average score in Dschang was 5.1, with a median of 6. The minimum was 0 (1.2% of households had consumed no food in the 12 food categories over the last 24 hours) and the maximum was 12 (0.6% of households had consumed foods from all 12 food groups in the previous 24 hours). Most households had consumed three to eight food groups.

The HDDS provides a window into the nutritional value of the foods widely consumed by Dschang households. Higher household dietary diversity does not necessarily mean better nutrition if the foods eaten are less nutritious or are related to health problems, as is the case with sugar, which has been linked to obesity and diabetes (Legwegoh & Hovorka, 2016). Most households in Dschang had consumed foods made with oil, fat, or butter (81%), fish or shellfish (61%), and grains (58%). Milk and dairy products (22%), eggs (19%), and meat and poultry (18%) were the least widely consumed. About 42% of households had consumed sugar/honey. The high frequency reported for oil, fat, or butter could indicate a public health problem; however, the typical meal compositions for people in Dschang commonly use local oils such as palm for cooking. Sugar/honey is consumed in tea or coffee and with other pastries, which are customs originating with Cameroon's dual colonial heritage of French and English.

Urban Food System

A view of the food system in Dschang emerged from a series of questions about where and by what means households accessed food (Legwegoh et al., 2020). Households in Dschang overwhelmingly rely on periodic open-air markets to purchase their food. B Market in the city centre is open every day, but is busiest on "market days" (every eighth day is a Big Market Day, and the middle day between Big Market Days is a Small Market Day). C Market (Tsimfem) is vacant except on market days, when it becomes a vibrant centre of commerce (ADEID, 2011). Nearly all households (91%) buy food at open markets and three in every five households (60%) do so on a weekly basis. About one in four households reported that open markets were their sole food source. The sources of the top-10 foods purchased in the previous month further illustrate the dominance of open markets: most households usually buy eight of the 10 most widely purchased foods at markets, including a range of fresh foods (fresh/cooked vegetables 64%, eggs 62%, and fresh fruits 62%); groceries (cooking oil 80%, sugar 55%); and dry foods (rice 76%, dried fish 75%, pasta 69%). White bread is usually purchased at small shops (by 69% of households) and meat at a butchery (by 54%).

Urban and peri-urban agriculture is equal in importance to open-air markets. Only 24% of the surveyed households do not produce any of the food they consume. A similar proportion produce food for the household in both rural and urban areas (23%), 36% produce food only in urban areas, and 18% produce food only in rural areas. Two in every five households that produce food exclusively or partially in rural areas do so for their own consumption and almost three-quarters of these households own the land on which they produce food. Over three-quarters of those involved in urban agriculture cultivated food on their own housing plots. About one in every five produced food within their neighborhood but outside their residential property (either on unbuilt land or on rented land near their homes). The practice of growing food on balconies or terraces, on other urban land and on riverbeds, is less common. Among non-producing households, the most common reason for not producing food was that they have no land to cultivate. Almost half the households keep livestock in the city. Of these, 88% raise chickens, followed by pigs

(43%). Other livestock included ducks, rabbits, goats, sheep, guinea pigs, geese, and turkeys. More than half the households that do not keep livestock said they do not have the land to do so, making this the most important barrier to urban livestock rearing.

About half of households in Dschang received food transfers, making this socially and culturally embedded food source a vital part of the food system. More than three-quarters (77%) of transfer-receiving households receive them from a rural source and 31% from an urban source. Relatives were far more likely to be the source of food transfers than friends: 96% of transfer-receiving households received transfers from relatives and nine percent from friends. The most common sources of transfers are rural relatives (64% of transfer-receiving households), followed by urban relatives (29%). Very few households received food from rural or urban friends. Tubers and plantains were the most common items, received by 61% and 60% of beneficiary households, respectively. About half of these households received maize and potatoes. Other important foods transferred included oil, various vegetables, and fish. Many households received several of these foods, with 15% receiving five or more types. The foods most likely to be received from an urban source were fish, rice, beef or *canda* (cow skin), and oil. Those most likely to be received from a rural source were potatoes, maize, tubers, and plantains. The primary difference is that urban-source transfers tend to be *purchased* by the senders while rural-source foods are *cultivated*. Almost all recipients agreed that these food remittances were important or very important to the household, but only a few said they were "essential for our survival." In a context of price fluctuations in the markets, and the fact that staples such as cassava, cocoyams, plantains, and yams are becoming increasingly expensive, food transfers are likely to become even more important for urban household food security.

Notably absent as main sources of the most popular foods in Dschang are supermarkets and street vendors. The only shop named *supermarché* sells dry groceries, beverages, and other non-food items. Fewer than one in 10 households access food from supermarkets, possibly including international chain retailers in the major cities of Douala and Yaoundé or in smaller supermarkets in nearby Bafoussam (a city of about a million people, 55 kilometers away). The lack of supermarket chains speaks to the importance of the local, informal, and traditional food chains to support the needs of Dschang's growing population. It also contributes to the literature on food system transition in secondary cities, which expects a transition from traditional and informal food sources to greater integration in global commodity chains and a more prominent role for supermarkets. The case study of Dschang appears to be similar to Heather Mackay's (2019) examples of two Ugandan secondary cities where supermarkets were only used sporadically and mainly for specific items. Jane Battersby and Vanessa Watson (2019) found that the formal and informal and the local and national scales were more integrated than previously assumed, such that further analysis of the food chains supplying Dschang's informal markets could reveal more complex integration with international commodity markets than first appears. The ostensibly informal food transfers from urban relatives in other cities are one possible pathway.

Household Food Production and Food Security

The survey data provides some insight into the role of agricultural activities in reducing household vulnerability to food insecurity. The key limitation is that the cross-sectional data does not reveal any change over time that could demonstrate an impact, but it does reveal some relationships between the two characteristics. Households that produce crops are less food secure on average (mean HFIAS 8.1) than households that do not produce any of their own food (mean HFIAS 7.5). They also have less diverse diets (mean HDDS 4.9 versus 5.8). There are relatively small differences between households that raise livestock and households that do not raise livestock in terms of HFIAS (7.9 and 8.1, respectively). Households that raise livestock have less dietary diversity than households that do not raise livestock (4.8 and 5.4, respectively). While the data does not provide evidence associating urban agriculture with food security, it is likely that producing households would be worse off without the food they produce. The results suggest that, for many households, the food they produce is not sufficient to protect them against hunger and poverty.

The Months of Adequate Household Food Provisioning (MAHFP) evaluation tool captures monthly access to food over the course of the previous year and identifies which months in the year households faced difficulty in accessing food. This tool has been used in similar studies to observe the impact of the agricultural cycle on households' food security over the course of a year (Battersby & Watson, 2019; Crush & Frayne, 2010). The MAHFP score is calculated as the number of months out of 12 that the household had an adequate food supply. The average score for Dschang was 10.6 and the median was 11. The lowest score was 5, and six percent of households had a score below 9. Almost one-third (32%) had a score of 12.

The impact of the agricultural seasons on food security is evident in the months when the greatest number of households did not have an adequate food supply (Fig. 14.1). In March, April, and May, during the planting season for most staple crops, 23%, 37%, and 17% of households reported inadequate food access. The most widely inaccessible foods were root vegetables and foods made from them. January was another month when a high percentage of households had difficulty accessing food (25%), but for most households experiencing difficulties in January the reason was a lack of cash rather than the agricultural cycle (93%). The most inaccessible food in January was meat (84%). Notably, the HDDS categories of "sugar/honey," "foods made with oil, fat or butter," and "other" foods were inaccessible to only a few households, during the agricultural lean season and in January when household cash reserves are low. This suggests that the least nutritious foods are the most likely to be available year-round, perhaps pointing to a nutritional transition trend where the food system makes less nutritious food more easily accessible.

The overall picture that emerges from the data is an urban environment with high rates of food insecurity, which is often characterized by infrequent but severe events at the household scale, such as an absolute

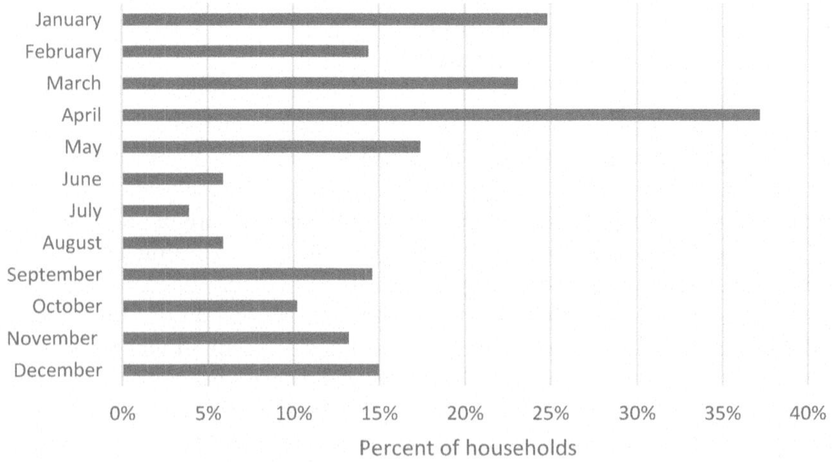

Fig. 14.1 Monthly pattern of food inaccessibility

lack of food for a short period of time. The three most important sources of food are periodic open-air markets, subsistence agricultural activities, and food transfers from relatives. Our previous work in Southern Africa compared the high rates of food security among poor households in Blantyre, Malawi, relative to poor households in Gaborone, Botswana, against the backdrop of expectations that a wealthier, more "developed" country such as Botswana would have more food-secure cities. Our analysis found that a robust set of sources from commercial and non-commercial sources created a resilient entitlement portfolio that reduced poor urban households' vulnerability to food insecurity (Legwegoh & Riley, 2014; Riley & Legwegoh, 2014). These conclusions do not appear to resonate in Dschang, where households use diverse means of accessing food and yet most are severely food insecure. The case of Dschang suggests that there are additional factors shaping vulnerability in Dschang, perhaps related to differences in Cameroon's food system, differences in secondary versus primary cities or differences in how Africa's urban food systems have changed over a decade. The differences in these case studies are far too complex to arrive at a definitive explanation, but the remaining section of our chapter explores three specific factors that might be limiting the effectiveness of households' multiple-source food security strategy. The study presented in this volume by Yanick Kamga demonstrates a fourth key factor, climate change, which is not addressed in this chapter.

Political Economic Factors of Vulnerability

Daniel Maxwell (1999) conducted a political economic analysis of urban food security in sub-Saharan Africa and found it to be a politically invisible issue. He argued that the problem of food access for urban residents in Accra and Kampala was discursively overshadowed by the politically salient issues of rural food production and urban infrastructure. A recent review found that urban food security continues to be overlooked in much of the global development policy literature, although there is an increasing number of clusters of interest and action (Crush & Riley, 2019). According to Maxwell (1999), urban food insecurity is symptomatic of the political economic changes that resulted from structural adjustment programs in the 1980s and 1990s, specifically the paradoxical reduction of state responsibility for social welfare just as many countries were adopting democratic political institutions and neoliberal approaches to economic development. The three specific factors explored here—the state's undermining of local governance capacity in Cameroon; the political crisis caused by the national government's crackdown on anglophone opposition; and the loss of food from Dschang due to regional market integration—demonstrate the enduring salience of political economy for understanding urban food insecurity.

The Political Organization of Local Governance

Councils constitute local government in Cameroon. They are made up of elected officials who are supposed to manage local development, including improving the living conditions of their inhabitants and driving socio-economic development. However, due to political challenges linked to imbalance in political and financial power held by the central government, councils generally lack the required resources and are not sufficiently empowered to carry out these functions (Bang, 2013; Kofele-Kale, 2011; Takwa et al., 2020; Yombi et al., 2019). As in the case of Dschang, the lack of proper functional infrastructure, even at the city council office, illustrates that councils are inefficient administrative units, which lack a strong economic base to drive development. Given the challenge in or the lack of devolution of revenue from central to local governments, and the lack of political accountability to the local population, one could understand why the daily priorities of residents such as food access do not appear to drive urban leadership.

Consequences of the Anglophone Crisis

Political conflict is an important factor in the loss of food entitlements in general, although its impact in the context of secondary African cities has been less researched. The government crackdown on protests by the anglophone minority in 2016 ignited a crisis in the Northwest and Southwest Regions of Cameroon, which has escalated into a full-fledged conflict and triggered mass displacements, especially in rural areas (UN-OCHA, 2021). Dschang is close to the border between Cameroon's anglophone and francophone regions and is therefore affected by the civil war and nearby military violence. There were an estimated 679,000 IDPs in Cameroon as of January 2020 (ACAPS, 2020). While the exact number in Dschang is not known, anecdotal evidence during the survey implementation in December 2017 suggested that there were many displaced people. Most were from the neighbouring Lebialem division, one of the worst affected regions, which shares indigenous cultural similarities with Dschang. There had been longstanding mobility between the two areas prior to the current political instability, and for many communities in Lebialem, Dschang is more accessible than urban centers in their own region.

Many of these people are absorbed by local households, for example when children are sent from conflict areas to attend school in a peaceful area and stay with relatives. The presence of IDPs likely contributed to the high level of household food insecurity in Dschang by stretching the resources of local households hosting guests and through the introduction of vulnerable displaced households. The conflict has also disrupted agricultural production and trade in areas of the Northwest Region that are important to Dschang's local food system, although precise data is not available.

Food System Integration as a Food Source for Neighboring Countries

In an informal interview with a leading city official in 2017, the official stated that the top reason for food insecurity in Dschang is that most of the food produced there is exported to major consumer markets in Cameroon's larger cities and other cities in West and Central Africa. His analysis aligns with the available data about the regional food networks (and the trucks we observed on the highway, spilling over with fresh produce). The Dschang area, like most agricultural regions in Cameroon, has become the breadbasket for most of the Central African region and parts of Nigeria (ADEID, 2011). A combination of oil-rich economies that have lower agricultural productivity, such as Equatorial Guinea, or large populations, such as Nigeria, purchase major quantities of agricultural staples, from tubers to fresh vegetables, that are exported daily. Food price data shows a major difference in the average price from one region to another. For example, between 2010 and 2016 in Libreville, Gabon, a kilogram of cassava was US$1.13, while in Douala, Cameroon, it was US$0.29. Between 2010 and 2016, the average price for a kilogram of plantains in Libreville was US$1.23; in Douala it was US$0.43 (FAO, 2017; Legwegoh & Fraser, 2017). The COVID-19 crisis, which led to regional border closures and limited mobility in the region, has also had an impact on the extent of food exports from Cameroon. It is estimated that more than 800,000 tonnes of tomatoes produced annually for markets in Nigeria, Equatorial Guinea, Gabon, Chad, Congo and the Central African Republic had to be dumped in Cameroonian markets, sending prices crashing from between 4,000 FCFA (Central African francs) and 5,000 FCFA during peak harvest season to 1,000 FCFA and 2000 FCFA (Foka-Nkwenti et al., 2020). Local consumers might benefit in the short run from the closed borders, but the structural problem of food security for food-producing regions, in an international market with highly uneven geographies of wealth, will continue to exact a toll on the viability of local food systems and household food security.

Conclusions

Further research is needed to understand the direct and indirect impacts of these linkages and to develop a theoretical understanding of secondary city food systems as part of the food production side of the equation within national and international urban hierarchies. The evidence suggests that households in Dschang are facing an increasingly uphill battle to meet their own food needs on a consistent basis. There are bright spots, such as the dietary diversity and the richness of the environment for food production. Nonetheless, contextual factors beyond the control of households and city leaders limit the effectiveness of efforts to build a food-secure city.

References

ACAPS. (2020, January 21). *Cameroon: Escalation of the anglophone crisis*. ACAPS short note. https://reliefweb.int/sites/reliefweb.int/files/resources/20200121_acaps_short_note_escalation_of_the_anglophone_crisis_cameroon_0.pdf

Action pour un Développement Equitable, Intégré et Durable (ADEID). (2011). *Plan communal de développement de Dschang*. https://www.pndp.org/documents/PCD_DSCHANG1.pdf

Bang, H. (2013). Governance of disaster risk reduction in Cameroon: The need to empower local government. *Jàmbá: Journal of Disaster Risk Studies, 5*(2), 1–10.

Battersby, J., & Watson, V. (Eds.). (2019). *Urban food systems governance and poverty in African cities*. Routledge.

Coates, J., Swindale, A., & Bilinsky, P. (2007). *Household Food Insecurity Access Scale (HFIAS) for Measurement of Household Food Access: Indicator Guide (Version 3)*. FHI 360/FANTA.

Crush, J., & Frayne, B. (2010). *The Invisible Crisis: Urban Food Security in Southern Africa* (Urban Food Security Series No. 1). African Food Security Urban Network (AFSUN).

Crush, J., & Riley, L. (2019). Rural bias and urban food security. In J. Battersby & V. Watson (Eds.), *Urban food systems governance and poverty in African cities* (pp. 42–55). Routledge.

Food and Agriculture Organization (FAO). (2017). *GIEWS Food Price Monitoring and Analysis (FPMA) Tool*. Food and Agriculture Organization.

Foka-Nkwenti, C., Nguendo Yongsi, H., Mpeh, N., & Nganou-Mouafo, M. (2020). COVID-19 and food insecurity in Cameroon. *GSC Advanced Research and Reviews, 5*(2), 111–117.

Fon, N. (2011). *The Right to Adequate Food in Cameroon. The Second and Third Periodic Report (Art. 1–15) of Cameroon to the Committee on Economic, Social and Cultural Rights*. Cameroon: Community Initiative for Sustainable Development (COMINSUD), Cameroon Movement on the Right to Food (CAMORIF) and National Committee World Day of Prayer-Cameroon. Parallel Report.

Government of Cameroon. (2005). *Enquête sur L'emploi et le Secteur Informel au Cameroun, Phase 1: Enquête sur l'Emploi, Rapport Principal*. National Institute for Statistics.

Kofele-Kale, N. (2011). Local governance under Cameroon's decentralization regime: Is it all sound and fury signifying nothing? *Commonwealth Law Bulletin, 37*(3), 513–530.

Krishna Bahadur, K., Legwegoh, A., Therien, A., Fraser, E., & Antwi-Agyei, P. (2018). Food price, food security and dietary diversity: A comparative study of urban Cameroon and Ghana. *Journal of International Development, 30*, 42–60.

Legwegoh, A., & Fraser, E. (2017). High food prices in urban Cameroon: Coping strategies and suggested policy actions. *Development in Practice, 27*(8), 1064–1077.

Legwegoh, A., & Hovorka, A. (2016). Exploring food choices within the context of nutritional security in Gaborone Botswana. *Singapore Journal of Tropical Geography, 37*, 76–93.

Legwegoh, A., Kamga, Y., Riley, L., Martin, L., & Njukeng, P. (2020). *Food Security in Africa's Secondary Cities: No. 3. Dschang, Cameroon* (Urban Food Security Series No. 29). African Food Security Urban Network (AFSUN).

Legwegoh, A., & Riley, L. (2014). Food, place, and culture in urban Africa: Comparative consumption in Gaborone and Blantyre. *Journal of Hunger and Environmental Nutrition, 9*(2), 256–279.

Mackay, H. (2019). Food sources and access strategies in Ugandan secondary cities: An intersectional analysis. *Environment and Urbanization, 31*(2), 375–396.

Maxwell, D. (1999). The political economy of urban food security in sub-Saharan Africa. *World Development, 27*, 1939–1953.

Nickanor, N., Kazembe, L., & Crush, J. (2019). *Food Security in Africa's Secondary Cities: No. 2. The Oshakati-Ongwediva-Ondangwa Corridor, Namibia* (Urban Food Security Series No. 28). African Food Security Urban Network (AFSUN).

Riley, L., Chilanga, E., Zuze, L., & Joynt, A. (2018). *Food Security in Africa's Secondary Cities: No. 1. Mzuzu, Malawi* (Urban Food Security Series No. 27). African Food Security Urban Network (AFSUN).

Riley, L., & Legwegoh, A. (2014). Comparative urban food geographies in Blantyre and Gaborone. *African Geographical Review, 33*(1), 52–66.

Sneyd, L. (2013). Wild Food, prices, diets and development: Sustainability and food security in urban Cameroon. *Sustainability, 5*, 4728–4759.

Swindale, A., & Bilinsky, P. (2006). *Household Dietary Diversity Score (HDDS) for Measurement of Household Food Access: Indicator Guide (Version 2).* FHI 360/FANTA.

Takwa, C., Babila, F., & Teno, M. (2020). Dynamics of local government finance in Cameroon: An assessment of local councils in the Fako Division. *Asian Journal of Economics, Business and Accounting, 17*(2), 54–72.

Temgoua, E., Ntangmo Tsafack, H., Njine, T., & Serve, M. (2012). Vegetable production systems of swamp zone in urban environment in West Cameroon: Case of Dschang City. *Universal Journal of Environmental Research and Technology, 2*(2), 83–92.

United Councils and Cities of Cameroon (UCCC). (2021). *Dschang*. http://www.cvuc-uccc.com/national/index.php/en/component/content/article/158-association/carte-administrative/ouest/menoua/458-dschang

UN-Department of Economic and Social Affairs, Population Division (UN-DESA). (2018). *World Urbanization Prospects: The 2018 Revision, Online Edition*. https://population.un.org/wup/

UN-Office for the Coordination of Humanitarian Affairs (UN-OCHA). (2021, February 28). *Cameroon: North-West and South-West* (Situation Report No. 28). https://reliefweb.int/sites/reliefweb.int/files/resources/sitrep_nwsw_february_2021.pdf

World Bank. (2020). *Population Estimates and Projections.* Data Catalog. https://datacatalog.worldbank.org/dataset/population-estimates-and-projections

Yombi, G., Chouaïbou, M., & Agoume, L. (2019). Decentralized territorial communities and implementation of public policies: The case of Cameroon. In B. Açikgöz (Ed.), *Public economics and finance.* IntechOpen, ch. 4.

CHAPTER 15

Non-timber Forest Products in Cameroon's Food System and the Impact of Climate Change on Food Security in Dschang

Yanick Borel Kamga

Introduction

The world's reservoir of forest genetic resources, especially in tropical areas, is threatened by a series of human and natural factors, such as population pressures and climate change. The destruction of forest ecosystems is increasingly restricting the range of possibilities for generations of current and future users in rural and urban communities (Bele et al., 2011; Nkem et al., 2010; Tieminie et al., 2021). Globally, the area of tree cover declined by 10% from 2001 to 2020, with 94% of the loss in Africa due to shifting agricultural uses (WRI, 2021). The forests of Cameroon, one of the major components of the Congo Basin forests, do not escape this reality. From 2002 to 2020, Cameroon lost 3.7% of its

Y. B. Kamga (✉)
Department of Plant Biology, Faculty of Science, University of Dschang, Dschang, Cameroon

L. Riley and J. Crush (eds.), *Transforming Urban Food Systems in Secondary Cities in Africa*,
https://doi.org/10.1007/978-3-030-93072-1_15

humid primary forest (Global Forest Watch, 2021), the type of forest that provides the greatest density of biodiversity and cannot easily be regenerated (Marchese, 2015). There are many edible fruit species in Cameroon's forests (Eyog-Matig et al., 2006). There is also a vibrant trade in other foods such as game, caterpillars, snails, mushrooms, leaves, bark and seeds.

The loss of biodiversity might outpace efforts to catalogue the varieties of foods and their harvest methods and uses, since almost all forest landscapes in Cameroon are affected by climate variability and climate change. Sustainable urbanism implies sustainable use of natural resources, including forest resources, and research into the linkages between urban food systems and forest ecosystems is an important contribution to sustainability efforts. The changes to Cameroon's forest ecosystems due to shifting weather patterns are already being observed by people living and working in the forests (Kamga, 2020; Nkem et al., 2010; Tieminie et al., 2021). The potential impact of climate change on Cameroon's urban food systems makes it an urgent need to understand these linkages.

Many plant and animal products other than timber—non-timber forest products (NTFP)—support a food system through which rural and indigenous communities harvest food from the forests and sell it to traders, who then supply the food to consumers in urban and rural markets across the country. The majority of the population of Dschang uses NTFPs, which provide them with food and medicinal products. In addition, through various informal channels, these products offer employment opportunities and generate income for the many players involved in their exploitation and marketing. At the national level, the economic dimension of NTFPs is practically unknown because of the informal nature of the trade and the lack of research into these dimensions of the food system. At the international level, NTFP trade exists between the Republic of Cameroon and many African and European countries. Unfortunately, no official statistics exist on the trade between Dschang and other cities. The sale of NTFPs exploited from different plant formations allows indigenous and local populations, and even those living in urban centres, to obtain the income necessary to meet different social obligations, plan for their futures and access adequate food for their families.

The following sections explore the consumption, trade and production of NTFP food products. Each section is based on a different research study and the relevant methodological background is embedded in each section. The next section demonstrates the importance of NTFPs in the diets of households in Dschang. It is followed by results from a survey of small-scale food traders in Dschang, with a focus on *safou* sellers. The subsequent section presents findings from ethnobotanical research with communities that harvest NTFPs for sale, with a focus on bitter kola. The chapter ends with a reflection on the threat climate change poses to the food system and to the sustainability of cities such as Dschang.

Urban Consumers of NTFP Foods

Cameroon is an urban country and UN-DESA (2018) anticipates that by 2050 three in four Cameroonians will live in urban areas. Urban food systems research is lagging far behind the growing need for baseline data to identify opportunities to improve the resilience and sustainability of Cameroon's urban dwellers. Recent urban food research in Cameroon has focused on urban and peri-urban agriculture (Gwan & Kimengsi, 2020, p. 19; Karg et al., 2020; Ngome & Foeken, 2012; Prain et al., 2010; Tata et al., 2019), and the impact of food prices on dietary change (KC et al., 2018; Legwegoh & Fraser, 2017), with little detailed study of informal food-trading systems. There has been some work establishing the importance of "wild foods" to urban food systems and household food security, which documents the popularity and widespread availability of many NTFP foods in urban Cameroon (Sneyd, 2013). A common theme in the literature is that the rapid expansion of urban populations and built environments, coupled with anaemic food policies that rely on imports rather than investment in agriculture, threatens the sustainability of Cameroon's development. The threat to urban food systems posed by climate change and forest ecosystem loss adds a new element to this conversation.

The information in this section is drawn from original research conducted in Dschang, Cameroon, in December 2017 that examined multiple dimensions of food security and the food system from the household scale (Legwegoh et al., 2020). The in-person survey was conducted with 964 households across the population of Dschang Municipality, which includes a gradation of characteristically urban, peri-urban and rural environments. The survey included questions about food consumption, food access, livelihoods, agriculture and the consumption of indigenous foods (Legwegoh et al., 2020).

Dschang is the urban centre for a small agricultural area in Menua District, West Region. The urban area has a population of about 200,000, although the influx of internally displaced people from neighbouring North West Region due to political violence and the circular migration of students and staff attached to the local university mean that the actual population at any given time is much higher. While about one in four household members was born outside of Dschang, within the highly mobile age range of 15–44, about 40% were born elsewhere (usually in another city in Cameroon). Dschang's food system operates mainly through traditional periodic markets, with the main central market (Marché B) operating daily, with much higher activity on Big Market Day (every eighth day) and Small Market Day (four days after Big Market Day) (Republic of Cameroon, 2011). Subsistence agriculture is a key component of Dschang's food system (Temgoua et al., 2012). Two in five households produce food in a rural area for their own consumption. Food trading is also an important source of income: when asked to list all sources of income for the household, the informal production and sale of fresh produce was the most widespread income source (29% of households) and 16% had income from the informal sale of fresh produce they had not produced. The informal food system is an important source of food and income for most residents of Dschang, where only 16% rely exclusively on formal income sources and 24% do not produce any of the food they consume.

The large population of migrants from other regions of Cameroon shapes the food culture in Dschang, although it continues to be shaped as well by the multitude of foods specific to the environment and the Bamiléké cultural traditions of the region. The survey included a list of 20 indigenous foods developed by the local research team to reflect what one might expect to find locally (Table 15.1). Many NTFP foods are consumed by a large number of households. *Njansang*, a seed from the *Ricinodendron heudelotti* tree (CIFOR, 2021a) and bush onion, a fruit from the *Afrostyrax lepidophyllus* tree, whose seed is also used as a spice and whose leaves and bark are used for medicinal purposes (Moukette et al., 2015), were both consumed by two-thirds of households. Half of all surveyed households consumed *safou*, also known as "bush butter", the fruit from the *Dacryodes edulis* tree that provides nutritious oil and fatty acids (CIFOR, 2021b). Bitter kola, the nut from the *Garcinia kola* (*G. kola*) tree (Kamga et al., 2019), was consumed by 42% of households. These foods are cooked and consumed together as traditional meals. This is the case, for example, of the mixture made from potatoes, bananas,

Table 15.1 Indigenous food consumed by the households surveyed in Dschang

Indigenous foods	*Number of households*	*Percentage of households*	*NTFP (Y/N)*
Njansang	635	66.0	Yes
Bush onion	625	65.0	No
Spices	525	54.5	Yes
Safou	481	50.0	Yes
Honey	425	44.2	Yes
Bitter kola	406	42.2	Yes
Raffia palm	388	40.3	Yes
Rondelle	341	35.4	Yes
Okok/Eru	325	33.8	Yes
Cola nut	289	30.0	Yes
Bushmeat	288	29.9	Yes
Mushrooms	223	23.2	Yes
Mbongo	207	21.5	Yes
Bush mango	133	13.8	Yes
Pèpè	101	10.5	Yes
Forest snails	94	9.8	Yes
Fruit of Raffia palm	90	9.4	Yes
Termites	73	7.6	Yes
Green grasshopper (Ngoh)	69	7.2	No
Caterpillar	7	0.73	Yes

NB Multiple responses

beans and red oil (palm), commonly known as *tchoumtom* and *tchoukolo* in the area around Dschang.

A variety of local ingredients, spices (*njansang*, rondelles), condiments (fresh or dried) or bush onion are widely used in the preparation of different dishes. This is the case with fish broth or forest snails, accompanied by rice and/or plantain (ripe or not) and sweet banana. Other examples are *Folong*, a vegetable prepared with peanuts with smoked fish, and *N'dolé*, a vegetable cooked with meat and/or smoked fish and accompanied by rice, plantain, cassava, macabo, sweet potato and couscous (mixture of simple corn flour cooked at 100 °C). Other foods identified in the locality that are eaten directly or over the embers include bush mango, cola nuts, bitter kola, honey and *safou*. They play an important role for more than 100 million people in the sub-region because they are a source of food as well as medicine. Regular consumption of these products can also correct micronutrient deficiencies (FAO 2011).

Figure 15.1 shows that some of the indigenous foods are consumed on a frequent basis. Bush onion and *njansang* were consumed on a weekly basis by the majority of households. *Safou* was consumed by half of all households, but rarely on a weekly basis. Other foods such as mushrooms, green grasshoppers and caterpillars were less popular and when households consumed them, it was usually just once a year. The indigenous foods in this survey were mainly purchased, with some exceptions. These included bush mango (*Irvingia gabonensis*), which is grown and

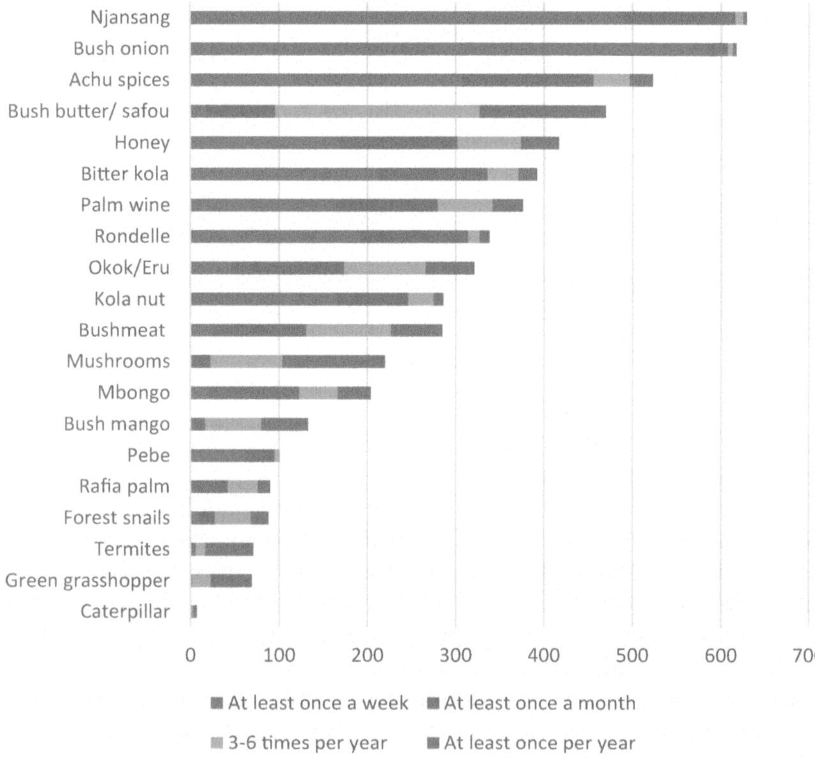

Fig. 15.1 Frequency of consumption of selected indigenous foods

collected in rural areas by 13 and 15% of consuming households, respectively; cola nuts, which are grown and collected in rural areas by 20% and 18% of consuming households, respectively, and bushmeat, grasshoppers and termites, which tend to be collected about as frequently as they are purchased by consuming households. Because most of these products are purchased, the frequency of consumption depends on their availability on the market, which in turn depends on the seasonality of different foods. Foods that are consumed by several households weekly or semi-weekly are available almost all year on the market. This availability would be linked to the ecological conditions of each production basin and the variety of ecosystems in which the trees can grow. The *Dacryodes edulis* tree that produces *safou* grows in different regions, with fruit maturing in different months in different parts of the country. This staggering of seasons could explain the consistent availability on the markets of large urban centres and therefore the fruit's accessibility to households. The high frequency of NTFP food consumption demonstrates that forests are a basic food source for households living in Dschang.

Safou and the Informal Food System in Dschang

The household food security survey was followed by a survey of small-scale food-trading businesses (fewer than five employees) in Dschang in August 2019 (Kamga et al., 2021). The survey questions addressed issues about the location of vendors, demographic characteristics, enterprise characteristics (structure, practices and business environment), sanitation and food production patterns. The survey included 854 trading businesses representing a distribution of types of businesses in line with a rapid survey conducted as a framing exercise (Kamga et al., 2021). The survey was implemented in 21 locations across Dschang's urban area, with the largest share (25%) conducted at the main food market (Marché B), followed by the informal residential areas of Foreké (15%) and Foto (14%). Three-quarters of the enterprises were owned and operated by one person, with the remaining 25% having between one and five employees. About two-thirds of the respondents were women and one-third were men. The majority of businesses (73%) operated without paying any kind of licensing or market fee (Kamga et al., 2021).

The survey included an open-ended question about what foods the businesses sold, in which each respondent named up to five foods that were the main items they sold and responded to a series of questions pertaining to each food. There was a wide variety of foods, with the most popular being bread (20%), rice (17%) and peanuts in various forms (15%). The most widely offered NTFP food was *safou,* sold by 39 traders (4.6%). Other NTFP foods sold were *eru* (11%), bitter kola (seven percent), cola (seven percent), *njansang* (six percent), *fruits noirs* (two percent), rondelles (two percent) and snails (two percent). These findings demonstrate that many NTFP foods are available in Dschang, confirming the findings of the household survey. They also put into perspective the scale of the NTFP foods relative to cultivated foods in Dschang's informal system. The small percentage of small-scale food traders selling each of these products illustrates that they are a small part of the informal food system relative to other cultivated and processed foods.

A closer look at the types of businesses selling the most popular NTFP food, *safou*, helps to paint a picture of the marketing of this product. Owners of businesses selling *safou* were 83% female and nine percent had primary education or less, reflecting a demographic with statistically fewer opportunities for formal sector employment than the average person in Dschang, and hence more economic vulnerability. Some characteristics that made these businesses more "informal" were that only six percent were paying fees of some kind, 75% were located in roadside stalls and another nine percent had no fixed place of operation. Each business owner named their usual source for *safou*, and 48% sourced it from a market in Dschang, 23% from a wholesaler, 20% from a rural market, and three percent each from the following sources: harvested by the respondent; purchased directly from the harvester; and purchased at a market in another city. These sources show that in almost all cases the supply chain for *safou* consumed by households in Dschang has multiple intermediaries between the harvester and the consumer.

The livelihoods of people engaged in this supply chain—including harvesters, wholesalers and retailers—are affected by the health, biodiversity and productivity of Cameroon's forests. As discussed below in more detail, rural communities living in and near the forests witness the signs of climate change in their daily lives. Yet there has been little research on the perceptions of climate change in urban communities in Cameroon. The food trader survey in Dschang found that 78% of respondents were aware of climate change. The top source of information about climate change was the change people observed in their environment (71%). According to about 64% of those who knew of climate change by observing a change in their environment, they observed dropping temperatures, rising

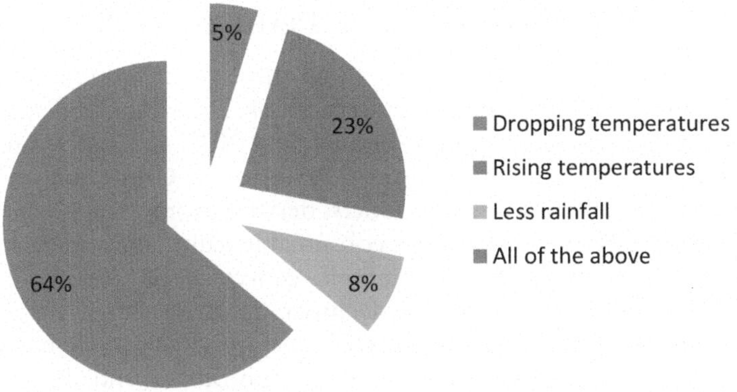

Fig. 15.2 Proportion of the main effect of climate change in Dschang Area

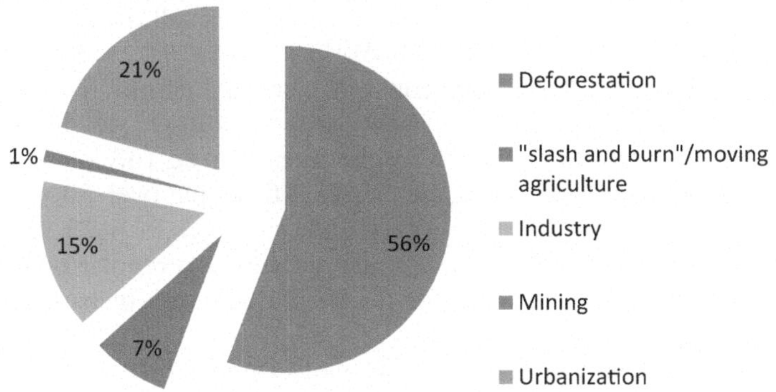

Fig. 15.3 Perception of the main human activity causing climate change

temperatures and less rainfall (Fig. 15.2). The majority of respondents (56%) believed that the main cause of climate change is deforestation, followed by urbanization (21%) (Fig. 15.3). This result is in part a reflection of the cultural importance of forests for most Cameroonians and in turn the cultural importance of forest products for rural and urban Cameroonians alike. The following section focuses on the rural communities whose livelihoods are arguably the most directly affected by the depletion of forest biodiversity. They play an important role in the connection between forests and urban consumers of NTFPs.

The Exploitation and Trade of *G. kola*

This section provides information about the source communities for NTFPs found in Cameroonian cities such as Dschang. The section is based on ethnobotanical research conducted in 2017 and 2018 in markets and villages in Central and Eastern Regions of Cameroon into the harvesting and marketing of products derived from the *G. kola* tree (Kamga et al., 2019). *G. kola* yields consumable fruit, nuts and bark and is endemic to the humid lowlands rainforest ecosystem of West and Central Africa, from Sierra Leone to the Democratic Republic of the Congo (Adebisi, 2004; Kouame et al., 2016). The nuts have high nutritional values and are widely used as a stimulant, aphrodisiac and digestive aid (Alaje et al., 2014).

In investigating the socio-cultural and economic importance of *G. kola* and practical information about harvesting and marketing practices, interviews were conducted in the markets of Lomié (a small town adjacent to the Dja Faunal Reserve, a UNESCO World Heritage Site, located in East Region), and Makénéné, a large town in Central Region (Kamga et al., 2019). The semi-structured interviews lasted about 25 to 40 minutes each. The 81 people interviewed were mostly male (63 men and 18 women), distributed into three age classes: 18 people over 50 years old, 56 people aged 25 to 50 and seven people younger than 25 years old. The criterion for selection of the 81 respondents was regular or seasonal exercise of the exploitation and/or marketing activities of bark or seeds of *G. kola* (Kamga et al., 2021).

The NTFP markets in the small towns of Lomié and Makénéné supply a variety of NTFPs for domestic trade and for export. The trade in *G. kola* products is mostly informal, in the sense that there is little regulation or government oversight of the market. No organizational or professional structure was identified during the study, in part because the taxation of NTFPs in Cameroon is poorly developed in comparison with that of timber. In this market, transportation is the limiting factor, as producers deal with both the law of supply and demand and the law of supply and transportation. If there is less supply, transportation becomes scarce, and if transportation is scarce, supply decreases and prices rise. This scenario causes a dislocation of the system and production capacity, which benefits only a narrow set of stakeholders who have a supply at hand near a consumer market with low supply (Ruiz Pérez et al., 2000). When the commercial sector presents a good interconnection with a regular transport network, the producers become interested, and this can be beneficial for all the actors involved.

The challenges and opportunities facing producers vary geographically. In the area served by Makénéné Market, most of the trees are parts of plantations that have been preserved by farmers (Kamga, 2020). In this area, trees are closer to the villages (two to 3.5 hours' walk). In the area served by Lomié Market, the trees are mainly in the wild forest reserve and farther from villages (five to 12 hours' walk). The distance to the harvest site is the key limiting factor, considering the return walk requires carrying loads of up to 40 kg of fresh bark, fruit or seeds. Harvesters need to have adequate time and physical strength to engage in harvesting. Much of what is collected is used locally, especially if markets are difficult to access and prices are low. In the area around Lomié Market, only 23% of harvesters harvested for commercial purposes, while the figure was 75% in the Makénéné area.

The income resulting from the sale of *G. kola* products, especially the *G. kola* seeds, varies according to the market and the various categories of people involved in the distribution chain. Generally, seeds are free in the farms or forest and harvesters benefit from 100% of the sale price, often after investing considerable time and energy to harvest and process the fruits into nuts. Income generated from the sale of *G. kola* is used for subsistence and other household expenses. The *bayam-salams* are market women who operate across Cameroon, often as intermediaries between harvesters and other traders or consumers (Fonjong, 2004).

Respondents estimated that the profit margin for *bayam-salams* was 30–40%, and up to 60% when they store the products until there is a scarcity in the market. There was a geographical difference in the gender of marketers. In the Makénéné Market area, 82% of traders were women, whereas in the Lomié Market area most of the traders were men. The men trading bitter kola in the Lomié Market area were mainly hunters, and several people indicated that the scarcity of bitter kola pushed most of the *bayam-salams* to focus on other NTFPs, including foods found in Dschang such as bush onion and *njansang*.

The harvesting practices have varying degrees of effect on the health of trees. The harvesting techniques applied by respondents were leaf removal (62%), followed by barking and uprooting (58%), felling (28%) and ring removal of the bark (19%). Most of the fruit and seeds were collected from the ground or the tree canopy. Collecting on the tree does not affect the survival or regeneration of species. Collecting from the ground can be dangerous for the dynamics of the species if all the products are harvested. Felling trees or uprooting certain species reduces the productive potential of the forest in the short term and can even affect the specific richness, if it becomes intensive for certain categories of species. The intensity of exploitation depends on the domestic and commercial demand for the product. The impact of logging on the structure and composition of the forest is closely linked to the intensity of the harvest and could lead to the disappearance or death of individual trees (Avana-Tientcheu et al., 2018; Guedje, 2002). This strong pressure can lead to a reduction in the availability of products on the market and, consequently, could drastically reduce the income of all actors involved in the value chain, in particular households that harvest the products. This pressure increasingly risks the future livelihoods of the people who depend on these NTFPs (Kamga et al., 2019; Loubelo, 2012; Ndoye et al., 1999).

Climate change poses a major threat to biodiversity, as species that cannot adapt or relocate risk extinction (Díaz et al., 2019). In addition, many ecosystem services will be affected by climate change, and individuals and communities that rely on NTFPs will be impacted. Disturbances in the development schedule of *G. kola* trees, such as drop of buds, aborted fruits and lack of flowering, were observed by the local population after the flowering periods (Kamga, 2020). They attributed these events to climate change, noting that during some years it is colder than the normal range and in those years the trees (which are vulnerable to frost) bloom late; in some cases, not at all. Despite the average global increase in temperatures due to climate change, the effect of climate change in Cameroon is a cooler climate that is less favourable to fertilization and pollination. This situation will lead to a drastic drop in production, and a concomitant drop in the income of the actors involved in the value chain (producers/farmers, harvesters, transporters, sellers and so on), as mentioned by the local population in these forest regions (Kamga, 2020; Kamga et al., 2019). As well, the scarcity or even the absence of some climate-affected NTFPs, such as bitter kola nuts, pushes the local populations to turn to other, more abundant NTFP species. A number of NTFP fruits are now widely available in urban markets: *Afrostyras lepidophyllus*; *Pentaclethra macrophylla*; *Ricinidendron heudelotii*; *Carapa procera*; *Monodora myristica*; *Baillonella toxisperma* (whose tasty pulp is eaten fresh); *Coula edulis* (whose almonds are eaten fresh or cooked); *Tetrapleura tetraptera*; *Irvingia gabunensis*; *Dacryodes edulis*; and *Cola nitida* (Kamga et al., 2019).

Many ecosystem services are affected by the effects of climate change. This is the case with the supply of forest (Dale et al., 2010) and agricultural resources (Howden et al., 2007, p. 6). Tropical African forests provide the livelihoods of around 100 million people, directly for 60 million rural people and indirectly for 40 million urban dwellers living near forest environments, by providing them with food, medicine, fuel, fibre and non-timber products. These products also serve social and cultural functions (Mayaux et al., 2013). These ongoing climate change effects pose new challenges for biodiversity conservation, as they strongly influence environmental factors and many vital ecosystem services for local populations. Indeed, food systems are highly dependent on biodiversity and a considerable proportion of medicinal drugs is directly or indirectly of biological origin. For example, tropical forests provide an array of medicinal plants used for health care. Eighty percent of people in developing countries depend on traditional medicines derived from plants and trees. The change in the rhythm of the seasons due to climate change also has consequences on the flowering of plants that are needed for foraging bees to make honey. An in-depth study remains necessary to assess the effects of climate change on household vulnerability. Sustainable planning will require a broader awareness of the role of forests in Cameroon's food system and a fuller picture of the threats facing forests. There has been an overemphasis on the threat that forest communities pose (e.g. through bark harvesting and shifting agriculture), as the impacts of these practices are small and manageable compared to the looming threat of climate change.

Conclusion

The popular perception in Cameroon that deforestation is the main cause of climate change reflects the importance of forests for Cameroonian cultures and livelihoods. Cameroonians are perhaps more keenly aware than others of the importance of forests, which are home to more than 80% of the earth's biodiversity and represent one of the last refuges for many animal and plant species. It is estimated that 27,000 animal and plant species disappear each year due to climate change. This loss of biodiversity, which can be irreversible, cuts off humanity from invaluable services and resources; deforestation is therefore a disaster for humans as well as for other species. As noted by Mekou Youssoufa Bele et al. (2011, p. 369): "In developing countries such as Cameroon, forest resources are frequently the principal assets of the poor, and the most proximate opportunity for food security and poverty alleviation". Indigenous foods such as *njansang*, bush onions, *safou* and bitter kola are frequently consumed by the majority of households in urban areas such as Dschang. The commercialization of these products in local markets provides incomes that contribute to the livelihood of those involved in trading activities.

Climate change is a new challenge for biodiversity conservation, as it threatens the availability of NTFPs as well as agricultural activities. Conservation-oriented exploitation is possible with knowledge about sustainable levels of exploitation, uses and potential commercialization of NTFPs. This chapter combines research in markets and households in the secondary city of Dschang with communities and households in rural forest areas in Eastern and Southern Regions of Cameroon to trace the vital importance of forests for urban food security. It should inspire further research with communities in forested and urbanizing areas engaged in the exploitation, commercialization and consumption of NTFPs. Further work is required to develop sustainable guidelines for forest exploitation that would benefit all communities and raise awareness of the practical importance of forest ecosystems.

References

Adebisi, A. (2004). A Case Study of *Garcinia kola* Nut Production-to-Consumption System in J4 Area of Omo Forest Reserve, South-West Nigeria. In T. Sunderland & O. Ndoye (Eds.), *Forest Products, Livelihood and Conservation (Case Studies of Non-Timber Forest Product Systems) Volume 2—Africa* (pp. 115–132). Centre for International Forestry Research.

Alaje, D., Owolabi, K., Olakunle, T., Oluoti, O., & Adetuberu, I. (2014). Nutritional, minerals and phytochemicals composition of Garcinia cola (Bitter cola) and Aframomummelegueta (Alligator pepper). *Journal of Environmental Science, Toxicology and Food Technology, 8*(1), 86–91.

Avana-tientcheu, M., Momo Solefack, M., Kamga, B., & Nguetsop, V. (2018). Vulnerability assessment of *Gnidia glauca* (Thymelaeaceae) exploitation, traditional uses and domestication potential in the community forest of Kilum-Ijim, North Western Cameroon. *Journal of Ecology and the Natural Environment, 10*(8), 182–191.

Bele, M., Sonorin, O., Sonwa, D., Nkem, J., & Locatelli, B. (2011). Forests and climate change adaptation policies in Cameroon. *Mitigation and Adaptation Strategies for Global Change, 16*, 369–385.

Carvalho, S., Brito, J., Crespo, E., & Possingham, H. (2010). From climate change predictions to actions: Conserving vulnerable animal groups in hotspots at a regional scale. *Global Change Biology, 16*(12), 3257–3270.

Center for International Forestry Research (CIFOR). (2021a). *Njansang: Ricinodendrenheudelotti.* CIFOR fact sheet. https://www.cifor.org/publications/pdf_files/factsheet/Njansang_eng.pdf

Center for International Forestry Research (CIFOR). (2021b). *Food of the forest.* CIFOR feature. https://www.cifor.org/feature/foodfromforest/

Dale, V., Tharp, M., Lannom, K., & Hodges, D. (2010). Modeling transient response of forests to climate change. *Science of the Total Environment, 408*(8), 1888–1901.

Díaz, S., Settele, J., Brondízio, E., Ngo, H. T., Guèze, M., Agard, J., Arneth, A., Balvanera, P., Brauman, K., Watson, R. T., Baste, I. A., Larigauderie, A., Leadley, P., Pascual, U., Baptiste, B., Demissew, S., Dziba, L., Erpul, G., Fazel, A., & Vilá, B. (2019). *Summary for policymakers of the global assessment report on biodiversity and ecosystem services of the Intergovernmental Science—Policy Platform on Biodiversity and Ecosystem Services.*

Eyog-Matig, O., Ndoye, O., Kengue, J., & Awono, A. (Eds.). (2006). *Les fruitiers forestiers comestibles du Cameroun.* International Plant Genetic Resources Institute.

Food and Agriculture Organization (FAO). (2011). *State of the World's Forests 2011.* Food and Agriculture Organization.

Fonjong, L. (2004). Challenges and coping strategies of women food crops entrepreneurs in Fako Division Cameroon. *Journal of International Women's Studies, 5*, 1–17.

Global Forest Watch. (2021). *Cameroon.* Global Forest Watch Country Statistics. https://www.globalforestwatch.org/dashboards/country/CMR/

Guedje, N. (2002). *La gestion des populations d'arbres comme outil pour une exploitation durable des produits forestiers non-ligneux: l'Exemple du Garcinia lucida (Sud Cameroun)* (Thèse de doctorat). l'Université Libre de Bruxelles.

Gwan, A., & Kimengsi, J. (2020). Urban expansion and the dynamics of farmers' livelihoods: Evidence from Bamenda, Cameroon. *Sustainability Journal, 12*(5788).

Howden, S., Soussana, J.-F., Tubiello, F., Chhetri, N., Dunlop, M., & Meinke, H. (2007). Adapting agriculture to climate change. *PNAS, 104*(50), 19691–19696.

Kamga, Y. (2020). *Ethnobotanique, diversité floristique, variabilité morphologique et anatomique dans les facies à Garcinia kola Heckel (Clusiaceae) des forêts denses du Sud-Cameroun* (Thèse de doctorat). School of Science and Technology, l'Université de Dschang.

Kamga, Y., Legwegoh, A., Riley, L., Wagner, J., Lékéufack, M., & Njukeng, P. (2021). *Food Security in Africa's Secondary Cities: The Informal Food Sector in Dschang, Cameroon*. African Food Security Urban Network (AFSUN).

Kamga, Y., Nguetsop, V., Anoumaa, M., Kanmegne, G., Momo Solefack, M., & Ngueguim, J. (2019). *Garcinia kola* (Guttiferae) in tropical rain forests: Exploitation, income generation and traditional uses, in the east and central regions of Cameroon. *Journal of Pharmaceutical, Chemical and Biological Sciences, 7*(1), 13–27.

Karg, H., Derechsel, P., Dittrich, N., & Cauchois, A. (2020). Spatial and temporal dynamics of croplands in expanding West African cities. *Urban Agriculture and Regional Food Systems, 5*(1), 16.

KC, K., Legwegoh, A., Therien, A., Fraser, E., & Antwi-Agyei, P. (2018). Food price, food security and dietary diversity: A comparative study of urban Cameroon and Ghana. *Journal of International Development, 30*(1), 42–60.

Kouame, N., Aké, C., Mangara, A., & N'guessan, K. (2016). Analyse de l'intérêt socio-économique des grains de *Garcinia kola* Heckel (Clusiaceae) dans la commune de Koumassi (Abidjan), Côte d'ivoire. *International Journal of Biological and Chemical Science, 10*(6), 2587–2595.

Legwegoh, A., & Fraser, E. (2017). High food prices in urban Cameroon: Coping strategies and suggested policy actions. *Development in Practice, 27*(8), 1064–1077.

Legwegoh, A., Kamga, Y., Riley, L., Lekeufack, M., & Njukeng, P. (2020). *Food Security in Africa's Secondary Cities: No. 3. Dschang, Cameroon* (Urban Food Security Series No. 29). African Food Security Urban Network (AFSUN).

Loubelo, E. (2012). *Impact des produits forestiers non ligneux (PFNL) sur l'économie des ménages et la sécurité alimentaire: cas de la République du Congo* (Thèse de doctorat: Ecole Doctorale). Sciences Humaines et Sociales de l'Université Rennes 2.

Marchese, C. (2015). Biodiversity hotspots: A shortcut for a more complicated concept. *Global Ecology and Conservation, 3*(1), 297–309.

Mayaux, P., Pekel, J.-F., Desclée, B., Donnay, F., Lupi, A., Achard, F., Clerici, M., Bodart, C., Brink, A., Nasi, R., & Belward, A. (2013). State and evolution of the African rainforests between 1990 and 2010. *Philosophical Transactions of the Royal Society B: Biological Sciences, 368*(1625).

Moukette, B. M., Pieme, C., Nya Biapa, P., Ama Moor, V., Berinyuy, E., & Ngogang, J. (2015). *Afrostyrax lepidophyllus extracts* exhibit in vitro free radical scavenging, antioxidant potential and protective properties against liver enzymes ion mediated oxidative damage. *BMC Research Notes, 8*, 344.

Ndoye, O., Ruiz Pérez, M., & Eyebe, A. (1999). Non-wood forest product markets and potential degradation of the forest resource in Central Africa:

The role of research in providing a balance between welfare improvement and forest conservation. In T. Sunderland, L. Clark, & P. Vantomme (Eds.), *Non-wood forest products of Central Africa: Current research issues and prospects for conservation and development* (pp. 183–206). Food and Agriculture Organization.

Ngome, I., & Foeken, D. (2012). "My garden is a great help": Gender and urban gardening in Buea Cameroon. *GeoJournal, 77*, 103–118.

Nkem, J., Fobissie, B., Idinoba, M., Somorin, O., Ndoye, O., & Awono, A. (2010). Shaping forest safety nets with markets: Adaptation to climate change under changing roles of tropical forests in Congo Basin. *Environmental Science and Policy, 13*, 498–508.

Prain, G., Karanja, N., & Lee-Smith, D. (Eds.). (2010). *African Urban Harvest: Agriculture in the Cities of Cameroon, Kenya, and Uganda*. IDRC.

Republic of Cameroon. (2011). *Plan Communal de Développement de Dschang*. Action pour un Développement Equitable, Intégré et Durable (ADEID). https://www.pndp.org/documents/PCD_DSCHANG1.pdf

Ruiz Pérez, M., Ndoye, O., & Eyebe, A. (2000). La commercialisation des produits forestiers non ligneux dans la zone de forêt humide du Cameroun. *Arbres, Forêts et Communautés Rurales, 19*, 29–35.

Sneyd, L. (2013). Wild food, prices, diets and development: Sustainability and food security in urban Cameroon. *Sustainability, 5*, 4728–4759.

Tata Ngome, P., Shackleton, C., Degrande, A., Nossi, E., & Ngome, F. (2019). Assessing household food insecurity experience in the context of deforestation in Cameroon. *Food Policy, 84*, 57–65.

Temgoua, E., Ntangmo Tsafack, H., Njine, T., & Serve, M. (2012). Vegetable production systems of swamp zone in urban environment in West Cameroon: Case of Dschang City. *Universal Journal of Environmental Research and Technology, 2*(2), 83–92.

Tieminie, R., Loh, C., Tiguhong, J., Nghobouche, M., Mandiefe, P., & Tieguhong, M. (2021). Non-timber forest products and climate change adaptation among forest dependent communities in Bamboko forest reserve, southwest region of Cameroon. *Environmental Systems Research, 10*(20), 1–13.

UN-Department of Economic and Social Affairs (UN-DESA). (2018). *World Urbanization Prospects: The 2018 Revision*. United Nations Department of Economic and Social Affairs.

World Resources Institute (WRI). (2021). *Global Forest Review: Indicators of Forest Extent*. https://research.wri.org/gfr/forest-extent-indicators/forest-loss

CHAPTER 16

Accessibility of Sanitary Facilities Among Food Sellers in African Secondary Cities: Implications for Food Safety and Urban Planning Policies

Emmanuel Chilanga and Liam Riley

Introduction

The rapid growth of secondary cities across sub-Saharan Africa has often outpaced the capacity for local governments to ensure the safety of food consumed by their urban populations. Foodborne illness is a major public health hazard that is directly related to the effectiveness of food safety policies, compliance and enforcement (Hoffmann et al., 2019). The

E. Chilanga
McGill University, Montreal, QC, Canada

L. Riley (✉)
Balsillie School of International Affairs, Wilfrid Laurier University, Waterloo, ON, Canada
e-mail: lriley@balsillieschool.ca

L. Riley and J. Crush (eds.), *Transforming Urban Food Systems in Secondary Cities in Africa*,
https://doi.org/10.1007/978-3-030-93072-1_16

World Health Organization (WHO) estimates that one in every 10 people suffered from foodborne illnesses in 2010 (Sharif et al., 2018). Foodborne diseases are caused by contaminated food or drink, most often from foods containing biological hazards such as bacteria, viruses and parasites (Vipham et al., 2020). African countries have reported a disproportionately high number of children and adults who suffer from foodborne diseases, such as diarrhea and cholera, compared to other countries in the world (Pires et al., 2021).

This chapter contributes to the literature on food safety practices in African secondary cities through two case studies of urban food systems: in Dschang, Cameroon; and in Mzuzu, Malawi. Both cities were studied as part of the African Food Security Urban Network—Food, Urbanization, Environment and Livelihoods project (AFSUN-FUEL), which included a survey of small-scale food traders. The food trader survey followed a household survey conducted in both cities in 2017 that found high rates of dependence on informal and open markets within the food systems (Legwegoh et al., 2020; Riley et al., 2018). Whereas Mzuzu has one large supermarket (of the South African Shoprite chain), Dschang has no similar establishment. Formal sector supermarkets provide some advantages for government safety regulators, such as indoor facilities, a fixed address and potential legal accountability, which are inconsistent factors among small-scale informal traders, making enforcement of safe food practices relatively difficult (Boatemaa et al., 2019). In many African countries, municipal governments of secondary cities have less capacity than primary cities to enforce food safety bylaws (Giroux et al., 2021). There are also gaps in the provisioning of facilities for safe hygiene in most secondary cities, which can negate the best efforts of vendors to protect their customers (Smit, 2018).

Food Safety in Mzuzu's Food System

Mzuzu is the third-largest city, after Lilongwe and Blantyre, and a major urban center in Malawi's Northern Region. Mzuzu was declared a city in 1985 and covers 147 square kilometers at an altitude between 1,200 and 1,370 meters above sea level. In 2018, the city had a population of 220,000, settled across 15 political wards (Manda, 2019). Food safety and hygiene is a known public health concern among residents of Mzuzu city. Current studies demonstrate that a high proportion of products sold in the city are contaminated. For instance, a study by Jazimoni Lazaro,

Fanuel Kaputi and Rochelle Holm (2019) found that almost 100% of fresh fish sold in the open markets of the city were contaminated with *Salmonella spp* bacteria. Another study found that 87% of fresh vegetables sold in the open markets were contaminated with *Escherichia coli* (Holm et al., 2017). The high burden of diarrheal infection in the city may be attributed to poor sanitary services that overstretch the budget of public hospitals (Holm et al., 2018).

Despite the food safety hazards, the majority of Mzuzu residents depend on open-air markets and vendors for their regular food supply. Riley et al., (2018, p. 35) found that, in the previous year, 67% of households had purchased food at the Main Market (partially covered), 57% at Vigwagwa Market (uncovered) and 46% from street vendors. About half (54%) had purchased food at supermarkets (54%), but only 25% purchased food at a supermarket weekly or daily. A study of informal settlements in Lilongwe found even higher rates of patronage of street vendors: all households had purchased food from sellers at least once per week in the previous year and 45% of households averaged five times per week (Chilanga et al., 2017). There appear to be strong linkages between informality, poverty and foodborne illness risk in Malawian cities.

The Government of Malawi recognizes that food safety and sanitation is a policy priority, human rights and social welfare issue, and that policies promoting food safety are a pathway to the attainment of the Sustainable Development Goals (Magamba et al., 2017). Unfortunately, the country has limited food safety and sanitation capacity due to limited infrastructure, food safety knowledge and data (Morse et al., 2018). Political corruption scandals in recent years, which have contributed to urban food insecurity and eroded public trust in governments (Riley & Chilanga, 2018), have further compounded the challenges for governments seeking to ensure food safety for urban residents. These factors are currently undermining social and public health workers to support best food safety and market hygiene practices, such as use of sanitary facilities and personal hygiene.

Food Safety in Dschang's Food System

Dschang has grown rapidly since the establishment of the University of Dschang in 1994 (Mboua et al., 2021). While accurate demographic data is not available, the 2020 extrapolated data suggests that the city is home to more than 200,000 people (Legwegoh et al., 2020). Food safety and sanitation is a topical public health and policy issue in Dschang. Empirical studies in the city have found that a large proportion of domestic water that is supplied and used in the city is not safe for human consumption in accordance with WHO water quality standards (Emile, 2011). Lack of potable water has been linked to a high prevalence of waterborne outbreaks that exacerbate morbidity and mortality (Fusi-Ngwa et al., 2014). Other studies attributed the high prevalence of intestinal parasites in children to consumption of salads that are watered by fecal-contaminated water (Ntangmo Tsafack et al., 2012). A national study that assessed the quality of meat and meat products produced in Cameroon found that most foods of animal origin were contaminated with pathogenic bacteria and aflatoxins (Pouokam et al., 2017).

Informal traders are critical actors in Dschang's food system. There are two major markets in the city center: Marché A specializes in non-food items and services and Marché B is the food market. Marché B operates daily, with variable numbers of vendors and customers. Every eighth day is Big Market Day (*ngam*) and the market area swells with traders and customers. The fifth day between Big Market Days is Small Market Day (*metah*), while the days in between are attended only minimally by vendors and customers (Fouepe et al., 2017). Other markets located in residential areas also follow the market day pattern, mainly accommodating rural traders and customers (ADEID, 2011). The household survey conducted in 2017 found that nearly all households (91%) had purchased food at open markets in the past year and for about one in four households open markets were their sole food source. There is no major supermarket outlet in Dschang and only 10% of households had purchased food at a supermarket in the past year, perhaps at a supermarket in another city (Legwegoh et al., 2020). Bread and meat were the only popular foods to be purchased, and these purchases were made mostly at places other than open-air markets (small shops and butcheries, respectively) (Legwegoh et al., 2020).

Given the widespread use of open-air markets, an understanding of sanitary practices among food vendors in Cameroon is critical in order to inform food and sanitation policies and programs (Montet et al., 2021). Research in Fako found that many food traders were not following hygienic food handling practices and thus caused bacterial contamination of vegetable products (Akoachere et al., 2018). In Yaoundé, it was observed that most street food vendors had compromised personal hygiene standards, with up to 45% of the vendors carrying feco-orally transmissible parasites (Blaise, 2014).

There is a dearth of studies exploring the predisposing factors of poor hygienic practices among food traders in secondary cities of Africa. There are also too few studies situating the issue in a social context that links food safety to food security and governance in African secondary cities. To address this gap in knowledge, and to provide a window onto the international differences in food safety, our study examined factors that are associated with access to sanitary facilities, with a focus on toilets, potable water, waste disposal and food inspections to reduce the risk of foodborne pathogens spread by food traders in Mzuzu and Dschang. The focus on vendors' access to both information about food safety and hygiene-promoting facilities is also relevant to the public health problems that have emerged in the context of COVID-19, since many of the challenges of raising awareness and changing routine behaviors are common to both campaigns.

Methods

A multistage descriptive cross-sectional study design was used to select a representative sample of diverse food traders in each city. Prior to this study, municipal governments in Mzuzu and Dschang did not have lists of all food traders, which posed a challenge in drawing a representative sample size. To address this problem, our survey teams mapped out all food traders in targeted areas of each city, using a rapid survey that recorded GPS coordinates and types of businesses, providing a basis for understanding the relative prevalence of different types of small-scale food trading businesses in each of the targeted neighborhoods. These maps were the basis for sampling targeted percentages of different types of small-scale food sellers in different areas of each city, achieving the objective of studying food trading activities in commercial and residential areas.

The survey was conducted in Mzuzu in April 2019 and in Dschang in August 2019. The survey questionnaire was adapted from the survey instrument developed by the Hungry Cities Partnership (HCP) (Chikanda et al., 2020). It captured information related to vending location, demographic and enterprise characteristics (structure, practices and business environment), food production patterns and sanitation. The sanitation questionnaire was in addition to the HCP questionnaire and was based on a validated questionnaire used in previous studies in urban areas of developing countries (Abdul-Mutalib et al., 2012; Aluko et al., 2014). It gathered information relating to access to sanitation resources, knowledge and practices. The survey questionnaire was designed on kobotoolbox.org and loaded onto GPS-enabled tablets using OpenDataKit software.

The Kolmogorov–Smirnov and Shapiro–Wilk tests were used to determine the normality of the distribution of the data. These include starting capital, average sales, number of employees and age. Bivariate and multivariate logistic regressions were performed to examine significant predictors of access to basic hygienic resources, practices and knowledge. The data was not normally distributed, so the nonparametric Mann–Whitney U test, Kruskal–Wallis test and Spearman's correlation were applied. The multicollinearity of explanatory variables obtained a variance inflation factor of 5,143, which indicated independence among the explanatory variables both at the individual and the cluster level. Consequently, a fixed-effects model was used to account for the clustering effect. The results of the multivariable analysis have been reported as crude and adjusted odds ratios with a 95% confidence interval (CI). A *p*-value of less than 0.05 was considered statistically significant. The data was analyzed using an IBM Statistical Package of Social Sciences for Windows, version 23.0 (IBM Corporation, Armonk, NY).

Sociodemographic Characteristics of Traders

Sociodemographic and sanitation data for Mzuzu ($n = 497$) and Dschang ($n = 848$) selected food traders are presented in Table 16.1. Most food traders in both cities were women: 67.6% in Mzuzu and 63.7% in Dschang. There was significant difference in education attainment between the cities. In Mzuzu, most food traders (47.3%) had a primary school education, while in Dschang, the majority of traders (62%) had a secondary education. It was also observed that fewer food traders in Mzuzu (3.2%) had post-secondary education qualification compared to Dschang (15.9%). In both cities, a significant number of food traders did not pay city taxes: 72.2% in Mzuzu and 74.1% in Dschang. About 36% and 66% of food traders in Mzuzu and Dschang, respectively, operated in a permanent structure.

Table 16.1 Sociodemographic characteristics of food traders in Mzuzu and Dschang

Variable	*Mzuzu*	*Dschang*
Sample size	n = 497	n = 848
Gender		
Male	161 (32.4%)	307 (36.2%)
Female	336 (67.6%)	541 (63.7%)
Business type		
Permanent	179 (36.0%)	558 (65.8%)
Temporary	298 (59.9%)	113 (13.3%)
Mobile vendor	18 (4.0%)	175 (20.6%)
Pay city rates		
Yes	138 (27.8%)	220 (25.9%)
No	354 (72.2%)	628 (74.1%)
Education level		
No formal education	19 (3.8%)	41 (4.8%)
Primary	235 (47.3%)	145 (17.1%)
Secondary	224 (45.1%)	526 (62.0%)
Post-secondary	16 (3.2%)	135 (15.9%)

Access to Sanitary Resources among Food Vendors

Table 16.2 summarizes the findings in terms of access to hygiene facilities and information about hygienic practices. Regarding access to facilities, the study found that 22.7% of food traders in Mzuzu and 76.7% in Dschang reported having limited access to toilet facilities. More than half (56.5%) had limited access to potable water at their main place of business in Mzuzu and about three-quarters (76.8%) of traders in Dschang reported the same. Almost half (48.3%) of the traders had difficulties accessing waste disposal facilities in Mzuzu, while 56.2% in Dschang reported having challenges of accessing waste disposal facilities. It was reported that 45.9% and 81% of food traders in Mzuzu and Dschang, respectively, had limited access to handwashing facilities at their business premises.

Regarding access to information, almost all the food traders in Mzuzu (97.4%) and two-thirds in Dschang (66.3%) reported having no formal food safety and hygiene training. It was reported that 69.2% and 74.6% of food traders in Mzuzu and Dschang, respectively, had never attended a food safety public awareness campaign. In terms of food and premises inspection, it was reported that only 19.3% of the food traders had been inspected in Mzuzu, while 27.2% of food traders had been inspected in

Table 16.2 Access to sanitary resources in Mzuzu and Dschang

Variable	*Mzuzu*	*Dschang*
Access to toilet facility		
Yes	383 (77.1%)	198 (23.3%)
No	113 (22.7%)	650 (76.7%)
Access to water facility		
Yes	213 (42.9%)	197 (23.2%)
No	281 (56.5%)	651 (76.8%)
Access to waste disposal		
Yes	257 (51.7%)	371 (43.8%)
No	240 (48.3%)	477 (56.2%)
Attended food safety training		
Yes	13 (2.6%)	286 (33.7%)
No	484 (97.4%)	562 (66.3%)
Attended food safety campaign		
Yes	153 (30.8%)	215 (25.4%)
No	344 (69.2%)	633 (74.6%)
Business facility inspected		
Yes	96 (19.3%)	231 (27.2%)
No	399 (80.3%)	617 (72.8%)
Available hand washing facility		
Yes	267 (53.7%)	161 (19.0%)
No	228 (45.9%)	687 (81.0%)
Food safety knowledge		
Yes	228 (45.9%)	511 (60.3%)
No	267 (53.7%)	337 (39.8%)

Dschang. Regarding food safety knowledge, 45.9% of the participants in Mzuzu and 60.3% in Dschang correctly answered most of the knowledge section questions, such as the effects of eating semi-cooked food, the role of insects in spreading food poisoning, rapid multiplication of harmful bacteria at room temperature and use of leftover food.

Correlates of Access to Sanitation Resources

Univariate logistic regression analyses suggest that women food traders were less likely to report having access to toilet facilities in Mzuzu (OR = 0.21, $p < 0.001$) and Dschang (OR = 0.41, $p < 0.002$). Traders that had permanent business spaces were more likely to report having access to a toilet facility in Mzuzu (OR = 2.13, $p < 0.000$; and in Dschang (OR = 2.16, $p < 0.000$) than mobile food traders. Similarly, food traders who were operating at a permanent structure were more likely to have access to a waste disposal facility in Mzuzu (OR = 1.25, $p < 0.000$) and in Dschang (OR = 2.10, $p < 0.000$) than mobile food traders.

Food traders in Mzuzu who pay city rates were more likely to report having access to a toilet facility (OR = 2.41, $p < 0.01$) and waste disposal (OR = 2.19; $p < 0.03$, respectively). Similar odds ratios were observed in Dschang (OR = 2.24, $p < 0.01$; and OR = 2.39, $p < 0.02$, respectively). Food traders who reported that they were inspected in the past 12 months were more likely to have access to a toilet, water and waste disposal facilities than their peers that were not inspected in Mzuzu (OR = 2.74, $p < 0.01$; OR = 1.32, $p < 0.03$; OR = 2.81, $p < 0.03$, respectively) and in Dschang (OR = 2.19, $p < 0.01$; OR = 1.22, $p < 0.02$; OR = 2.41, $p < 0.03$, respectively). In both cities, there were no significant differences in the odds of reporting access to sanitation facilities based on education attainment or access to food safety training.

In multivariable analysis (Table 16.3), the odds of reporting having access to toilet facilities was much lower among women than men in Mzuzu (OR = 0.18, $p < 0.001$) and in Dschang (OR = 0.14 $p < 0.001$). Traders who operated in permanent structures were more likely than mobile traders to report having access to toilet facilities in Mzuzu and Dschang (OR = 2.03, $p < 0.000$; OR = 2.11, $p < 0.000$ respectively). Similarly, food traders who had permanent structures were more likely than mobile traders to have access to waste disposal facilities (OR = 1.19, $p < 0.000$; OR = 1.25, $p < 0.000$ respectively). Similar differences were observed among trades that pay city rents or not, and also among the trades whose business facilities were inspected or not. Those that paid city rates, and those that had their facilities inspected, had a higher likelihood of accessing sanitary facilities such as water, toilet and waste disposal.

Discussion

The findings of this study reveal that the majority of food traders in Mzuzu and Dschang cities are constrained from accessing critical sanitary facilities such as toilet, waste disposal and water facilities. These findings have numerous implications to gender equality, public health and urban planning.

First, even though most food traders in Mzuzu and Dschang are women, women face significant gender discrimination in access to toilet facilities. The results add further evidence to observations that women

Table 16.3 Adjusted odds ratios (95% CI) for factors associated with access to food safety and practice resources

Variables	*Model 1* *Access to toilet facilities*		*Model 2* *Access to water*		*Model 3* *Access to waste disposal*	
	AOR-Dschang	*AOR-Mzuzu*	*AOR-Dschang*	*AOR-Mzuzu*	*AOR-Dschang*	*AOR-Mzuzu*
Gender						
Male	1	1	1	1	1	1
Female	0.14***	0.18***	0.52	0.16	1.23	1.09
Business type						
Mobile trader	1	1	1	1	1	1
Permanent structure	2.11***	2.03***	1.34**	1.41**	1.15**	1.19**
Business tenure						
Do not pay rent	1	1	1	1	1	1
Pay city rent	2.21**	2.26**	1.14**	1.12*	2.19*	2.12*
Business inspected						
No	1	1	1	1	1	1
Yes	2.34**	2.66**	1.22*	1.21*	2.31*	2.69*
Food safety education						
No	1	1	1	1	1	1
Yes	1.35	1.29	1.23	1.19	1.32	1.24
Food safety campaigns						
No	1	1	1	1	1	1
Yes	1.16	1.12	1.17	1.11	1.52	1.41

*$p < 0.05$, **$p < 0.01$, ***$p < 0.00$

in the Global South who work in informal market spaces have inadequate access to toilet facilities (Schmitt et al., 2018). "Toilet insecurity" refers to a situation when individuals do not have access to safe and usable toilets, and it is increasingly recognized as a serious indication of—and contributor to—gender inequality (Caruso et al., 2017). Inadequate access to sanitary facilities by women working as food traders is linked to their low productivity and poor health outcomes. It is also an infringement of their rights to an adequate standard of living, which includes access to sanitary facilities. From the food systems standpoint, it also creates a hazard for consumers whose food safety would be enhanced if women vendors' toilet insecurity were systematically addressed. Our study therefore contributes to the advocacy for a decent work environment for women in the informal urban sector in Mzuzu and Dschang, where there is urgent need for sanitary facilities.

Second, our study suggests that for sanitation knowledge and education programs to be effective tools to improve food safety practices among food traders, they need to be supported by sanitary infrastructures and inspection services. This point resonates with Warren Smit's (2018, p. 99) argument that public education and bylaw enforcement are important, "but it is arguably even more important to ensure that the necessary infrastructure and facilities are available." It has been observed that there were no significant statistical differences in utilization of sanitary resources between food traders that had formal food hygiene knowledge compared to those that had limited knowledge. The finding supports studies in developing countries that found that provisioning of sanitary facilities increases hygienic practices among food traders with higher education as compared to those with basic education attainment (Abdi et al., 2020; Rietveld et al., 2016).

The pathways in which food inspection is associated with access to sanitary facilities should be interpreted with caution. Potential mediating and moderating factors could weaken the observed association in Mzuzu and Dschang. For instance, city planners usually inspect food in formal markets where there are well developed and planned marketing infrastructures, leaving out the majority of food traders operating in informal markets, on roadsides and from home (Lazaro et al., 2019; Temgoua et al., 2009). Therefore, food safety regulatory frameworks in these cities should be inclusive, as leaving out informal food traders may have negative implications for the management of foodborne outbreaks, which are linked to unhygienic standards in many developing cities (Imathiu, 2017).

Finally, the findings of the study suggest that city planners can adopt system approaches to increase access to sanitary facilities, which can, in turn, improve food safety practices among food traders in Mzuzu and Dschang. System approaches consider multifaceted interrelationships between connected systems and allow for the integration of knowledge across a broad array of disciplinary and policy domains (Rietveld et al., 2016). In these cities, there is a need for collaborative efforts among social service delivery sectors, such as departments of water, infrastructure, public health and urban governing institutions to ensure that food traders have access to sanitary facilities. These planning activities can gain traction by linking their objectives to several relevant Sustainable Development Goals (Ilieva, 2017), almost all of which are connected in some way to the intersecting concerns with health, economic development, gender equality, urban sustainability and fair working conditions that surface in discussions of food safety in African cities. Building on the observation that food safety in cities relies on regulations at national and international levels (Boatemaa et al., 2019), there is ample potential to tie local efforts to promote food safety with larger-scale development efforts.

Conclusion

In developing countries, small-scale urban food traders play a significant role in food systems, as they are in proximity to the majority of low-income consumers. Unfortunately, these traders are often linked to poor food hygiene practices that compromise public health safety. In Mzuzu and Dschang, access to sanitary facilities such as toilets, water and waste bins are a problem of public concern. The intersection of social and structural factors such as gender, socio-economic status and city management influences access to sanitary facilities, with women facing extra challenges. To address these challenges, a systems approach has been proposed whereby diverse stakeholders across multiple levels of governance should work together to improve access to sanitary facilities.

References

Abdi, A., Amano, A., Abrahim, A., Getahun, M., Ababor, S., & Kumie, A. (2020). Food hygiene practices and associated factors among food handlers working in food establishments in the Bole Sub City, Addis Ababa, Ethiopia. *Risk Management and Healthcare Policy*, 13, 1861–1868

Abdul-Mutalib, N.-A., Abdul-Rashid, M.-F., Mustafa, S., Amin-Nordin, S., Hamat, R., & Osman, M. (2012). Knowledge, attitude and practices regarding food hygiene and sanitation of food handlers in Kuala Pilah, Malaysia. *Food Control, 27*(2), 289–293.

Action pour un Développement Equitable, Intégré et Durable (ADEID). (2011). *Plan communal de développement de Dschang*. https://www.pndp.org/documents/PCD_DSCHANG1.pdf

Akoachere, J.-F., Tatsinkou, B., & Nkengfack, J. (2018). Bacterial and parasitic contaminants of salad vegetables sold in markets in Fako Division, Cameroon and evaluation of hygiene and handling practices of vendors. *BMC Research Notes, 11*(1), 100.

Aluko, O., Ojeremi, T., Olaleke, D., & Ajidagba, E. (2014). Evaluation of food safety and sanitary practices among food vendors at car parks in Ile Ife, southwestern Nigeria. *Food Control, 40*, 165–171.

Blaise, N. (2014). An assessment of hygiene practices and health status of street-food vendors in Yaoundé, Cameroon. *International Journal of Tropical Disease & Health*, 1153–1170.

Boatemaa, S., Barney, M., Drimie, S., Harper, J., Korsten, L., & Pereira, L. (2019). Awakening from the listeriosis crisis: Food safety challenges, practices and governance in the food retail sector in South Africa. *Food Control, 104*, 333–342.

Caruso, B., Clasen, T., Yount, K., Cooper, H., Hadley, C., & Haardörfer, R. (2017). Assessing women's negative sanitation experiences and concerns: The development of a novel sanitation insecurity measure. *International Journal of Environmental Research and Public Health, 14*(7), 755.

Chikanda, A., Crush, J., & Tawodzera, G. (2020). Urban food security and south-south migration to cities of the global south. In J. Crush, B. Frayne, & G. Haysom (Eds.), *Handbook on urban food security in the global south* (pp. 261–281). Edward Elgar.

Chilanga, E., Riley, L., Ngwira, J., Chalinda, C., & Masitala, L. (2017). *Food insecurity in informal settlements in Lilongwe, Malawi*. African Food Security Urban Network (AFSUN). Urban Food Security Series No. 25.

Emile, T. (2011). Chemical and bacteriological analysis of drinking water from alternative sources in the Dschang Municipality, Cameroon. *Journal of Environmental Protection, 2*(5), 620.

Fouepe, G., Kana, J., Kengmo, G., & Mfewou, A. (2017). Socioeconomic and technical characteristics of broiler and laying hens layers in peri-urban and urban areas in the city of Dschang, West Cameroon. *International Journal of Agricultural Economics, 2*(4), 110.

Fusi-Ngwa, C., Besong, E., Pone, J., & Mbida, M. (2014). A cross-sectional study of intestinal parasitic infections in children in ghettoed, diverse and affluent communities in Dschang, West Region, Cameroon. *Open Access Library Journal, 1*(9), 1–14.

Giroux, S., Blekking, J., Waldman, K., Resnick, D., & Fobi, D. (2021). Informal vendors and food systems planning in an emerging African city. *Food Policy*, 103, 101997

Hoffmann, V., Moser, C., & Saak, A. (2019). Food safety in low and middle-income countries: The evidence through an economic lens. *World Development*, 123, 104611

Holm, R., Mwangende, J., Tembo, M., & Singini, W. (2017). Bacteriological quality of fresh produce and link to water and sanitation service access from informal markets in Mzuzu, Malawi. *Environment, Development and Sustainability, 19*(6), 2487–2497.

Holm, R., Kamangira, A., Tembo, M., Kasulo, V., Kandaya, H., Gijs Van Enk, P., & Velzeboer, A. (2018). Sanitation service delivery in smaller urban areas (Mzuzu and Karonga, Malawi). *Environment and Urbanization, 30*(2), 597–612.

Ilieva, R. (2017). Urban food systems strategies: A promising tool for implementing the SDGs in practice. *Sustainability, 9*(10), 1707.

Imathiu, S. (2017). Street vended foods: Potential for improving food and nutrition security or a risk factor for foodborne diseases in developing countries? *Current Research in Nutrition and Food Science Journal, 5*(2), 55–65.

Lazaro, J., Kapute, F., & Holm, R. (2019). Food safety policies and practices in public spaces: The urban water, sanitation, and hygiene environment for fresh fish sold from individual vendors in Mzuzu, Malawi. *Food Science & Nutrition, 7*(9), 2986–2994.

Legwegoh, A., Kamga, Y. B., Riley, Martin, L., & Njukeng, P. (2020). *Food security in Africa's secondary cities: No. 3. Dschang, Cameroon*. African Food Security Urban Network (AFSUN). Urban Food Security Series No. 29

Magamba, K., Matumba, L., Matita, G., Gama, A., Singano, L., Monjerezi, M., & Njoroge, S. (2017). Aflatoxin risk management in commercial groundnut products in Malawi (sub-Saharan Africa): A call for a more socially responsible industry. *Journal of Consumer Protection and Food Safety, 12*(4), 309–316.

Manda, M. (2019). *Understanding the context of informality: Urban planning under different land tenure systems in Mzuzu city, Malawi*. PhD thesis, School of Architecture, Planning and Geomatics, Faculty of Engineering and the Built Environment, University of Cape Town

Mboua, P., Siakam, C., & Mabo, N. (2021). Impact of the resumption of classes on the mental health of students of the Faculty of Letters and Social Sciences of the University of Dschang, in the context of Covid 19. *Journal of Affective Disorders Reports* 5, 100147

Montet, D., Brabet, C., Ray, R., & Galindo, S. (2021). *Mycotoxins in food and beverages: Innovations and advances part 1*. Routledge and CRC Press.

Morse, T., Masuku, H., Rippon, S., & Kubwalo, H. (2018). Achieving an integrated approach to food safety and hygiene: Meeting the sustainable development Goals in sub-Saharan Africa. *Sustainability, 10*(7), 2394.

Ntangmo Tsafack, H., Temgoua, E. and Njine, T. (2012). Physico-chemical and bacteriological quality of the vegetable watering water in the Dschang Town, Cameroon. *Journal of Environmental Protection*, 3(8A)

Pires, S., Desta, B., Mughini-Gras, L., Mmbaga, B., Fayemi, O., Salvador, E., Gobena, T., Majowicz, S., Hald, T., Hoejskov, P., Minato, Y., & Devleesschauwer, B. (2021). Burden of foodborne diseases: Think global, act local. *Current Opinion in Food Science, 39*, 152–159.

Pouokam, G., Foudjo, B., Samuel, C., Yamgai, P., Silapeux, A., Sando, J., Atonde, G., & Frazzoli, C. (2017). Contaminants in foods of animal origin in Cameroon: A one health vision for risk management "from farm to fork". *Frontiers in Public Health*, 5

Rietveld, L., Siri, J., Chakravarty, I., Arsénio, A., Biswas, R., & Chatterjee, A. (2016). Improving health in cities through systems approaches for urban water management. *Environmental Health, 15*(1), S31.

Riley, L., & Chilanga, E. (2018). Things are not working now': Poverty, food insecurity and perceptions of corruption in urban Malawi. *Journal of Contemporary African Studies, 36*(4), 484–498.

Riley, L., Chilanga, E., Zuze, L., & Joynt, A. (2018). *Food security in Africa's secondary cities: No. 1, Mzuzu, Malawi. African food security urban network.* African Food Security Urban Network (AFSUN). Urban Food Security Series No. 27

Schmitt, M., Clatworthy, D., Ogello, T., & Sommer, M. (2018). Making the case for a female-friendly toilet. *Water, 10*(9), 1193.

Sharif, M., Javed, K., & Nasir, A. (2018). Food borne illness: Threats and control. In A. Holban & A. Grumezescu (Eds.), *Foodborne diseases* (Vol. 15, pp. 501–523). Academic Press.

Smit, W. (2018). Urban governance in Africa: an overview. In C. Ammann, & T. Förster (Eds.), *African cities and the development conundrum* (pp. 55–77). Brill-Nijhoff

Temgoua, E., Ngnikam, E., & Ndongson, B. (2009). Drinking water quality: Stakes of control and sanitation in the town of Dschang, Cameroon. *International Journal of Biological and Chemical Sciences*, 3(3), Article 3

Vipham, J., Amenu, K., Alonso, S., Ndahetuye, J.-B., Zereyesus, Y., Nishimwe, K., Bowers, E., Maier, D., Sah, K., Havelaar, A., & Grace, D. (2020). No food security without food safety: Lessons from livestock related research. *Global Food Security*, 26, 100382

CHAPTER 17

Migrant Remittances and Household Food Security in Mzuzu, Malawi

Anil Dhakal

Introduction

Food insecurity is a growing concern, in particular among the urban poor residing in secondary African cities (Battersby & Watson, 2019; Frayne et al., 2010; Mackay, 2019). The ability to access food in an urban context largely depends on the availability of cash income. The cash-based nature of the urban economy requires urban households to depend mostly on the market for accessing food and other basic necessities (Kimani-Murage et al., 2014; Tacoli, 2017). As such, people living in poor urban conditions with precarious income streams are particularly vulnerable to fluctuations in the price of food and related market shocks (Mohiddin et al., 2012). In Malawi, studies have documented that the price of maize reaches its highest point at the end of the lean season, in

A. Dhakal (✉)
Department of Geography and Environmental Management, University of Waterloo, Waterloo, ON, Canada
e-mail: adhakal@uwaterloo.ca

L. Riley and J. Crush (eds.), *Transforming Urban Food Systems in Secondary Cities in Africa*,
https://doi.org/10.1007/978-3-030-93072-1_17

February, and that income shocks due to this price increase hit the poor hard, especially low-income urban dwellers (Harttgen et al., 2016; Sassi, 2015). Urban households with irregular income are more likely to be food insecure, which means that remittances received by migrant-sending households contribute to income stability and food security. They are the added source of income that can be used for meeting day-to-day needs and planning for future income security through investments. Most remittance-receiving households in low-income urban areas use a large proportion of the remitted income for food (Crush, 2013). However, research on migrant remittances and their implications for household food security and economic stability in the long term is crucial but very little explored (Choithani, 2017; Crush, 2013; Crush & Caesar, 2017).

This chapter uses the New Economics of Labor Migration (NELM) approach to explore the linkages between migrant remittances and household food insecurity in the sending community. The NELM approach is based on three main assumptions: relative deprivation, investment and insurance (Atuoye et al., 2017; Taylor & Taylor, 1999). Due to relative deprivation, households make migration decisions to improve household welfare through remittances (Stark, 1991; Taylor & Taylor, 1999). Similarly, remittances can be the source of investment in income-generating activities to expand household income streams; likewise, remittances can act as insurance in times of crisis or food price hikes (Atuoye et al., 2017; King & Collyer, 2016). The NELM approach is considered most relevant to study migration from low-income households because it uses qualitative interviews and household surveys and positions households as the central unit of research (Castles, 2009; de Haas, 2010). This approach facilitates an assessment of the degree to which remittances help migrant households to have continuous access to food without disruption, even in crisis situations.

Context of Malawians Migrating in Southern Africa

The history of international labor migration in the African continent dates back to the European colonial period, when the mining and farming industries established by Europeans required cheap labor from across the continent (Long & Crisp, 2011). Malawi (then Nyasaland), specifically Northern Malawi, has a long history of labor migration; most of its migrant workers traveled to South Africa to work in the mines through formal mechanisms (Ndegwa, 2015; Oucho, 1995). The formal migration of Malawians started in the late nineteenth century, coming to an end in the late twentieth century (Chirwa, 1997). In the early decades of the formal migration process, the migration was deeply rooted. The labor migration of men was sometimes even characterized as a major export of the country, as in an article published in the early 1940s, which noted that "the chief export of Nyasaland in the past fifty years has been men" (Read, 1943, p. 606). Since then, labor migration to the Southern African Development Community (SADC) countries and beyond has been the tradition in Malawian communities (Ndegwa, 2015; Oucho, 1995). Within SADC, South Africa is the major destination of labor migration from Malawi and is dominated by undocumented and informal workers, especially in domestic work and informal businesses (Banda, 2019).

The World Bank estimates that Malawians sent about US$181 million in 2018 to their households back home, which was much higher than the US$800,000 remitted in 2000 (World Bank, 2020). Similarly, remittances as a share of GDP rose from 0.065% in 1994 to 2.57% in 2018 (World Bank, 2020). This figure is low when compared with other remittance-receiving low- and lower-middle-income African countries such as Nigeria, Kenya, Mali and Zimbabwe (Ratha, 2021). However, remittances carry great importance in the livelihoods of the receiving households, as the transfer reaches directly to the household level. These statistics also hide much of the informal remittances and the remittances of goods.

Methodology

This research mainly employs qualitative methods based on the data collected at the household level. It focuses on the households living in informal settlements of Mzuzu that have at least one family member working in neighboring SADC countries. A combination of purposive and snowball sampling was applied to select migrant households and key informants. In-depth semi-structured interviews were conducted with the migrants' household representatives/heads to have a nuanced understanding of peoples' lived experiences. A total of 42 migrant household representatives/heads and 37 key informants from different sectors/organizations working with migrant communities or having expertise in the migration and development and food security issues were also interviewed. Out of 42 migrant households, 41 had at least one family member who had migrated to South Africa—including three overlapping households that had an additional family member who had migrated to Tanzania, Zambia or Zimbabwe as well. The remaining one household had a member working in Tanzania only. Furthermore, to better understand the food security situation of migrant-sending households, the Household Food Insecurity Access Scale (HFIAS) and Months of Adequate Household Food Provisioning (MAHFP) indicators were also captured. The HFIAS captures the household access to food in the previous four weeks, based on nine indicators from which Household Food Insecurity Access Prevalence (HFIAP) is calculated (Coates et al., 2007). The HFIAP categorizes food access components into four categories: 1 = food secure; 2 = mildly food insecure; 3 = moderately food insecure; and 4 = severely food insecure (Coates et al., 2007). The household food security status presented in the following sections refer to HFIAP. The MAHFP captures the status of the household food supply in the past 12 months (Bilinsky & Swindale, 2010). The household food supply status mentioned in the following sections refer to MAHFP.

The data were analyzed using the general inductive approach: finding the core meaning in the text; identifying themes most relevant to research objectives; and description of the most important themes (Thomas, 2006). NVivo 12 software was used to help with coding. The final step included the interpretation of the entire analysis (Creswell et al., 2003). The qualitative methodology revealed many nuanced perceptions of the effect of remittances on individuals' experiences of food insecurity. These perceptions fall into the four topics discussed in the following sections: remittances invested in maize production; the consequences of lost remittances; the improvement of dietary quality with remittances; and how remittances create stability. The discussion of these topics is followed by reflections on the findings through a NELM analysis.

Remittances Support Maize Production

In the Malawian context, having enough stock of maize at home is considered the hallmark of food security (Kerr, 2014; Smale, 1995). A 2017 household food security survey found that one in three surveyed households in Mzuzu cultivates maize, which was by far the most widely cultivated crop (Riley et al., 2018). Many urban dwellers (about 38%) were involved in urban agriculture: about 20% were engaged in livestock rearing, and about one-third of households cultivated crops at rural farms (Riley et al., 2018). Interestingly, the survey found that households involved in farming in the rural and urban areas had better food security status and lived poverty index (Riley et al., 2018). About 37% of respondents who did not grow food cited lack of access to agricultural inputs as a barrier (Riley et al., 2018). Residents of the informal settlements in Mzuzu were also found to engage in cultivating maize, growing vegetables and rearing poultry and pigs.

These survey findings aligned with the rate of urban agriculture in this qualitative study. Out of 42 migrant households interviewed, 19 reported using remittances for farming, mainly for buying fertilizer and hiring farm workers (Table 17.1). When they could not cultivate, they would generally receive remittances to buy maize. In responding to a question about the importance of remittances in her household's access to food, the participant, a widow in her late sixties who was living with a granddaughter, observed:

> At first, it was hard for me to buy food. But now they send money each and every month for me to buy food. (Before) I wouldn't even afford to buy fertilizer. Of course, this year I didn't cultivate maize, but they have already started sending money for me to buy maize at the market. So, I don't have food problems right now (Interview 8, June 30, 2018, Mzuzu, Malawi).

Table 17.1 Sectors of remittance use by migrant households

Sectors (sub-themes)	*No. of households (42)*	*No. of references*
Food	30	30
Education	19	20
Farming and farm inputs	19	21
Utilities and taxes	17	17
Housing	8	9
Investment	7	10
Clothes	6	6
Health care	4	5
Rent	4	4

Note The table was generated based on the frequency of the sub-themes mentioned by migrant households as per the NVivo 12 coding

The respondent had three household members working in South Africa. Her household was food secure and had a food supply for 12 months. She reported that the rental income of MWK40,000 (US$56) from two houses built using remittances, and the monthly transfers of between MWK30,000 and 70,000 (US$43–98) received from her three migrant family members made it easier to run her household of two.[1] While answering the importance of remittances in her livelihood, she mentioned living an easy life as she was having bread and tea for breakfast and did not have to worry about lunch and dinner. Her case indicates that the extended sources of income as a result of direct investment of remittances in building houses, along with regular remittance support and a small family, helped in having better access to food.

Similarly, a male participant in his mid-twenties had experienced improved access to food because of the regular remittance support from his father working in South Africa. His household was food secure and had adequate food supply for 11 months. When comparing the situation before and after migration, he observed, "[Before] you know it was possible that maybe some other days you could have maize flour to cook *nsima*, but you don't have money to buy relish. But now we are able to find everything for our meal" (Interview 6, June 26, 2018, Mzuzu, Malawi). His family was receiving monthly remittances of between MWK70,000 and 100,000. He reported using remittances to buy fertilizer as well.

Remittances have improved households' access to farm inputs, specifically chemical fertilizer. Many respondents stated that the use of chemical fertilizer is very important for a good harvest; however, they could not afford fertilizer without remittances. The average retail fertilizer price in the country remained around MWK20,113 in 2018 (Africa Fertilizer n.d.). The price is very high when compared to the minimum wage rate, which remained at MWK962 per day in 2018 (MWK28,860 per month) (WageIndicator, n.d.). Moreover, some migrant households were also using remittances for hiring agricultural pieceworkers—locally known as *ganyu*. A female respondent in her early seventies put the importance of remittances this way: "At first, we were buying maize at the market, but now we have our own maize, and we use remittances to buy fertilizer and pay agricultural expenses. … We buy six bags of fertilizer each year

[1] In mid-2018, the exchange rate was approximately US$1 = MWK715.

using remittances" (Interview 39, August 10, 2018, Mzuzu, Malawi). Her family had been cultivating for over a decade on borrowed land in Ekwendeni, a small town situated 20 kilometers north of Mzuzu. Her household was food secure and did not face any shortages of supply over the previous 12 months.

These narratives support the conclusion that remittance-receiving households, having remittances as an additional source of income, experienced increased access to food as these households used remittances to buy not only food, but also farm inputs for a better yield. Such findings also indicate that agriculture is still a livelihood strategy for many low-income households. The finding may be indicative of the peri-urban nature of Mzuzu—a city that shares common characteristics with other cities of sub-Saharan Africa, such as the challenges of urban poverty, rapid urbanization, clean drinking water, sanitation and health facilities—where a significant number of households still depend on agriculture to support their livelihood (Githira et al., 2020; Kita, 2017; Riley et al., 2018). Such observations further support the importance of remittances for low-income urban households in their struggle to achieve food security.

Without Remittances, Food Insecurity Worsens

For most migrant households, the frequency and the volume of remittances ensured improvements in their access to food even when they were faced with increasing food prices. However, some households with zero or nominal remittances from their family members had to face more challenges in accessing food. An example is a case of a female respondent in her mid-fifties. She took a loan of MWK60,000 to support her son for passport and transport. He went to South Africa in 2014. Instead, the burden of repaying the loan deteriorated her household's ability to access food. While responding to her present situation regarding food access in comparison to the situation before 2014, she said:

> There is a change, because, at first, the food at the house was just okay, but because I borrowed some money to send this boy [the migrant] for transport, and now I am still paying back the money. The way I am buying food now and the way I was buying at first is different. Now, I am struggling to buy food because the other money is being used to pay back the loans. (Interview 3, June 20, 2018, Mzuzu, Malawi)

Her household was severely food insecure. Likewise, the household had food supply for only seven months. For similar households, the added burden of paying back the loan and lack of support from the migrant could make households vulnerable to food insecurity. To this end, a female respondent in her late sixties, who had a migrant son in South Africa, raised two points. First, she did not see any significant difference in her food security situation from the nominal remittance of MWK20,000 (US$28) she had received from her son in the previous 12 months. Second, increasing food prices were the main problem for her family's restricted ability to access adequate food. Her household had a severe food insecurity scale and had access to sufficient food to meet her household needs for only seven months. While responding to the reason for the inadequate supply for the months from August to December, she said, "The food we get [from the farm] is not enough to take us from January to December," indicating the main staple food they produced (maize) was not enough to feed the family for the whole year (Interview 32, July 31, 2018, Mzuzu, Malawi).

The cases of migrant households such as those mentioned by the respondents in interviews 3 and 32 may be exceptional. However, even the migrant households that are solely dependent on remittances for food access may fall into the critical food insecurity situation once they stop receiving support from their migrant family member(s). In the Malawian context, a study on the effects of household income composition and occupational mobility on household welfare and poverty conducted between 2010 and 2013 found that diversified income sources result in improved household welfare in the urban areas (Benfica et al., 2018). Therefore, motivating and providing necessary support to migrant households to engage in income-generating activities that ensure long-term food security to the migrant households is essential. A similar observation was made by Atuoye et al. (2017) in their study in Ghana. The authors found that remittances were not enough to make households fully food secure and hence suggested developing an alternative livelihood strategy through the promotion of small enterprises and self-employment activities.

Remittances of Food and Money Support Healthier Diets

This study found that remittances support households' access to healthier diets. Some households were able to add nutritious food to their daily dietary intake due to the remittances of food and money. Such was the experience of a female respondent in her mid-twenties whose husband had been working in South Africa since 2014. She was living with her three-year-old daughter in a grass-thatched house in one of the informal settlements of Mzuzu and did not have any source of income except remittances. While replying to a question about the changes in her ability to access food, she said, "Now, we eat a good diet. We manage to buy a tray of eggs and buy cooking oil. Before, I couldn't manage that" (Interview 16, July 7, 2018, Mzuzu, Malawi). However, her household was mildly food insecure as she couldn't afford to eat her desired food in the last two weeks due to the lack of resources; she received remittances just four days before the day of the interview. For her household, the time immediately before receiving remittances was critical to access desired food. However, she experienced a significant improvement in her household's overall access to food.

The food packages sent by migrants also carry great significance in improving household food security. Some migrant households received food packages that lasted for many months. Generally, migrants send cooking oil, sugar and Cremora (milk powder) from South Africa in December for the year-end festive season. In another interview, a female respondent in her mid-sixties, who had two family members in South Africa, said:

> Milk, sugar, soap, body oil, cooking oil—like those things I received in December 2017 and they ended last month [June 2018]. He [migrant son] sends those once a year.... We can say it is more than MWK50,000 [in monetary value]. Because these things are cheap in South Africa but if you buy here, they cost more. He tells me that "if I send you money to buy groceries there, means it's a lot of money. So it's better I should just send them from here." (Interview 17, July 7, 2018, Mzuzu, Malawi)

Interestingly, some households that received such packages multiple times a year mostly did not need to buy those items in Mzuzu. For example, another respondent, a divorced woman in her mid-forties, reported receiving food items such as cooking oil and sugar three to four times a year. When responding to a question about whether her household needs to buy those items at the local market, she said, "Most of the times these things are sufficient to meet the household requirements" (Interview 18, July 16, 2018, Mzuzu, Malawi). Another female respondent in her late fifties also reported receiving food packages and groceries (specifically, soap, oil and milk) four times a year (Interview 41, August 17, 2018, Mzuzu, Malawi). Moreover, the above excerpt from interview 17 indicates that migrants prefer to send food packages as the items are cheaper in South Africa. Furthermore, it may also minimize the chances of remittances being used for other, less important, things by the family member. Overall, remittances sent in food and money helped remittance-receiving households with improved access to food.

Overreliance on remittances can create its own vulnerability. Households that lack sources of income beyond remittances have to compromise their food intake if they fail to receive support on time. The case of a respondent who was living with a three-year-old daughter (interview 16) suggests the need to have an alternative source of income in addition to remittances to better ensure household food security. Studies in South African cities have also shown that households with irregular income are more likely to be food insecure (Frayne & McCordic, 2015). Remittance-dependent households going through such situations are more likely in a migration system that is dominated by low-skilled and irregular migration, which ultimately determines the volume and frequency of support from migrants to their families back home.

Remittances Create Stability

Another improvement found due to remittances was fully or partially reduced incidence of skipping meals. Skipping meals is one of the coping strategies practiced by many poor households across Africa (Ngidi & Hendriks, 2014; Stella et al., 2015; Tsegaye et al., 2018). This coping strategy was also adopted by many migrant households in Mzuzu in difficult times. The respondent of another migrant household, a male in his late forties, was the area chief of one of the settlements chosen for the study. He had a son in South Africa who was sending support to him every two to three months. While responding to the changes in his family's access to food, he said:

> There is a great change right now.... Earlier, I was using the money I received as a pension but at present, as he went there and [is] sending something, that something is helping us. We are buying extra food and the other things which were lacking before. ... First, we didn't have breakfast and rice. But now we do [have] breakfast and also eat rice for breakfast. Now, we use rice for breakfast every day. In most cases, we use rice with tea instead of buying bread. Before we just possibly ate porridge, using common flour and sometimes without sugar. Now we don't have that problem (Interview 10, July 3, 2018, Mzuzu, Malawi).

His household belonged to the food secure category. In his case, the respondent had a regular pension income, he was also working part-time, and the family had remittance support from the son. These diverse income sources improved the household's food stability enough that family members no longer had to skip meals.

A similar improvement was experienced by another respondent, a male in his mid-twenties, but in reduced incidences of skipping breakfast. His household decided to send a second family member, the elder daughter, to South Africa, as they could not receive enough support from the first migrant (a younger son). When the daughter started sending remittances, the household noticed a considerable improvement in its access to food. When responding to a question about the changes observed due to the availability of remittance income, the respondent replied:

> At least now we are better off because if we talk about the most difficult times, we used to stay a whole week without breakfast, even a month, but now we can only stay without it maybe for two days only. ... Before, we would sometimes just mix Super Dip with water and eat. [Super Dip is one of many types of juice powder that is mixed with lots of water. People often drink it to deal with hunger. It costs about 50 to 60 kwacha per pack.] (Interview 35, August 8, 2018, Mzuzu, Malawi).

Sometimes, even if remittances were not received on time, having remittances as a support mechanism gave migrant households enough privilege to borrow money. A female respondent in her early fifties had experienced improved confidence to borrow money during difficult times. For instance, when there was no maize at home or when there was no money to bring maize to the maize mill for processing, she was sure that she could pay back a loan, knowing that she would receive remittances from her migrant son. She said:

> There is a big difference because, at first, I was only relying on fetching firewood. As you can see there is raining outside, [and] when it rains, I couldn't go and fetch firewood. It means I would stay maybe three, four days without food. But right now, because I receive remittances, even though they [migrants] skip some months, I do keep the money to buy maize. If I don't have money to go to the maize mill, I will borrow some from someone because I have a hope that I receive remittances later to cover the expenses (Interview 23, July 20, 2018, Mzuzu, Malawi).

Another effect of remittances on ensuring food stability was households' improved ability to buy food in bulk. In Malawi, food price generally goes down during the harvest season and up in the rainy season, and such a trend has been recurring for years (Sassi, 2015). So urban dwellers having enough resources prefer to buy during the harvest season, which starts in April and continues until June. For example, the participant who was living with a three-year-old daughter and had remittances as only the source of income observed her increased ability to buy maize for a whole year because of remittances. In her words:

> I buy food [maize] for a whole year now. Before, we were buying just for a month. Back then, we didn't have money to buy maize for the whole year. But now, we manage to buy. In the village, the price of the maize remains the same throughout the year. But here, you find maize cheap in the harvest season and high during the rainy season. So we buy for a whole year [during the harvest season] (Interview 16, July 7, 2018, Mzuzu, Malawi).

This was similar to the experience of another respondent, a female in her mid-thirties whose husband was working in South Africa. She had a timber shop at the local Vigwagwa Market. When responding to a question about the changes in her household's ability to access food, she said:

> There is a difference because, at first, we were buying most of the things daily. But now we buy groceries, maybe some other groceries which can take us for many days. For maize, we buy it every month but ... when the

> month of June comes, we buy maize for a long period of time ... because the maize price goes high in November-December and the rainy season (Interview 34, August 8, 2018, Mzuzu, Malawi).

Another participant in her early forties also observed the improved capacity to buy in large quantities due to remittances. She said:

> Right now, because we don't usually go to the market, when we want to buy maize, we buy it for some months. We go to the market for relish only. Before, we used to go to the market and buy a few kilograms of flour, not maize. But now we buy many bags of maize that lasts for months (Interview 13, July 5, 2018, Mzuzu, Malawi).

While visiting informal food markets in Mzuzu, anyone can easily observe vendors selling food items such as rice, flour, beans, vegetables and oil in the lowest quantities or units possible. People from low-income categories tend to buy in small quantities and eventually pay more than those buying in large quantities. Furthermore, recent studies in Malawi have also documented that poor urban dwellers are vulnerable to food price fluctuations (Harttgen et al., 2016; Sassi, 2015). As such, buying maize during the harvest season not only saves money but also ensures enough supply of food for months. The excerpts presented above support the notion that households experience better food stability through an improved supply of food due to remittances.

NELM Theory and the Linkages Between Migrant Remittances and Household Food Security

The results indicate a complex association between migrant remittances and household food security. It is intertwined with many factors. The major ones are the informal nature of migration, which determines the volume and frequency of remittance support, and recurring food price volatility, which makes the urban poor vulnerable to food insecurity. Interestingly, the migration of a family member is one of the coping strategies adopted by many low-income households to overcome household food insecurity (Derribew, 2013; Khatri-Chetri & Maharjan, 2006). The findings of this study show that remittances helped improve the household food security situation of many migrant households; however, the support was not enough to completely lift them out of the condition of food insecurity.

The NELM approach facilitated the assessment of this study. The approach postulates that migratory decisions are made to increase household welfare through diversified income sources, to make investments in income-generating activities and to support households in times of shocks (Stark, 1991). The migrant households used remittances directly to access food and indirectly to make investments in agriculture—mainly for buying fertilizer and hiring farm laborers. The remittance investments in agriculture were the economic activity that the migrant households chose to improve their household welfare, as suggested by the NELM (de Haas et al., 2020). Such investment ultimately helped migrant households with improved food supply as well as income. Remittances were also found as insurance, a risk-aversion mechanism for households during difficult times (King & Collyer, 2016), as in the case of the migrant household in interview 23, where households could rely on remittances when suffering from reduced income sources due to economic and environmental shocks.

The findings indicate that remittances lead to only short-term improvements in access to food, unless remittances are used to expand household income sources. On the other hand, remittance investment in entrepreneurial activities is critical in the migration system, which is dominated by low-skilled and irregular migrancy. As such, this investigation suggests that achieving all four pillars of food security—availability, access, utilization and stability—will remain a distant prospect without appropriate policy interventions in the context of Mzuzu, where people are living with precarious income sources and have to go through recurring food price shocks. The NELM theory is helpful in addressing the situation, as it stresses creating a favorable investment environment to leverage remittance investments through economic policy reforms (Taylor & Taylor, 1999). Empirical evidence from Atuoye et al.'s (2017) recent study in Ghana also supports this conclusion. Their study documented the fact that remittances alone are not enough to make households completely food secure; hence, they suggested developing an alternative livelihood strategy through the promotion of small enterprises and self-employment activities.

Finally, the findings of this chapter should be understood in the context of international labor migration from informal settlements in Mzuzu, Malawi. These findings are also relevant in understanding the situation of migration and household food security in the context of growing urban poverty in the rapidly urbanizing cities of the developing world.

References

Africa Fertilizer. (n.d.). *Retail fertilizer prices*, https://africafertilizer.org/local-prices/#tab-id-2.

Atuoye, K., Kuuire, V., Kangmennaang, J., Antabe, R., & Luginaah, I. (2017). Residential remittances and food security in the Upper West Region of Ghana. *International Migration, 55*(4), 18–34.

Banda, H. (2019). *The dynamics of labour migration from Northern Malawi to South Africa since 1974*. PhD thesis, Faculty of Humanities, University of Witwatersrand

Battersby, J., & Watson, V. (Eds.). (2019). *Urban food systems governance and poverty in African cities* (1st ed.). Routledge.

Benfica, R., Squarcina, M., & De La Fuente, A. (2018). *Structural transformation and poverty in Malawi*. International Fund for Agricultural Development. IFAD Research Series No. 25

Bilinsky, P. and Swindale, A. (2010). *Months of Adequate Household Food Provisioning (MAHFP) for measurement of household food access: Indicator guide (version 4)*. FHI 360/FANTA

Castles, S. (2009). Development and migration or migration and development: What comes first? *Asian and Pacific Migration Journal, 18*(4), 441–471.

Chirwa, W. (1997). "No TEBA ...Forget TEBA": The plight of Malawian ex-migrant workers to South Africa, 1988–1994. *International Migration Review, 31*(3), 628–654.

Choithani, C. (2017). Understanding the linkages between migration and household food security in India. *Geographical Research, 55*(2), 192–205.

Coates, J., Swindale, A., & Bilinsky, P. (2007). *Household Food Insecurity Access Scale (HFIAS) for measurement of household food access: Indicator guide (version 3)*. FHI 360/FANTA

Creswell, J. W., Plano Clark, V., Gutmann, M., & Hanson, W. (2003). Advanced mixed methods design. In A. Tashakkori & C. Teddlie (Eds.), *SAGE handbook of mixed method in social and behavioral research* (pp. 209–240). Sage.

Crush, J. (2013). Linking food security, migration and development. *International Migration, 51*(5), 61–75.

Crush, J., & Caesar, M. (2017). Introduction: Cultivating the migration-food security nexus. *International Migration, 55*(4), 10–17.

de Haas, H. (2010). Migration and development: A theoretical perspective. *International Migration Review, 44*(1), 227–264.

de Haas, H., Castles, S., & Miller, M. (2020). *The age of migration: International population movements in the modern world* (6th ed.). Guilford Press.

Derribew, A. (2013). An assessment of coping strategies for drought induced food shortages in Fedis District, East Hararghe Zone, Ethiopia. *International Journal of Science and Research, 4*(1), 284–289.

Frayne, B., & McCordic, C. (2015). Planning for food secure cities: Measuring the influence of infrastructure and income on household food security in Southern African cities. *Geoforum, 65*, 1–11. https://doi.org/10.1016/j.geoforum.2015.06.025

Frayne, B., Pendleton, W., Crush, J., Acquah, B., Battersby-Lennard, J., Bras, E., Chiweza, A., Dlamini, T., Fincham, R., Kroll, F., Leducka, C., Mosha, A., Mulenga, C., Mvula, P., Pomuti, A., Raimundo, I., Rudolph, M., Ruysenaa, Sh., Simelane, N., Tevera, D., Tsoka, M., Tawodzera, G., & Zanamwe, L. (2010). *The state of urban food insecurity in Southern Africa*. African Food Security Urban Network (AFSUN). Urban Food Security Series No. 02

Githira, D., Wakibi, S., Njuguna, I., Rae, G., Wandera, S., & Ndirangu, J. (2020). *Analysis of multiple deprivations in secondary cities in sub-Saharan Africa,* https://www.ittransport.co.uk/projects/analysis-of-multiple-deprivations-in-secondary-cities-in-sub-saharan-africa/

Harttgen, K., Klasen, S., & Rischke, R. (2016). Analyzing nutritional impacts of price and income related shocks in Malawi: Simulating household entitlements to food. *Food Policy, 60*(1), 31–43.

Kerr, R. (2014). Lost and found crops: Agrobiodiversity, indigenous knowledge and a feminist political ecology of Sorghum and Finger Millet in Northern Malawi. *Annals of the Association of American Geographers, 104*(3), 577–593.

Khatri-Chetri, A., & Maharjan, K. (2006). Food insecurity and coping strategies in rural areas of Nepal: A case study of Dailekh District in mid western development region. *Journal of International Development and Cooperation, 12*(2), 25–45.

Kimani-Murage, E., Schofield, L., Wekesah, F., Mohamed, S., Mberu, B., Ettarh, R., Egondi, T., Kyobutungi, C., & Ezeh, A. (2014). Vulnerability to food insecurity in urban slums: Experiences from Nairobi, Kenya. *Journal of Urban Health, 91*(6), 1098–1113.

King, R., & Collyer, M. (2016). Migration and development framework and its links to integration. In B. Garcés-Mascareñas, & R. Penninx (Eds.), *Integration processes and policies in Europe. IMISCOE Research Series* (pp. 167–188). Springer

Kita, S. (2017). Urban vulnerability, disaster risk reduction and resettlement in Mzuzu city, Malawi. *International Journal of Disaster Risk Reduction, 22*(1), 158–166.

Long, K., & Crisp, J. (2011). *In Harm's way: The irregular movement of migrants to Southern Africa from the Horn and Great Lakes Regions.* UNHCR. Research Paper No. 200

Mackay, H. (2019). Food sources and access strategies in Ugandan secondary cities: An intersectional analysis. *Environment and Urbanization, 31*(2), 375–396.

Mohiddin, L., Phelps, L., & Walters, T. (2012). Urban malnutrition: A review of food security and nutrition among the urban poor. *Nutrition Works, 8*

Ndegwa, D. (2015). *Migration in Malawi: A country profile 2014*. International Organization for Migration. IOM Report

Ngidi, A., & Hendriks, S. (2014). Coping with food insecurity in rural South Africa: The case of Jozini, Kwazulu-Natal. *Mediterranean Journal of Social Sciences, 5*(25), 278–289.

Oucho, J. (1995). Emigration dynamics of Eastern African Countries. *International Migration, 33*(3–4), 391–467.

Ratha, D. (2021). Keep remittances flowing to Africa. In A. Ordu (Ed.), *Foresight Africa 2021* (pp. 17–19). Brookings Institution.

Read, M. (1943). Migrant labor in Africa and its effects on tribal life. *International Labour Review, 45*(6), 605–631.

Riley, L., Chilanga, E., Zuze, L., & Joynt, A. (2018). *Food security in Africa's secondary cities: No. 1. Mzuzu, Malawi*. African Food Security Urban Network (AFSUN). Urban Food Security Series No. 27

Sassi, M. (2015). The welfare cost of maize price volatility in Malawi. *Bio-Based and Applied Economics, 4*(1), 77–100.

Smale, M. (1995). "Maize is life": Malawi's delayed Green Revolution. *World Development, 23*(5), 819–831.

Stark, O. (1991). *The migration of labor*. Basil Blackwell.

Stella Wabwoba, M., Wanambacha Wakhungu, J., & Omuterema, S. (2015). Household food insecurity coping strategies in Bungoma County, Kenya. *International Journal of Nutrition and Food Sciences, 4*(6), 713–716.

Tacoli, C. (2017). Food (in)security in rapidly urbanising, low-income contexts. *International Journal of Environmental Research and Public Health, 14*(12), 1554.

Taylor, E., & Taylor, J. (1999). The new economics of labour migration and the role of remittances in the migration process. *International Migration, 37*(1), 63–88.

Thomas, D. (2006). A general inductive approach for analyzing qualitative evaluation data. *American Journal of Evaluation, 27*(2), 237–246.

Tsegaye, A., Tariku, A., Worku, A., Abebe, S., Yitayal, M., Awoke, T., Alemu, K., & Biks, G. (2018) Reducing amount and frequency of meal as a major coping strategy for food insecurity. *Archives of Public Health*, 76(1), Article 56

WageIndicator. (n.d.). *Minimum wages in Malawi with effect from 01–07–2017,* https://wageindicator.org/salary/minimum-wage/malawi/archive-before-2019/minimum-wages-in-malawi-with-effect-from-01-07-2017

World Bank. (2020). *Personal remittances, received (% of GDP)—Malawi*, https://data.worldbank.org/indicator/BX.TRF.PWKR.DT.GD.ZS?locations=MW

CHAPTER 18

Rural-Urban Migrants in Mzuzu's Informal Food Trading System

Lovemore Itai Zuze

Introduction

The urban food trading system needs much further research in order to tackle the challenges of urban food insecurity and urban poverty. Although there has been considerable work on urban food systems and urban food markets (FAO, 2018; Frayne et al., 2010; iPES-Food, 2017; Pothukuchi & Kaufman, 1999; Riley et al., 2018), there has been much less exploration of the relationship among different food traders and the implications of new entrant migrant food traders for the food marketing systems of secondary cities. Some scholars have explored the dualistic relationship that exists between formal and informal food traders and the implications for food governance and the creation of an enabling institutional environment that supports sustainable urban food systems

L. I. Zuze (✉)
Department of Food Security and Nutrition, University of Livingstonia (Laws Campus), Livingstonia, Malawi
e-mail: itai.zuze@unilia.ac.mw

L. Riley and J. Crush (eds.), *Transforming Urban Food Systems in Secondary Cities in Africa*,
https://doi.org/10.1007/978-3-030-93072-1_18

(Skinner & Haysom, 2016; Skinner et al., 2018). They warn against labeling informal systems as obstacles to sustainability (Gondwe et al., 2011). Most policymakers, however, associate informality with barriers to sustainable food systems and are of the view that formalizing the informal food trading system can improve urban food systems.

Throughout sub-Saharan Africa (SSA), rural-urban migration is reshaping the foodscape of secondary cities (Morgan & Sonnino, 2010). Small towns have always been spaces of connectivity between societies, in terms of trade, exchange of ideas and goods, and the establishment of food routes (Agergaard et al., 2019). As larger primate cities become overwhelmed, smaller urban centers are experiencing an unprecedented and revolutionary demographic transition (Haysom & Fuseini, 2019). The ongoing revolution has profound social and economic implications for urban food systems in secondary cities. Small cities are experiencing a surge in economically active populations who are becoming actively engaged in the urban food systems in attempts to secure income and livelihoods (Tacoli & Agergaard, 2017).

This chapter focuses on the case of the secondary city of Mzuzu in northern Malawi. Mzuzu is Malawi's third administrative city, located in the northern region of the country. The other two regions of the country are home to the political capital (Lilongwe in the central region) and the commercial capital (Blantyre in the southern region). Mzuzu has emerged as a fast-growing administrative city for the northern region, deliberately supported by decentralization policies meant to diffuse the high rate of urbanization in primate cities in Malawi. The Government of Malawi has therefore prioritized making the city an economic hub for the northern region. The government's proposed framework advocates local government systems and the spread of responsibilities from traditional primate cities to smaller cities across Malawi (Government of Malawi, 2013). By SSA standards, Mzuzu fits in the subnational secondary city category, which is mainly characterized by administrative functions for the central government (Roberts, 2014).

Mzuzu has 15 administrative wards and covers an area of 144 square kilometers within the municipal boundaries. The Government of Malawi National Statistical Office (2018) estimated Mzuzu's population at around 221,000 in 2018. The city had the highest rate of population increase (5.4% between 2008 and 2018) among all the urban areas in Malawi. The city's population accounts for around eight percent of

Malawi's urban population and 1.3% of Malawi's entire population. In the foreseeable future, there is potential for Mzuzu to become a corridor secondary city due to its proximity to neighboring Tanzania, which is one of the country's main trading partners and a gateway to the Asian markets. The city has a youthful demographic profile, with youth and children under 20 years of age constituting more than half the population. A high number of poor migrant households are food insecure and heavily reliant on the informal sector as their source of food. Rural-urban migration is a major contributor to the rise in the population of the city (Riley et al., 2018). Informality plays a major role in food supply and occurs in the context of a negligible investment environment and insufficient employment creation opportunities (Government of Malawi, 2018). These characteristics are familiar in most secondary cities in SSA (FAO, 2018; Haysom & Fuseini, 2019; Roberts, 2014).

Combatting food insecurity and urban poverty in secondary cities require a concerted effort and set of interventions appropriate to the nature and types of households inhabiting these cities, as well as the food systems and forms of food governance within them. The focus of this chapter is the social and economic interactions of rural migrants actively involved in the food market system and their possible impact on food security (access, distribution, availability) in shaping a sustainable urban food system. The local government has attempted to accommodate and acknowledge rural migrants as permanent urban settlers by expanding social service provision to match the growing urban population through the implementation of the priority areas one and seven of the National Urban Policy (Government of Malawi, 2019). While this approach has partially assisted in managing the ever-increasing urban population, there has been less understanding of the social and economic integration of rural migrants and their insertion into the urban economy and especially the food trading sector. This chapter aims to contribute to the literature by exploring the dynamic changes facing migrants in secondary cities through the lens of urban poverty and urban food market systems. It aims to understand the extent to which rural migrants are involved in food trading in comparison with their urban-born counterparts. It further explores how the integration of migrant rural food traders can build sustainable food market systems and reduce poverty in urban societies undergoing transition.

Methodology

The research presented in this chapter was undertaken as part of the African Food Security Urban Network—Food, Urbanization, Environment and Livelihoods project (AFSUN-FUEL), an interdisciplinary policy research initiative aimed at influencing the direction of policy on urban poverty and food security in Africa (FUEL, 2021). A city-wide survey of small-scale food traders was conducted in Mzuzu to gather data about the nature of the food system. The first stage involved the multiple-frame sampling method to map the spatial location units (area sampling) and the types of enterprises (list sampling) (Battersby et al., 2016; FAO 2015). The second phase was the administration of the questionnaire to 485 respondents with purposive representation of each trading enterprise type. The number of trading enterprises of each type interviewed was proportional to the number in the overall enterprise profile. The sampling was also organized such that half the participants were from the central business area of Mzuzu and the other half were from the residential areas. In practice, a larger number of food traders in the central business district of the city were interviewed than suggested by the profile data set.

The data collected during this research was cross-tabulated to identify interactions among variables of interest (test of independence). The Pearson Chi-square test (χ^2) of independence and the likelihood ratio chi-square test (G^2) were used to determine significant associations among variables. In cases where the Pearson Chi-square test requirements were violated (cells with expected count less than 5 being > 20%) (Cochran, 1954), the likelihood ratio (the Monte Carlo method for computing the significance level) was used. Apart from demographic traits, variables of interest included the area of origin of food traders (migrant status), the type of enterprise (based on the main trait of the food trader), the form of enterprise (formal or informal), their motivation for selling food products, and their business practices.

Profile of Food Traders and Their Enterprises

Female vendors constituted 67% of the respondents and the food trading system is also dominated by vendors who are married (77.5%). The data indicated a significant Pearson Chi-square association between gender and marital status (χ^2 (2, N = 485) = 7.319, p = 0.025). Female food vendors were more likely to be married than their male counterparts.

The majority of the food trading participants had attained at least some basic education, with only 3.7% having not attained any form of education (Table 18.1). The city exhibited a rich and multifaceted food trading sector, with migrant participants from across the country. The majority of food traders interviewed (64.3%) had been born in a rural area within Malawi, while 15.1% were from another city within the country, and 2.7% were from outside the country. Only 18% had been born in Mzuzu (Fig. 18.1). Among the respondents, 88% were full-time food traders with no alternative form of employment. The average age of the respondents was 36 and the majority of food traders had been in the business for one year. A typical trader owned only one business and was part of a family comprising five family members (Table 18.2).

The city food trading system is largely controlled by informal food traders. Out of the eight types of food traders identified within the city, only those operating market stalls (9.7%) and those operating small shops (20.8%) exhibit an element of formality in their trading enterprise. The rest (69.5%) are enterprises reflecting the informal side of food trading. The majority of respondents (64.3%) within the city were rural-urban migrants. However, less than half the migrants (46.2%) were operating one of the informal enterprise types, refuting the assumption that those who migrate from rural areas mainly practice informal food

Table 18.1 Enterprise owner demographic characteristics

Characteristic	*Variable*	*Frequency*	*Percent (%)*
Sex	Male	159	32.8
	Female	326	67.2
Marital status	Married	376	77.5
	Single	103	21
Education level	No education	18	3.7
	Primary education	232	47.8
	Secondary education	216	44.5
	Tertiary education	16	3.3
Employment before food vending	Yes	194	40
	No	290	60
Alternative employment besides vending	Yes	57	12
	No	427	88

Sample size = 485

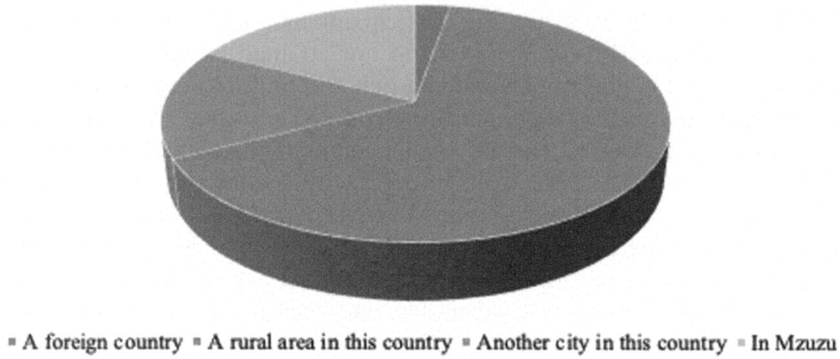

Fig. 18.1 Birthplace of enterprise owners

Table 18.2 Key Socio-economic Attributes

Household Socio-economic Factors	*Mean*	*Mode*	*Minimum Values*	*Maximum Values*
Age of enterprise owner (years)	35.8	30.36	17	75
Number of years in the business	8	1	0	48
Number of businesses owned	1.3	1	1	6
Number of businesses co-owned	0.3	0	0	5
Household members	4.75	4	0	14

trade. Furthermore, the lack of association between migrant status and type of enterprise is consistent with this assertion. Among the group that practices informal food trade, the majority were roadside vendors (48.7%).

Greater financial security for the family and the need for money to survive were the major motivational factors for establishing a food trading enterprise. Three-quarters of the traders highly rated greater financial security for the family, and 48% were food traders from rural areas. Nearly 90% of the traders cited the need for money to survive, with 56.7% being rural migrant food traders. There was a high likelihood ratio chi-square between migrant status and the motivation of wanting greater financial security for the family (G^2 $(16, N = 485) = 38.504, p = 0.001$), and marital status and wanting greater financial security for the family (G^2 (16, $N = 485$) = 24.203, $p = 0.002$). Migrant status and the need for money to survive also had a high likelihood ratio chi-square motivation factor (G^2 (16, $N = 485$) = 32.912, $p = 0.004$). Rural–urban migrant food traders were more likely to attribute greater importance to financial security and the need for money to survive as their main motivations for setting up a food enterprise. The main factors in choice of location for the food enterprises were places that had plenty of customers and closeness to traffic (G^2 (14, $N = 485$) = 21.002, $p = 0.021$ and G^2 (7, $N = 485$) = 15.349, $p = 0.043$, respectively).

Food Sourcing and Other Business Practices

Variables of interest were selected to investigate the business practices associated with migrants and the types of enterprise they run. The variables included the source of food products sold, storage capabilities of food traders, payment of city licenses, profit-saving culture, and access to finance.

The formal market in Mzuzu was a popular place for sourcing food products for sale. Most types of traders chose the formal market as their main source of products sold (48.9%), followed by sourcing products directly from farmers (17.5%) and from wholesalers (15.9%). Most traders were selling multiple products and the above picture is based only on the main food source that a trader was using. Among the food traders who sourced their products from the formal market, 31.3% were migrant food traders.

Most food traders who participated in the study either preferred not to divulge their storage facility (41.9%) or stored their products at home (41%). There was an association between food storage and enterprise type (likelihood ratio chi-square G^2 (56, $N = 485$) = 381.121, $p = 0.000$). Food enterprises that were informal in nature, such as vendors of hot cooked foods, mobile vendors, local restaurants, roadside vendors and home-based traders, preferred storing their products at home, whereas the formal enterprises (market stalls and small shops) were more reluctant to say where they store their products.

The majority of the food traders (77.1%) indicated that they do not pay any form of tax to the city authorities, with 44.3% of these being roadside vendors. There was, however, a Pearson Chi-square association between the type of business and the remitting of license fees to the city authorities

Table 18.3 Major sources of business start-up capital

Source of Capital	*Frequency*	*Percentage*
Personal savings	258	53.2
Monetary gift from a relative	43	8.9
Loan from relatives within Malawi	33	6.8
Loan from a micro-finance institution	26	5.4

(χ^2 (7, N = 485) = 102.489, p = 0.000). Informal enterprise owners were less likely to pay any form of tax compared to the owners of formal enterprises. Migrant status had no direct association with the payment of tax.

Many respondents within the city were unwilling to indicate their preferred way of storing their profits. Those who did revealed that their main mode of savings was keeping money at home (26.4%), followed by keeping the savings at a bank (14.6%). Of those who kept their earnings at home, 19.6% were migrants. Other saving channels included village banks, mobile money, and investing in stock. There was a significant likelihood ratio chi-square between the type of enterprise and the profit-saving preferences (G^2 (56, N = 485) = 146.116, p = 0.000). The saving culture is more likely influenced by the type of enterprise being operated.

In the city, food enterprise owners were most likely to engage in the food business using finance obtained from their own savings (Table 18.3). The second most popular source of capital was a money gift from a relative and the third was a loan from relatives within Malawi. This indicates a significant social investment among the food enterprise owners within the city. Despite a large number of participants indicating that they established their food enterprises from personal savings, there was no statistical association between preferential access to business finance and migrant status.

Assessing Rural-Urban Migrants' Role in the Food Trading System

The overall objective of the AFSUN-FUEL project for secondary cities is to produce data that can be used to advocate for the integration of food system planning with the development agenda for secondary cities. This effort is part of a broader fight against urban poverty framed through the lens of food, mostly focusing on the urban poor (Haysom & Fuseini, 2019). A major pillar of this notion is that bringing meaningful development to these up-and-coming urban centers will only be effective and sustainable with the urban poor residents as the main beneficiaries (Haysom, 2015). This chapter focuses on providing better insights into how participation in urban life by migrants influences their activities at the food marketplace. One trend evident in Mzuzu is rural-urban migration as a major driver of population growth within the city (Riley et al., 2018). This demographic transition has prompted significant active participation in the food economy by those identifying rural background origins and traits. The largest group of food traders interviewed were married female rural migrants engaged in full-time informal food trading. The participation of women in employment and income-generating food enterprises is a particularly important element in reducing poverty in the city (Jaglin, 2014).

The city has a relatively new informal food market system, with most enterprises established in the year prior to the survey. Most food enterprise owners have a single business that they run on a full-time basis. The food traders characteristically lack business diversity after establishing an enterprise and are not engaged in any form of business other than food trading. This suggests the existence of a group of enterprise owners who are economically poor and at the mercy of established formal food traders. Rural migrant food traders are part of the broader cohort of migrants flooding the city in search of economic opportunities, only to become third-party traders for established formal food traders.

Migration patterns and associated food marketing practices have prompted broader changes in urban food availability, in the process altering the relationship between formal and informal food enterprises (Tefft et al., 2017). At first glance, Mzuzu appears to be a city predominantly under the control of the informal sector. However, the food market sector within the city is actually dominated by more inconspicuous formal food enterprises. The large group of informal traders relies on the sourcing of their food products from the formal market. In essence, the majority of the food traders exhibit informal characteristics, but it is the formal food sector that provides the city with the majority of food products and therefore remains in economic control of the food trading system.

The 2017 Mzuzu household food security survey indicated that the city had a high food diversity score, but identified high food prices as one of the key issues affecting food access (Riley et al., 2018). The sourcing of food products by informal traders from formal sources within the city and not directly from farmers could be a reason for the high prices associated with food products. In Mzuzu, as elsewhere, access to food in urban areas is mainly enabled and constrained by the market rather than own production (Haysom & Tawodzera, 2018).

The unwillingness of most respondents to divulge where they keep their business profits suggests that the respondents are reliant on less secure methods. If they kept their profits in secure facilities, they probably would have said so. This, in turn, reflects the low use of formal money management institutions (such as banks) by food traders. The fact that most informal enterprise owners utilize their personal savings for establishing and running their business rather than using formal loans is because they perceive financial institutions as unfriendly to informal business.

Across the world, collection of revenue from informal enterprises has proved a daunting task for most central and local governments (Kundt, 2017). License paying by traders for the provision of services in the city was very limited among food traders, with no obvious mechanisms to tap into the potential revenue generation by this group. There is a widespread belief that payment for licenses is only for established formal entities. On the other hand, making informal traders pay city rates is viewed by established formal traders as acceptance of the operations of informal traders. The tolerance of informal food traders by the city authorities has met to date with mixed reactions by those operating formal enterprises, with some advocating their removal from the food market system. Mzuzu City Council experienced a two-year (2017–2019) boycott on market fee payment by members of the Mzuzu Vendors Association in a bid to force the council to remove informal traders, whom they accused of operating illegally.

Finally, informal food traders are attracted to doing business in areas that have plenty of customers and are close to traffic, which is indicative of their impermanent nature. With the city authorities providing storage facilities only for those conducting business in market stalls, the storage of food products is a further challenge and even a tool of exclusion for informal trading. On the other hand, the city's food market system has been established and is dominated by food traders who are highly motivated to secure greater financial security for their families. For them, informal work is a necessity rather than a choice. This motivation suggests that there is an opportunity to build a sustainable market system based on full-time employees who will serve market demand while their families stay in the city (Zogli et al., 2019).

Policy Directions for a Sustainable and Inclusive Food System

This chapter offers a view of secondary cities as spaces of connectivity for food and development in the context of high levels of rural–urban migration. It offers new insights on how different variables relate to the main participants in the urban food trading system. By exploring migrant status and business practices, it shows how the actions of small food enterprise owners shape the food market environment in a secondary city. This chapter demonstrates the importance of examining multiple variables to understand the broader food system. This is similar to Heather Mackay's (2019) intersectional analysis of food security in Uganda, with observations linking migrant status, gender, and age support interventions based on a more sophisticated understanding of food system actors.

The chapter also points to the need for a coherent local governance framework in which food enterprise owners would pay for some form of license for the services they use and that could raise city revenue. This would require a new, inclusive approach to informality that acknowledges and supports the informal trade and stimulates traders to purchase licenses. In turn, there is a need for greater understanding of the motivations and financial and governance challenges faced by migrant informal food traders as well as their positive contributions to the urban food economy.

In Mzuzu, the majority of informal food market participants are rural migrants, which have impacts on motivations for participation, types of food trading, sourcing strategies and other business practices, and economic status. It is widely accepted that rural populations will play a role in the success of growing secondary cities. This chapter argues that the economic participation of this social group in the food system will also benefit their urban counterparts. This is because many rural–urban migrants are finding ways to become more active and integrated into the formal sector. What is less clear is whether migrant status affects the types of enterprises that migrants can establish. Certainly, the majority of rural migrants engage in informal food trading and the food marketing system has networks of symbiotic relationships among informal traders who act as intermediaries between fixed formal markets and low-income consumers.

References

Agergaard, J., Tacoli, C., Steel, G., & Ørtenblad, S. (2019). Revisiting rural-urban transformations and small town development in sub-Saharan Africa. *European Journal of Development Research, 31*(1), 2–11.

Battersby, J., Marshak, M., & Mngqibisa, N. (2016). *Mapping the invisible: The informal food economy of Cape Town, South Africa* (Urban Food Security Series No. 24). African Food Security Urban Network (AFSUN).

Cochran, W. (1954). The combination of estimates from different experiments. *Biometrics, 10*(1), 101–129.

Food and Agriculture Organization (FAO). (2015). *Handbook on master sampling frames for agricultural statistics frame development, sample design and estimation*. Food and Agriculture Organization.

Food and Agriculture Organization (FAO). (2018). *The role of cities in the transformation of food systems: Sharing lessons from Milan pact cities*. Food and Agriculture Organization.

Frayne, B., Pendleton, W., Crush, J., Acquah, B., Battersby-Lennard, J., Bras, E., Chiweza, A., Dlamini, T., Fincham, R., Kroll, F., Leduka, C., Mosha, A., Mulenga, C., Mvula, P., Pomuti, A., Raimundo, I., Rudolph, M., Ruysenaar, S., Simelane, N., Tevera, D., ... Zanamwe, L. (2010). *The state of urban food insecurity in Southern Africa* (Urban Food Security Series No. 2). African Food Security Urban Network (AFSUN).

FUEL. (2021). *Food, urbanization, environment and livelihoods*. http://www.afsun.org/fuel-project/.

Gondwe, J., Feng, G., & Ayenagbo, K. (2011). Planning for sustainability in Malawian cities: A conceptual analysis of the missing links. *International Journal of Human Sciences, 8*(2), 699–715.

Government of Malawi. (2013). *Guidebook on the local government system in Malawi*. Ministry of Local Government and Rural Development.

Government of Malawi. (2018). *Malawi population and housing census*. National Statistical Office. Preliminary Report.

Government of Malawi. (2019). *National urban policy*. Ministry of Lands, Housing and Urban Development.

Haysom, G. (2015). Food and the city: Urban scale food system governance. *Urban Forum, 26*(3), 263–281.

Haysom, G., & Fuseini, I. (2019). *Governing food systems in secondary cities in Africa* (Consuming Urban Poverty Working Paper No. 10). University of Cape Town.

Haysom, G., & Tawodzera, G. (2018). Measurement drives diagnosis and response: Gaps in transferring food security assessment to the urban scale. *Food Policy, 74*, 117–125.

iPES-Food. (2017). *What makes urban food policy happen? Insights from five case studies*. International Panel of Experts on Sustainable Food Systems. iPES-Food Report, http://www.ipes-food.org/_img/upload/files/Cities_full.pdf.

Jaglin, S. (2014). Regulating service delivery in southern cities: Rethinking urban heterogeneity. In S. Oldfield & S. Parnell (Eds.), *The Routledge handbook on cities of the global south* (pp. 434–447). Routledge.

Kundt, T. (2017). *Opportunities and challenges for taxing the informal economy and subnational taxation*. Institute of Development Studies. K4D Emerging Issues Report.

Mackay, H. (2019). Food sources and access strategies in Ugandan secondary cities: An intersectional analysis. *Sage Journals, 31*(2), 375–396.

Morgan, K., & Sonnino, R. (2010). The urban foodscape: World cities and the new food equation. *Cambridge Journal of Regions, Economy and Society, 3*(2), 209–224.

Pothukuchi, K., & Kaufman, J. L. (1999). Placing the food system on the urban agenda: The role of municipal institutions in food systems planning. *Agriculture and Human Values, 16*(2), 213–224.

Riley, L., Chilanga, E., Zuze, L., & Joynt, A. (2018). *Food security in Africa's secondary cities: No. 1 Mzuzu, Malawi* (Urban Food Security Series No. 27). African Food Security Urban Network (AFSUN).

Roberts, B. (2014). *Managing systems of secondary cities*. Cities Alliance/UNOPS.

Skinner, C., & Haysom, G. (2016). *The informal sector's role in food security: A missing link in policy debates* (HCP Discussion Paper No. 6). Hungry Cities Partnership.

Skinner, C., Reed, S., & Harvey, J. (2018). *Supporting informal livelihoods in public space: A toolkit for local authorities*. WIEGO and Cities Alliance.

Tacoli, C., & Agergaard, J. (2017). *Urbanisation, rural transformations and food systems: The role of small towns* (IIED Working Paper). International Institute for Environment and Development.

Tefft, J., Jonasova, M., Adjao, R., & Morgan, A. (2017). *Food systems for an urbanizing world: Knowledge product* (Working Paper No. 122014). World Bank Group.

Zogli, L., Lawa, E., Dlamini, B., & Martin, K. (2019). Exploring the motives for operating in Ghana's informal slum sector. *Acta Universitatis Danubius, Economica, 15*(6), 24–35.

CHAPTER 19

Rent as Ransom: Lodging and Food Security in Gweru, Zimbabwe

Miriam Grant

Introduction

Life was going well for Chidiwa[1] when we first interviewed her in 1993. Her husband, who worked for the army, came home to visit on weekends and earned a healthy monthly wage of Z$1,000. Their rent for lodging (for a family that included four children) and their food each consumed 20% of their monthly household income. Chidiwa described their links with their rural home in Rusape as very strong; Chidiwa and her family sent money monthly and other goods annually, and in exchange, they received "mealies" (sweet corn) after the harvest.

[1] A pseudonym, Chidiwa means "adorable" in Shona.

M. Grant (✉)
Department of Community, Culture and Global Studies, University of British Columbia Okanagan, Kelowna, BC, Canada
e-mail: miriam.grant@ubc.ca

L. Riley and J. Crush (eds.), *Transforming Urban Food Systems in Secondary Cities in Africa*,
https://doi.org/10.1007/978-3-030-93072-1_19

However, in 1995, the family was in crisis. By then Chidiwa had had another baby, but unfortunately in the meantime her husband had died, leaving her with five children to support. Although there was no indication of the cause of her husband's death, given the high prevalence rates of HIV/AIDS within the armed forces, one could not help but be concerned with Chidiwa's health, as well as that of the younger children. The household income dropped by 88%, and the meagre $120 earned by selling doilies and some of their furniture was spent entirely on their monthly rent. Given that rural homes were usually established through the husband, it is not surprising that these ties had become quite weak, with no exchange of money, goods or mealies. Chidiwa and her five children were solely dependent on urban relatives for food, and social welfare was paying for school fees. Chidiwa's only hope was that her husband's pension would come through.

This chapter will explore the vulnerability and resilience of 100 lodgers in the medium-sized city of Gweru, Zimbabwe, in a crucial three-year period, from 1993 to 1995. It seeks to understand the elements of food security from the vantage point of insecure housing during a period of growing tumult, with structural adjustment-driven retrenchments and the removal of food subsidies. The chapter examines social and economic rural–urban linkages within translocal families, modes of livelihood, household fluidity, and rent and food increases, and presents issues around vulnerability and resilience. The translocal lens allows us to examine multidirectional and overlapping networks that facilitate migration and transactional flows (Greiner & Sakdapolrak, 2013). The focus of a case study on Gweru helps us to unpack how secondary urbanism plays a role in food security. The chapter presents insights into the everyday lives of lodgers with respect to the challenges of poverty and hunger, and the gendered aspects of vulnerability and resilience.

Insights into food, housing and livelihood insecurity during the 1993–1995 period establish a baseline of resilience and vulnerability right at the start of the IMF/World Bank-sponsored Economic Structural Adjustment Programme (ESAP). However, this landscape was merely the precipice of a period of accelerating food poverty and food insecurity. More recently, Godfrey Tawodzera and Easther Chigumira (2019) declared that food poverty and food insecurity in Zimbabwe had reached epidemic proportions. Between January and March 2017, close to 4.1 million people out of a total population of 16 million needed food assistance (Tawodzera & Chigumira, 2019).

Tawodzera et al. (2016) trace the impact of ESAP, the Fast Track Land Reform Program, and Operation Murambatsvina (Restore Order), all of which negatively impacted urban livelihoods and exacerbated household food insecurity. Their research revealed that 83% of households in Harare's poorest areas experienced food insecurity. Thus, the ESAP-driven economic shocks to urban livelihoods and to the costs of food and basic services for the 1993–1995 period covered by this research was only the precursor to an accelerating, pervasive economic crisis that was particularly devastating for the urban poor.

The Early 1990s Zimbabwean Context: Structural Adjustment, Drought and HIV/AIDS

In what ostensibly should have been an exciting, prosperous second decade of independence, things went wrong from the start. The early 1990s saw a series of events that would set the stage for disaster within the fledgling country.

Beginning in 1991–1992, ESAP laid the groundwork for extreme economic hardships in Zimbabwe. These came in the form of massive formal sector retrenchments and unpredictable, sudden and frequent increases in the cost of basic food and services (Grant, 2007). The official unemployment rate was 40% and real wages fell by 33% between 1990 and 1997. By 1997, inflation had reached 45%, some 46% of all households had incomes below the food poverty line and 42% of urban households were poor (Grant, 2007, p. 79). Unfortunately, since 1998 Zimbabwe has been in a spiral of what Patrick Bond and Masimba Manyanya (2003) describe as "nearly uninterrupted economic chaos". Two other concurrent events add to the perfect storm of accelerating deterioration.

Just as ESAP was imposed, Zimbabwe experienced a major drought in 1991–1992. The direct impacts on food security included widespread crop failure, drops in agricultural production and rising food prices through 1993. Due to this situation, the government postponed the ESAP-driven removal of food subsidies on staples such as maize meal, bread and sugar. Instead, in 1991 the government brought in a wide range of successful drought-relief strategies, which included direct food relief and food for work programmes (Marquette, 1997, p. 1145).

The success of the drought-relief programme, coupled with productive harvest seasons 1991 and 1992, led to the removal of food subsidies in

1993. While maize meal costs remained stable, the price of bread and sugar in real terms rose by 40% and 50%, respectively, in 1993 (Marquette, 1997). The lethal forces of rising food prices, inflation, retrenchments and diminished real wages and incomes severely impacted food security for the urban poor. The 1996 Poverty Assessment Survey identified food insecurity as the most common factor for households that had plummeted below the poverty line in 1994–1995, with an overwhelming 74% of all poor households and 76% of all very poor households reporting food shortages (Marquette, 1997, p. 1146).

The second critical event was the rapidly expanding HIV/AIDS pandemic, with its devastating social, economic and demographic impacts. Not only were breadwinners dying, but households were forced to use scarce resources for medicine and funerals, widows had to reconfigure households, and grandparents and other extended family members had to assume responsibility for AIDS orphans (Grant & Palmiere, 2003). By 1997, the World Health Organization data reported an HIV prevalence of 25.84% of the population (WHO, 1998). The impact of the HIV/AIDS pandemic forms part of the backdrop to this research. A couple of households in the study endure deaths and subsequent household reconfiguration, and there are also lodging individuals and households that simply disappear during this period.

This chapter is based on a longitudinal study situated in the secondary city of Gweru, located partway between the capital, Harare, and the second largest city of Bulawayo. Using the typological categorization featured in the Consuming Urban Poverty research project (Haysom in this volume), Gweru represents both a corridor secondary city and a subnational urban centre. As a corridor secondary city and with direct international links to South Africa and Mozambique, Gweru has a well-developed connectivity network (Sheppard, 2016) that enhances its economic development. As a subnational urban centre, it is the capital of Midlands province, an important industrial city and a vital agricultural service centre for a large hinterland. Gweru's 1992 population of 128,000 has since grown to 146,000 (World Population Review, 2021). Post-independence, Gweru has served as a major destination for rural–urban and interurban migrants in search of employment and an urban base. Similar to other Zimbabwean cities, Gweru could not meet the insatiable demand for housing that was driven by high growth rates and in-migration; by 1990, Gweru's official waiting list was 14,468 (Grant,

1996), but realistically the shortfall would have been much greater. Prohibitions on squatting, high costs of new housing, and acute shortages of formal sector rental shelter serve to funnel urbanites to lodging, that is, the informal, private rental of rooms and part-rooms as the only viable, affordable option (Grant, 2007).

The data for this chapter is based on a three-year longitudinal study of 100 lodging households in 1993–1995. This involved purposive, stratified sampling to capture lodgers in all four density areas: peri-urban, and low-, medium-, and high-density. Types of lodging shelter range from peri-urban cottages to rooms and part-rooms in private houses, to former servants' quarters, to overcrowded, poorly serviced conglomerate houses in high-density areas (Grant, 2007). With lodging, oral agreements replace formal leases. Although lodging can be arranged in a timely manner with the repurposing of rooms and part-rooms, it also means that rents can be raised arbitrarily and lodgers may be evicted with little notice, often when extended family members of landlords/ladies arrive for long-term stays. However, the fluidity of lodging makes it an ideal shelter choice for migrants, especially as economic and household situations change.

One of the challenges we faced in this study was in locating our respondents for follow-up interviews in 1994 and 1995, especially in light of the difficult economic environment and the propensity for lodgers to move frequently. Moves within Gweru were often driven by the need for lower rent, to be closer to work or markets for selling goods, and for different spatial needs, among other factors such as lack of upkeep and noise. Although our sample size diminished to 40 in 1994 and 25 in 1995, we were able to determine the movement patterns of almost 50% of the original 100.

Within this difficult context of retrenchments, inflation, increasing costs of food and the growing incidence of food insecurity, all against the backdrop of the high prevalence of HIV/AIDS, this chapter seeks to answer the following questions:

> What are the characteristics of the most vulnerable households and how do their circumstances change?
> How do households demonstrate resilience and combat food insecurity during this difficult time?

How do economic and social linkages and bonds with the rural and urban-based translocal family change and what role does this play in the food security of urban households?

Rents, Household Incomes and Modes of Livelihood

Although the sample size of 100 lodgers is in no way representative, the longitudinal aspect of this research allows insight into how the dynamic economic situation impacted ordinary households. In 1993, most lodging households were dependent on formal sector employment, with 56% engaged solely in the formal sector, 22% solely in the informal sector and 17% in both. Even though formal sector wages were, on average, 31% higher than informal sector wages, there was a substantial range in both sectors. Women's average incomes were considerably lower than men's average wages, at 55% in formal employment and 82% in informal income (Grant, 2007), emphasizing gender discrimination in employment, one of several key factors that determine household food security (Riley & Dodson, 2020). Altogether, the average household income in 1993 was Z$488. Almost three-quarters of households relied on just one mode of livelihood, and for 75% of these, this constituted formal sector employment, which made them vulnerable to retrenchment. Only one-quarter of 1993 households engaged in two modes of livelihood and a mere two households had three modes.

Many households struggle with the cash-intensive nature of urban life. Already in 1993, we interviewed 16 lodgers whose household expenses exceeded monthly incomes. Another 30% spent 75 to 100% of incomes on expenses and 34% used 50 to 74% for expenses, such that a full 80% of households used at least 50% of their incomes for expenses. When households have little to no savings and exist from pay cheque to pay cheque, the shock of a temporary layoff or a full retrenchment constitutes an immediate catastrophe. Rent, utilities and school fees were all fixed expenses. The only available options to reduce school fees were to pull children from school or send them to less expensive rural schools. When lodgers were asked how they reduce costs when needed, 65% stated that they cut back on food. While some never borrowed money, others borrowed from friends (50%), family (26%), the landlord/landlady (three%) and friends and family (three%).

Even though lodging rents in this medium-sized city were not excessively high, the cost of rent was the "ransom" that lodgers had to pay to remain in the city. During these three years, rent increased by 21% from Z$103 to Z$125; in 1993, lodgers spent, on average, 22% of monthly income on rent and this increased to 27% in 1995. Nationwide, the rising cost of utilities was the main driver for increased rents. The Central Statistical Office Consumer Price Index for 1993–1998, using 1990 as a base of 100, found that rents, rates and utilities across Zimbabwe increased by 219% by 1995 (Grant, 2007). In a smaller city such as Gweru, however, landladies and landlords likely had more personal relationships with their lodgers and may have been more sensitive regarding the ability of lodgers to pay increased rents. Most small-scale landladies and landlords depended on lodgers to help them make house payments and/or as a key—and sometimes solitary—source of income. In addition, there was strong competition for dependable lodgers and lodgers could easily move elsewhere for cheaper rents. Altogether, just over one-fifth (22%) of lodgers moved elsewhere in the city during the course of the study. One can speculate that this was in search of cheaper rents and perhaps for different space and location as household compositions changed and, in many cases, as modes of livelihood increased.

Resilient households were best placed to cope with increasing rents. For example, Lodging Household #11 (extended family) changed their modes from one wage (1993) to one wage and one self-employment in 1994, to three modes of self-employment in 1995 and increased household income by 317%. Since their rent only increased by 11%, the%age of income spent on rent declined from 16 to six%. By 1995, there was a generational shift as two adult daughters, one with a child, took over the urban lodging household and both parents moved permanently to their rural home. While their father was retired, their mother contributed by crocheting goods, which provided both households with some cash. All unemployed adult family members were based at the rural home, which indicates a recognition of the cash-intensive nature of urban residence. Lodging Household #11 exhibits what Malte Steinbrink (2009) termed translocal economic diversification and transmigration. The urban household diversified and transmigration entailed a shift of unemployed and retired members to the rural home, with goods to sell being remitted to the urban household. This is an example of what

Steinbrink and Hannah Niedenführ (2020) emphasize as the coordination of consumption, reproduction and resource use activities of a split urban–rural translocal family.

In 1993, even though only the husband and wife lived in town and all adult and younger children were at their rural home, Lodging Household #45 was a more vulnerable household since it depended on one mode of self-employment (sewing/doilies) and rent absorbed 28% of their income. However, by 1995, there were three modes, with one wage and two self-employment modes of doilies/sewing and working as a contractor. Not only did household income increase by 180%, but rent was free since the wife was now employed as a domestic servant.

In stark contrast, vulnerable nuclear Lodging Household #86 depended on one wage job in 1993, and rent absorbed 22% of income for the household of two adults, two youths and four children. In 1994 the husband lost his wage job, so that by 1995, the household income had declined by 88%. Even though only the husband and wife remained in the city, and both sold vegetables for an income, they were in debt and rent alone absorbed 115% of their monthly income. This, of course, was not sustainable and unless the couple were able to generate more income to pay off their debt and cover expenses, or unless transmigration brought a productive family member to the household, a move to their rural home would be inevitable.

The Spatiality of Lodging, Impacts on Food Security and the Rural–Urban Lifeline

It is helpful to note that the main focus of this research is lodging, and thus this chapter views food security through the lens of private rental, which in most situations is the "view from the room". Within the framework of David Harvey's "matrix of spatialities" (2019, p. 135), the absolute space of the lodging room (or part-room) restricts food storage and preparation. Any food acquired in bulk (usually from rural areas) must be stored in cramped quarters along with all other household items. Food is prepared in the same space in which the household sleeps and is usually cooked on a hotplate (unless situated in peri-urban quarters, which may lack electricity) or on a small paraffin stove. Only a small number of households had any access to refrigeration. The water source was either in the kitchen (in the case of lodging space shared with the landlord's/lady's family) or a standpipe in the yard of overcrowded lodging houses. In the latter case, extension cords that snake between rooms also present serious fire hazards. Thus, sanitation and safety aspects of absolute space are challenges that impact food storage, preparation and safety. Cecilia Tacoli (2020, p. 28) reminds us that "lack of adequate housing and sufficient living space are considerable obstacles to buying food in bulk and at lower cost". While the absolute space of lodging varied considerably, overcrowded, dilapidated lodging houses represented micro-slums where physical hazards and lack of hygiene created unsafe living environments, inclusive of food preparation and storage, thus contributing to food insecurity.

Another drawback to the absolute space of lodging is lack of access to yards and gardens. This access can depend on the type of lodging shelter, the number of households on the plot and the generosity of the landlord/landlady. If one lodges on a commercial lodging property, you may have to share with 10 to 12 other households, and either there is no room for a garden, or the yard is hard as cement. Even if you are the only lodger living with a family, they may claim exclusive use of the garden. The cost of water may also be a factor as to whether it's economically feasible to grow vegetables and mealies. One lodger complained that their landlord charged extra for every person in their household, including children, to pay for the alleged extra water they consumed as a household, and that was without access to a garden.

Thus, in this case study, lodgers who were given access to a garden were rare exceptions. In 1993, only seven of the 100 lodgers were allowed to grow food; six of these urban farmers lived in high-density areas and one lived in a peri-urban area. One high-density lodger had a fowl run and made a living by selling chickens. Overall, most lodgers had to buy their food and, when possible, supplement this with mealies from their rural-based translocal family, depending on the harvest and the strength of those social ties.

Ninety-three percent of lodgers had a rural home, and thus the majority were involved in translocal networks that involved movement of people and resources (Greiner & Sakdapolrak, 2013). Gweru's position in the urban hierarchy and its attractiveness as a destination for job seekers are evident from the fact that there are at least 30 different locations of rural Zimbabwean homes among this group of lodgers. In 1993, for those who identified as having a rural home, 40% felt they had strong to very strong linkages, in contrast to almost half (48%) who declared just moderate links and 12% with weak to very weak links. Understandably, most lodgers whose rural homes were at least 300 kms away reported moderate links. Other than that small group, the strength of social capital (Lin, 2001) with the translocal family appeared to surpass the importance of distance. For example, many with nearby rural homes had weak to moderate links and many with more distant rural homes had very strong links.

In 1993, lodging households used, on average, 36% of their monthly incomes for food, with a wide range from seven to 120%. One-fifth of households used at least 50% of monthly incomes and four% spent at least 75% of monthly incomes. At that juncture, 28% had at least one household member—usually the wife—farming at their rural home and one lodger raised chickens within Gweru. Almost two-thirds (64%) of households received food remittances from their rural translocal family. At this point in time, no respondents mentioned receiving food remittances from translocal family members in other urban areas, but this certainly may be more common now (Crush & Caesar, 2020).

Households who farmed spent slightly more—38% on average—for food, which may either reflect larger households or increased diversity in types of food consumed. It is not surprising that households that received food remittances spent significantly less on food, at an average of 22%. However, these households also had the expenditure of sending money and/or urban goods at least once a year.

By 1994, households in the study ($n = 40$) spent, on average, 26% of monthly income on food and 15% spent at least 40%. Most households had increased their incomes during this time. Forty% who had received mealies in 1993 continued to do so and another 27.5% who had not received mealies previously did so in 1994, for a total of just over two-thirds (67.5%) receiving mealies. More than half of those who now received mealies decreased the%age of income spent on food. The 1994 increase in households who now received mealies, and sometimes groundnuts, would also correlate to an increase in households where at least one person farmed, from 28% (1993) to 37.5% (1994). Transactional flows of food, cash, urban goods and the circulation of increased farming labour represent social and economic translocalization key survival strategies (Steinbrink, 2009).

Twenty percent of households who had previously received food remittances no longer did so. Many of these households may have decided that they didn't have enough resources to send money and urban goods to their rural-based family in exchange for some mealies. Of the eight households who no longer received food remittances, three were headed by widows. Most of these households showed resilience by increasing modes of livelihood and, in some cases, changing livelihood strategies. Two women were now going to Botswana monthly to sell goods and to bring back goods for sale, and one of these women had been farming the previous year. Other households took in lodgers, sold doilies and second-hand clothing, and a couple of households added a productive family member. Altogether, the increases in modes of livelihood and incomes, and increases in farming and receipt of mealies, meant that for 75% of households, the%age of monthly income spent on food decreased from 1993.

The remaining 1995 respondents ($n = 25$) averaged 23% of monthly income spent on food, with a range from six to 60%. Only three households spent in excess of 40% of their monthly income on food. By 1995, of the 21 lodgers who had rural homes, more than half (52%) had maintained strong links or had developed stronger ties than they had in 1993. Almost two-thirds (64%) of households still received mealies and also sometimes groundnuts, and one lodging household grew their own mealies since they lived in a peri-urban area. The proportion of households who had at least one member who farmed dropped from 37.5% (1994) to 32% in 1995.

Resilient Lodging Household #73—an extended family household with nine members—may help us uncover what may have been a shift in survival strategy for some. This household sent at least one member to farm in both 1993 and 1994, but no one in 1995. They still received mealies and groundnuts twice a year and remitted Z$100 monthly. Having added a productive translocal family member by 1995, they had increased their 1993 income by 300% and engaged in three livelihood modes.

Resilient, female-centred Lodging Household #22 did not have anyone farming over the three years, but reported that their translocal rural linkages evolved from moderate in 1993 to strong in 1995. While they originally remitted money and goods once or twice a year, by 1995 they remitted Z$300 monthly, received mealies three times a year and had sent one youth household member to live there, which would mean one less mouth to feed. The option to send children, youth and unemployed adult children to the rural home was a critical coping strategy.

For vulnerable Lodging Household #86 (male-centred, with eight members), linkages with the rural-based family also evolved from moderate in 1993 and 1994 to very strong in 1995. During the first two years, they received mealies, but not in 1995, and they reported that there had been a poor harvest. In addition, the frequency of their remittances of money and urban goods declined from seven to 12 times per year down to three to six times per year to nothing, and this reflected the household's income decline and increased vulnerability. In 1993, the wife had farmed, but by 1995, she remained in Gweru with her husband, where they both sold vegetables. Two adult daughters farmed and two youth and three children also lived at the rural home. The transmigration of the overall translocal household revealed a coordination strategy where the adult daughters could add to their parents' livelihood through vegetables, while youth and children would be less of an economic burden at the rural home.

Female-centred, vulnerable Lodging Household #24 consistently described linkages with her rural-based family as moderate. In 1993, she sent goods once a year but didn't receive any mealies. In 1994, this increased to sending goods three times a year and receiving mealies annually; by 1995, she remitted money and received mealies three times annually. No household member farmed or stayed at the rural home. With three adults, two children and two sources of income, the food remittances would have been critical to the household's survival. This supports Agnes Djurfeldt's (2021, p. 6) contention that in sub-Saharan Africa, "translocality is driven by survivalist motives and tied to increasing vulnerability".

Female-centred Lodging Household #70 faced survival challenges similar to thousands of other households during this time. In 1993, the wife had farmed in the rural areas, but not after that. By 1995, her brother had died and this meant that her sister-in-law, her sister-in-law's youth and three children had joined the household for a total of two adults, one youth and five children (note: the husband was absent and not discussed). The family received some social welfare for two of the orphaned children's school fees and survived on some self-employment and rent from two lodgers. They often bought a packet of mealies from the neighbours for Z$36 and spent Z$2 to get it ground; this would last for two weeks if they ate only twice a day, in the afternoon and evening. They described their family as "too big for the food [needed]".

The next section attempts to tie together the factors that influence levels of vulnerability and resilience during this challenging and rapidly changing period of time.

Resilience and Vulnerability for Lodgers in Gweru

Table 19.1 summarizes the types of strategies employed by the urban and rural translocal families over the three-year period. Diversification of income modes and the food security strategies impact each other, and most households engaged several strategies in both areas. Some lodging households that were highly vulnerable in 1993 became resilient. It is critical to appreciate that this environment was highly chaotic, with rapid changes in wage employment, and the presence of multiple, sudden shocks in the value of the currency and the costs of basic goods. These households had to pivot quickly and many changes were drastic. Overall, they were able to increase monthly incomes, decrease risk by diversifying modes of livelihood, and either spend proportionately less on food and/or strengthen their bonds with the rural-based translocal family such that remittances of food, cash and goods were increased. Urban lodging rents in Gweru did not increase significantly during those three years, and that would add to the economic security of most lodging households. And the fluidity of lodging allowed households to seek lodgings that met their needs regarding some or all factors such as cheaper rent, better location, quieter, less crowded, cleaner and better maintained.

Table 19.1 Translocal household strategies

Urban lodging household	Bonds between individuals, relations of exchange and cooperation transaction flows	Rural household
Send children/youth/elderly to rural home	→	
	←	Send productive family member
Send/increase cash and goods	→	
Send family members to farm	→	
	←	Send food transfers 1-3 times per year
Grow food in city		
Diversify livelihoods	←	Send goods to sell
Sell goods and services		
Reduce daily meals		
Move to cheaper lodging		
Combine households		

Source Adapted from Steinbrink (2009, p. 233)

On the other hand, vulnerable households that were unable to recover from major shocks such as retrenchment or a death in the family usually left the city for their rural homes, if this was an option. Just over one in 10 moved back to their rural homes permanently in either 1994 or 1995. For those without rural homes, and widows in particular, their fates were closely tied to the mercy of urban-based family members and, to a lesser extent, friends, neighbours and social welfare. Thus, their urban social networks were crucial for survival. In some cases, the landlord/lady allowed their tenant to stay without paying rent until their situation improved. Many households were "hanging by a thread" and their subsistence depended on selling low-value goods such as used clothing or doilies. Some single women, who saved face by claiming that they sold doilies for a living, were commercial sex workers who were ill by 1995—supported by parents and other extended family—with bleak days ahead. In other households, the main wage earner had fallen sick, adding another dimension of uncertainty and added costs of medicine.

The reinforcement of translocal rural family bonds was a key factor for many of those lodgers whose situations actually improved that source of staple foods from rural homes, the option to delegate family members to farm and live there, and increased ability to help with cash and urban good remittances reinforced the stability of many resilient lodgers during this period. Reciprocal remittances were also found to be prevalent among internal migrants in research on Windhoek, Namibia. Bruce Frayne (2001, in Crush & Caesar, 2020) found that 62% of urban households had received food remittances from their translocal rural family during the previous year. And overall, "levels of food insecurity in Windhoek were lower than predicted given pervasive poverty, high unemployment, a relatively small informal economy, and minimal urban agriculture" (Frayne, 2005a, 2005b, 2007, in Crush & Caesar, 2020, p. 296).

Another extensive food security study centred on Epworth, a peri-urban, mainly informal settlement with a population of 167,462 in 2012 on the outskirts of Harare (Tawodzera in this volume). Tawodzera and Chigumira (2019) found that remittances (rural–urban: 13.5%; intra-urban: 4.8%; and international: 4.4%) played a minor role in overall annual food sources. With the majority of sample households relying on casual work and informal wage work in an unhealthy environment that lacked access to water, sanitation and electricity for most residents, food poverty and limited dietary diversity were acute (Tawodzera & Chigumira, 2019).

Due to the process of deindustrialization, formal sector retrenchments, sudden increases in food and basic services and the resultant shift towards reliance on informal work and self-employment, the 1993–1995 period was only the start of acute challenges for low-income urbanites.

As a secondary city, Gweru provided a locale that was more easily accessed and negotiated than larger cities, and where there was some (perhaps only temporary) buffer from high costs in larger cities such as the capital, Harare. By 1995, "the money did not go far" and lodging households depended on their own resilience and on translocal urban and rural social networks to increase food security and economic diversification in the chaotic economic and social environment that was—and unfortunately still is—Zimbabwe.

This case study reflects the "view from the room" as a lens through which to examine the interstices between low-income housing, livelihoods, food security and translocal processes and strategies. In secondary cities across the Global South, millions of households rent a room and the physical attributes of that room and its immediate environment have impacts on overcrowding, physical security, access to water and sanitation, ability to store food, hygiene and safety around food preparation and storage, and, in rare cases, access to yards for growing or raising food. Whether the room is in an overcrowded lodging house, or a shack in a slum situated far from work, as Tacoli (2020) notes, these non-income factors contribute to malnutrition and food insecurity, with a high price being paid by women. Shelter—space, environment and locale—is a critical component of the complex issue of food security for the urban poor.

References

Bond, P., & Manyanya, M. (2003). *Zimbabwe's plunge*. Weaver Press.

Crush, J., & Caesar, M. (2020). Food remittances and food security. In J. Crush, B. Frayne & G. Haysom (Eds.), *Handbook on urban food security in the global south* (pp. 282–306). Edward Elgar.

Djurfeldt, A. (2021). Translocal livelihoods research and the household in the Global South: A gendered perspective. *Journal of Rural Studies* (online).

Frayne, B. (2005a). Rural productivity and urban survival in Namibia: Eating away from home. *Journal of Contemporary African Studies, 23*(1), 51–76. https://doi.org/10.1080/0258900042000329457

Frayne, B. (2005b). Survival of the poorest: Migration and food security in Namibia. In L. Mougeot (Ed.), *Agropolis: The social, political and environmental dimensions of urban agriculture* (pp.32–50). IDRC.

Frayne, B. (2007). Migration and the changing social economy of Windhoek Namibia. *Development Southern Africa, 24*(1), 91–108. https://doi.org/10.1080/03768350601165918

Grant, M. (1996). Vulnerability and privilege: Transitions in the supply pattern of rental shelter in a mid-sized Zimbabwean city. *Geoforum, 27*(2), 247–260.

Grant, M. (2007). Lodging as a migrant economic strategy in urban Zimbabwe. *Development Southern Africa, 24*(1), 77–90.

Grant, M., & Palmiere, A. (2003). When tea is a luxury: The economic impact of HIV/AIDS in Bulawayo, Zimbabwe. *African Studies, 62*(2), 213–240.

Greiner, C., & Sakdapolrak, P. (2013). Translocality: Concepts, applications and emerging research perspectives. *Geography Compass, 7*(5), 373–384.

Harvey, D. (2019). *Spaces of global capitalism: A theory of uneven geographical development*. Verso.

Haysom, G. (2021). Understanding secondary city typologies: A food governance lens. In J. Crush & L. Riley (Eds), *Transforming food systems in African secondary cities*. Palgrave Macmillan.

Lin, N. (2001). *Social Capital: A theory of social structure and action*. Cambridge University Press.

Lohnert, B., & Steinbrink, M. (2005). Rural and urban livelihoods: A translocal perspective in a South African context. *South African Geographical Journal, 87*(5), 95–103.

Marquette, C. (1997). Current poverty, structural adjustment, and drought in Zimbabwe. *World Development, 25*(7), 1141–1149.

Ministry of Public Service, Labour and Social Welfare (MPSLSW). (1996). *1995 poverty assessment study survey*. Social Development Fund Unit. MPSLSW Preliminary Report.

Minot, N. (1994). *Trends in social indicators*. SDA Advisor Monitoring and Implementation Unit. MPSLSW Working Document.

Riley, L., & Dodson, B. (2020). The Gender-Urban-Food Interface in the Global South. In J. Crush, B. Frayne & G. Haysom (Eds.), *Handbook on urban food security in the global south* (pp. 218–232). Edward Elgar.

Sheppard, E. (2016). *Limits to globalization*. Oxford University Press.

Steinbrink, M. (2009). Urbanisation, poverty and translocality: Insights from South Africa. *African Population Studies, 23*(Suppl), 220–252.

Steinbrink, M., & Niedenführ, H. (2020). Translocal livelihoods: New perspectives in livelihood research. In M. Steinbrink & H. Niedenführ. *Africa on the move* (pp. 35–52).

Tacoli, C. (2020). Food (In)Security in rapidly urbanizing, low-income contexts. In J. Crush, B. Frayne, & G. Haysom (Eds.), *Handbook on Urban food security in the global south* (pp. 23–33). Edward Elgar.

Tawodzera, G., Riley, L., & Crush, J. (2016). *The return of food: Poverty and urban food security in Zimbabwe after the Crisis* (Urban Food Security Series No. 22). African Food Security Urban Network (AFSUN).

Tawodzera, G., & Chigumira, E. (2019). *Household food poverty in Epworth, Zimbabwe* (Urban Poverty Working Paper No. 8). University of Cape Town. Consuming.

Tawodzera, G. (2021). The role of the informal sector in Epworth's Food System. In J. Crush & L. Riley (Eds), *Transforming food systems in African secondary cities*. Palgrave Macmillan.

World Health Organization (WHO). (1998). *Report on the global HIV/AIDS epidemic: The Evolving picture region by region*. WHO.

World Population Review. (2021). *Zimbabwe population 2021 (Live)*. https://worldpopulationreview.com/countries/zimbabwe-population

Index

L. Riley and J. Crush (eds.), *Transforming Urban Food Systems in Secondary Cities in Africa*,
https://doi.org/10.1007/978-3-030-93072-1

E

F

G

M

N

O

P

The manufacturer's authorised representative in the EU is Springer
Nature Customer Service Centre GmbH, Europaplatz 3, 69115 Heidelberg,
Germany. If you have any concerns regarding our products, please
contact ProductSafety@springernature.com

Printed and bound by CPI Group (UK) Ltd, Croydon, CR0 4YY
30/09/2026
02225661-0012